"十二五"普通高等教育本科规划教材

环境化学

（双语教学用）

弓爱君　刘杰民　王海鸥　编著

化学工业出版社
·北京·

This book includes five chapters, including Introduction, Atmosphere Environmental Chemistry, Water Environmental Chemistry, Soil Environmental Chemistry and Disasters and Environmental Protection. This book focused on atmosphere, water and biosphere, systematically illustrates the behavior of pollutants, the techniques and theories of easing and removing them. The section of biosphere combines the atmosphere, water and soil, thus it is not an independent chapter. Disasters and environmental protection are specifically included in the chapter 5 for the sake of enhancing the environmental awareness of modern constructors.

In order to improving the teaching effect, there are key points in every chapter and questions to be considered at the end. Meanwhile, the teaching board has already published an exercise book *Environmental Learning and Practicing Collection* for bilingual teaching.

The book can be used as the bilingual teaching material of Environmental Chemistry courses for undergraduates in environment, chemistry and biology majors. It can also be used as a reference book for postgraduates in related majors and environmentalists.

本书共五章，包括绪论、大气环境化学、水环境化学、土壤环境化学、灾害与环境保护。本书以大气、水、土壤三大圈层为主线，系统阐述了污染物的行为特征、减缓或消除污染的方法和原理。生物圈层的内容融入大气、水和土壤圈层，没有单独立章。为了增强新世纪建设者的环境意识，特将灾害与环境保护纳入本书的第五章。

为了提高本书的教学效果，每章正文开始前编有“本章要点”，每章末尾配有思考题。同时，我们教学组已经出版了与本教材配套的《环境化学习题集》(双语教学用)。

本书可作为高等学校环境、化学、生物等专业本科生的《环境化学》(双语) 课程教材，也可作为相关专业研究生及环境工作者的参考书。

图书在版编目 (CIP) 数据

环境化学 (Chemistry of the Environment)/弓爱君，刘杰民，王海鸥编著. —北京：化学工业出版社，2015.10 (2022.9 重印)

ISBN 978-7-122-24899-2

Ⅰ.①环… Ⅱ.①弓… ②刘…③王… Ⅲ.①环境化学-高等学校-教材-英文 Ⅳ.①X13

中国版本图书馆 CIP 数据核字 (2015) 第 187869 号

责任编辑：杨 菁 王 婧　　文字编辑：李锦侠
责任校对：王 静　　装帧设计：关 飞

出版发行：化学工业出版社 (北京市东城区青年湖南街 13 号 邮政编码 100011)
印　　装：北京七彩京通数码快印有限公司
787mm×1092mm 1/16 印张 15½ 字数 387 千字 2022 年 9 月北京第 1 版第 2 次印刷

购书咨询：010-64518888　　售后服务：010-64518899
网　　址：http://www.cip.com.cn
凡购买本书，如有缺损质量问题，本社销售中心负责调换。

定 价：56.00 元

Preface

The development of human society is becoming increasingly fast while various environmental problems appear noticeable than as past. Environment events are alarming people and stirring more attention from the public. Those events are promoting the flourishing development of environmental chemistry and make people clearly aware of the significance and necessity of sustainable development. Higher education is the principal base of fostering elites, spreading environmental knowledge and researching into the environmental issues. Thus, compelling this bilingual material for undergraduates has an indispensable.

Environmental Chemistry is one branch of science, which studies on the existence, characteristics, behavior, and impact as well as chemical principles and strategies to control pollutants. The book is divided into five chapters, which are Introduction, Atmosphere Environmental Chemistry, Water Environmental Chemistry, Soil Environmental Chemistry and Disasters and Environmental Protection. It systematically illustrates the definition, principle, method, problem and last development in bilingual way.

In writing the book, we have been aided greatly by bilingual teaching demonstration curriculum construction project and the textbook project of the "12 five-year plan" in USTB. We highly appreciate it.

Due to knowledge limitation, writing and English level, mistakes may occur in this book. We sincerely hope readers' feedback and correction.

Editor

Jul. 2015

前言

人类社会的发展日新月异，各种环境问题随之突显。环境事件给人们敲响了警钟，引起了人们对环境问题的高度重视，推动了环境化学的飞速发展，迫使人们清醒地认识到走可持续发展之路势在必行。高等学校是培养人才、传播环境知识、研究环境问题的重要基地。因此，编写本科生使用的环境化学中英文教材具有重要的意义。

环境化学是一门研究有害化学物质在环境介质中的存在、化学特性、行为和效应及其控制的化学原理和方法的科学。全书分为五章，即绪论、大气环境化学、水环境化学、土壤环境化学和灾害与环境保护，以中英文双语的形式系统阐述了环境化学的概念、原理、方法、问题及最新进展。

本教材的编写得到了教育部双语教学示范课程建设项目及北京科技大学“十二五”规划教材项目的大力支持，在此表示感谢！

限于学识、文字和英语水平，书中疏漏及欠妥之处在所难免，恳请读者批评指正。

编著者

2015 年 7 月

Contents

目录

Chapter 1
Introduction

Key Point: Definition of Environmental Chemistry; Classification of Environmental Chemistry; Research Content of Environmental Chemistry; Characteristics of Environmental Chemistry; Development of Environmental Chemistry; Environmental Pollution; Eight public nuisances in the world; Recent years' the world famous public nuisance; The main environmental problems; People's awareness of environmental problem; The solution to the key problems of the prevention and control of environmental pollution in China.

Section 1 What is Environmental Chemistry?

1. Definition of Environmental Chemistry

Environmental Chemistry is one branch of science, which studies on the existence, behavior, and impact of hidden harmful chemicals and finally reduces or eliminates them. Three purposes are laid on it. First, it is supposed to study the existence, chemical feature, behavior and impact of harmful chemicals in their specific ambient medium. Second, it researches into the chemical principle and method of controlling contaminant. Third, it enables scientists to find and solve environmental issues by applying basic theories and techniques. Environmental Chemistry is the core proportion of Environmental Science and also an instrumental branch of Chemistry the subject.

2. Classification of Environmental Chemistry

Environmental Chemistry basically includes Environmental Analytical Chemistry, Environmental Pollution Chemistry and Pollution Control Chemistry.

① Environmental Analytical Chemistry aims at studying how to apply modern scientific theories and advance experimental techniques to differentiate and measure the types, composition which includes state and structure, and content. On the account of the features of analyzing pollutants, like the various categories, the low density and severe interfere, scientists are usually required to introduce the measurement that possesses high flexibility, selectivity, accuracy and precision. The progressively integration of automatic constant monitor and satellite remote sensing technology into Environmental Analytical Chemistry enhances its ability to control pollutants in real time. Environmental Analytical Chemistry comprises organic analytical chemistry of environment and inorganic analytical chemistry of the environment.

② Environmental Pollution Chemistry mostly researches into the chemical behavior and changing law of pollutants in mediums during their movement, acceleration and transformation. This subject includes Atmosphere Environmental Chemistry, Water Environmental Chemistry, Soil Environmental Chemistry and Pollution Eco-Chemistry.

③ Pollution Control Chemistry is a science that exposes the separation and control of chemical pollutants. It focuses on studying pollution control and other chemical problems that may occur in the course of treatment. It also provides theories and methods for the treatment. Pollution Control Chemistry includes air pollution control chemistry, water pollution control chemistry and solid waste control chemistry.

3. Research Content of Environmental Chemistry

The research content of Environmental Chemistry includes five points. ①It examines the density level and state of harmful substance in the environment. ②It studies the origin, distribution, migration, transformation and fate as well as their various behaviors when dealing with different environmental mediums. ③It explores the mechanism and risk of the impact those harmful substances have on the environment, ecosystem and human body. ④It finds out the methods to ease and remove bad effects made by harmful substances. ⑤It seeks the solution to avoid the production of harm.

4. Characteristics of Environmental Chemistry

The characteristic of Environmental Chemistry is to study the mechanism and prevention methods for macro environmental phenomenon and change. The core is the research into chemical migration and reaction of pollutants in specific environment.

(1) The substances of research have a large quantity and their systems are complicated

The pollutants have various forms with large quantity and complex composition. Meanwhile, many kinds of pollutants may have chemical reactions and physical changes under different mechanisms in the environment, which makes the substances of Environmental Chemistry a big system with various configurations, changeable states and complex mechanisms.

(2) The density of pollutant is very low, usually with the order of μg/mL or μg/L

Pollutant has a very low density with the order of μg/mL or μg/L. However it is on the normal level when comparing with other compounds coexisting. For the precision of qualitative and quantitative analysis of these pollutants, modern analytical techniques are often required, for example, the Infrared Spectrum Technology, the NMR (nuclear magnetic resonance) technology, the GS-MS (Gas Chromatography and Mass Spectrometry) and the isotope tracer techniques.

(3) The study of Environmental Chemistry needs a wide span of time or space and high comprehensiveness

It is difficult to control chemical reactions in the environment. Thus, doing Environmental Chemistry study demands wide time and spatial span. It explores the distribution, migration, transformation and fate of pollutants in ecosystem, especially their acceleration, interaction, biological effect and so on.

5. Development of Environmental Chemistry

As the environmental problems being increasingly severe and the awareness of the public enhancing, new directions emerge from the research of Environmental Chemistry.

(1) Environmental Analytical Chemistry

The direction of Environmental Analytical Chemistry mainly focuses on the refinement and exploration of pretreatment techniques of environmental samples. The substances of analysis change from mostly inorganic compounds to organic compounds and organic metal compounds. The analyzing techniques of instruments are generally improved for developing the precision and reproducibility.

(2) Environmental Pollution Chemistry

Environmental Pollution Chemistry studies pollutants' chemical behavior and ecological effect when they taking shape, migrating, transferring and ending in the environments of atmosphere, water and soil. The subjects of Atmosphere Environmental Chemistry include atmosphere particles, acid deposition, atmospheric organic compounds, trace gases, ozone depletion and global warming, etc. Its spatial scale goes from inner air, city, and regional environment even to the whole world.

Atmosphere Environmental Chemistry firstly studies river, lake and reservoir. Estuary, bay, offshore area and groundwater pollution are all of the second concerns. It mainly explores the process, for instance, water interface chemical process, metal morphology transformation dynamics process, organic compounds chemical degradation process, metal and metalloid methylation process, etc.

Soil Environmental Chemistry mainly studies agrochemicals' migration, transformation and fate and its effects on soil and human body in the environment of soil. It includes degradation of organic pollutants in the soil, the release of greenhouse gases, pollutants' chemical reactions process between solid and liquid interface and chemical and biological remediation, etc.

Ecology Environmental Chemistry generally researches into the process of agrochemicals taking effects in the ecosystem, such as the environmental chemical behavior and ecological security of rare earth application.

(3) Pollution Control Chemistry

Currently, pollution control has two models, which are the traditional terminal control and pollution prevention and clean production. Terminal control has positive effects on the development of pollution control techniques and pollution treatment in each country, but it can only reduce or control pollutants emission, not prevent the production of them. Terminal control research includes the contents, which are atmosphere pollution control, water pollution control, solid waste control and resource analysis. The study in pollution control material and technology includes flocculants, ion exchanger, membrane material, catalyst and separation technology, etc.

On account of the limitation of terminal treatment, in the mid of 1980s, European countries proposed the transformation of control model from End-of-pipe Control to whole process control and also put forward several strategy views, including Pollution-Prevention, Cleaner-Production, Green Chemistry and Industrial Ecology. It aims at no harmful intermediate and sub product and builds inner circulation of waste and emission in order to achieve waste minimization, resource and energy maximization. It was one of the instrumental breakthroughs of environmental management strategy.

Section 2 Environmental Pollution Events

Environmental issue refers to various environmental recession and harmful effect on biosphere, mainly including acid rain, global warming, ozone hole, freshwater and other resources deficiency, and ecological degradation.

1. Environmental Pollution

Human factors destroyed ecosystem, led to the decrease of the quality of the environment, has affected people's normal life.

Pollution of the environment reflects in the following four aspects. First one is harmful substances pollution of the atmosphere, water, soil and animals and plants, and reaches the extent of the damage. Second one is improper interference and destruction of ecological system. Third one is that non-renewable resources are abused by excessive mining. Fourth one is that solid waste, noise, vibration, stench, radiation caused the damage to the environment. Factors that cause environmental pollution have three aspects of physical, chemical and biological, which caused by chemicals accounted for about 80% to 90%.

2. Eight public nuisances in the world

(1) Belgium Meuse Valley Smog Event

From December 1 to 5, 1930, excessive amounts of smog were discharged and could not spread from thirteen factories in Belgium Meuse Valley industrial district, which caused thousands of people there had chest hurting, coughing, tearing, sore throat and dyspnea. More than 60 people died within one week and then their relatives. It was the first record of atmosphere pollution in 20th century.

(2) America Los Angeles Chemical Smog Event

From 1940s, yellow smog started to emerge in the sky of Los Angeles in USA, which, at that moment, owned large amounts of vehicles. It was caused by photochemical smog and it harmed people's eyes, throat and lung and also brought symptoms like chest distress. A big scale of plants was affected, for example, the pine threes died and orange output dropped. In 1955, the number of victims reached 400 and it was the earliest atmosphere pollution due to automobile exhaust.

(3) Nora Smog Event

During October 26 to 31, 1948, it was smog constantly in Nora town, Pennsylvania, USA. It was just the plan where lay in major sulphuric acid plant, steel plant and zinc smelting plant. The smog released was trapped inside the valley and caused 6000 people eyes hurting, sore throat, nose running, and headache and chest distress. 20 of them soon died. This smog event was mainly caused by the fact that harmful substances like sulfur dioxide and metal particles attached on some suspended particulate matter and when people inhaled too much those in short time, there came the tragic.

(4) London Smog Event

From December 5 to 8, 1952, there remained high pressure above the sky of London. Thick fog and no wind. As currently, it was the coal-fired heating period, the ashes and moisture accelerated, which led to dyspnea, eyes hurting of many citizens. Within only four days, more than 4000 people died. In the later two months, 8000 people died as well. It was the biggest city smog event caused by coal firing in the 20th century.

(5) Yokkaichi Asthma Event

In 1955, the first petro-chemical integrated complex in Japan claimed to be listed in Yokkaichi. In 1958, the fishes started to have stingy petrol smell which made the local fishing industry drop. In 1959, Zhao Stone Petrol Company invested 18.6 billion yen to refinery in Yokkaichi. This city soon developed to be "petrol integrated city". However, exhaust fumes from petrol smelting began to cover the local air place. The thickness of smog was even 500m. The fact many floating matter inside the smog carrying harmful substances and metal ashes resulted many people get headache, sore throat, eyes hurting and vomiting. Since 1960, the local asthma patients exceedingly increased and some of them even could not stand the pain and committed suicide. Till the end of October in 1979, the patients with air pollution diseases reached the number of 775491 and the typical respiratory diseases include bronchitis, asthma, pneumonectasis and lung cancer.

(6) Itai itai disease

In the 1880s, modern strategy of management was applied to the Kamioka mine on the upstream of Toyama Prefecture, Japan. Kamioka mine dedicated in extraction and refinement of lead, zinc and mine. However, in the process of extracting and from the pile of slag, waste water with heavy metal like cadmium directly ran into the surroundings in the long period. As a result, the accumulation of heavy metal like cadmium happened in local paddy field, soil, river and sediment. Cadmium went through rice into human body and firstly resulted in kidney obstacle, gradually chondropathy. It also caused women severe pain in the middle of pregnant, lactation, endocrine disorder, which called Itai-itai disease. If the condition becomes more serious, victim might get multiple fractures and die with intolerable pain. From 1931 to 1968, 258 people were diagnosed and 128 of them died. Till December in 1977, another 79 died.

(7) Minamata Disease

Since 1949 in Minamata city in Kumamoto prefecture, Japan, a local nitrogenous fertilizer corporation began to produce vinyl chloride and vinyl acetate. As catalysts with Hg were needed in the process, large amounts of Hg then were released into Minamata bay without any sewage disposal. In 1954, there appeared an odd disease no one knew the cause in Minamata bay, named "Minamata disease". The victims were initially cats and human. The symptoms include numbness in the hands and feet, ataxia, deformation of hands and feet, metal disorder, scream till death. Through ten-year analysis, scientists confirmed Hg in the sewage from the factory was the cause. Hg was transferred into CH_3HCl after eaten by aquatic organisms. This matter harmed brain and other parts of body since migrating into animal body through fishes and shrimps and led to brain atrophy, cerebellum balance system destroyed. It was highly poisonous. In Japan, the victim eating poisonous fishes and shrimps with CH_3HCl reached the number of 100000. 283 people got poisoned and 60 of them died. Till 1991, 2248 people were diagnosed to be public nuisance patients, 1004 of which have already died. ChiSS Corporation compensated more than 10 billion yen. It has not been fully resolved for 40 years. It is a typical event of water pollution.

(8) Rice Bran Oil Event

In 1968, a cooking oil factory at Aichi prefecture in Kyushu, Japan, operated wrongly when producing rice bran oil so that heat carrier polychlorinated biphenyls (PCBs) used in deodorization process was mixed up with the oil. It caused cooking oil pollution. As the black poisonous oil was made into chicken feed at that moment, hundreds of thousands of chicken died in Kyushu, Shikoku and other regions. Then Kyushu University affiliated hospital consistently found patients poisoned by PCBs. Early stage symptoms are skin rash, black fingernails, skin pigmentation and conjunctival congestion. Then it transferred to liver failure and general muscle pain and if severe, it led to acute liver necrosis, hepatic coma till death. In 1978, 1684 people have been diagnosed.

3. Recent years' the world famous public nuisance

More than 200 serious pollution incidents happened throughout the world each year. Following 10 were the most remarkable among those pollution incidents.

(1) Seveso chemical pollution accident in Italy

On 1976 July 10, an insecticide factory explored in northern Italy, Seveso, which leads to highly toxic chemicals of dioxin pollution. Many people were poisoned and nearby residents were forced to move. In following years, the birthrate of local oaf increased greatly.

(2) The leakage of nuclear power accident at three mile island in America

On March 28, 1976, a meltdown occurred in a PWR at three mile island in America. Most of a reactor components burned and part of radioactive material leaked. The accident remained 36 hours, and left profound effect on people. It caused more than $ 10 billion economic losses.

(3) Mexico's liquefied gas explosion accidents

On November 19, 1984, the liquefied gas supply center of Mexico's state oil company exploded. It caused more than 1000 people dead, 400 injured and 30000 homeless.

(4) India Bhopal pesticide leakage accident

On December 3, 1984, in an insecticide factory of Union Carbide Corporation located in Bhopal city, a tank, which contained highly toxic chemical methyl carbimide, burst and leaked out. It caused 200000 victims and killed more than 2000 people.

(5) The leakage of nuclear power at Chernobyl in the former Soviet union

On April 26, 1986, in the Chernobyl nuclear power plant located in Kiev region, No. 4 reactor explored, which caused severe radioactive contamination. More than 100000 residents were evacuated. About 100 people were injured and 31 people dead.

(6) The Rhine river pollution accident

On November 1, 1986, the warehouse of Swiss chemical company out broke a fire. Large amount of poisonous chemicals flowed into the Rhine with water for fire extinguishing, which caused the most severe pollution of Western Europe in ten years.

(7) Environmental pollution caused by the gulf war

The gulf war, which out broke at the end of 1990, lasted 42 days. During that time, the fire of in the oil wells burned day after night. Up to now, it has been the biggest petroleum fire and marine oil pollution. Also, it is the most serious environmental pollution in the history, whose pollution level is higher than the nuclear leak of the chernobyl nuclear power plant. The environmental pollution caused by gulf war is a disaster, which affects the whole world.

(8) Sudan red pollution event

Sudan red (1) is a kind of red dyes, used to grace solvent, oil, wax, gasoline and polish shoes, and floor board. As a kind of industrial dye, it is forbidden to be used as food additive. On February 2nd, 2005, the first food company of UK (Premier Food) reported to British sanitation department that five tons of cayenne pepper imported from India contained Sudan red (1) and that it had been sold to many downstream grocer for chill pasta as flavoring . On February 18, UK's food standards agency confirmed the pollution, and traced grocer which used raw material supplied by Premier Food Company. It listed 575 kinds of food that contained Sudan 1 in four batches, warning consumer of the danger of those food in order to decrease the rate of cancer caused by Sudan. Those foods were forced to remove from the shelves under supervision before February 24. As for the production on the list of sale, it must be unconditionally return. British food standard agency said that it was the biggest food recall event in British history. After Britain issued food warning, China's General Administration of Quality Supervision, Inspection, and Quarantine released the Urgent Notice of Strengthening Inspection and supervision of Food for Sudan 1. It required checking food (especially imported food) sold in the mainland in order to prevent Sudan 1 being sold and eaten. On May 29, China formulated method for test Sudan 1 immediately and put into effect. Sudan has been found in many foods in China, including KFC's New Orleans roast chicken wings, spicy chicken fort, explosive chicken rice flower, and Heinz food, Guilin chill sauce, temple incense altar chili products, etc.

(9) Melamine-contaminated dairy products

On November 8, 2009, 14 babies suffered from kidney stone at same time in Min County, Gansu province, which drew public attention. Up to September 8, 2008, 59 cases of children with kidney stone had been found, some of who had grown into renal inadequacy. Besides, one child was killed. Those babies all ate Sanlu milk whose price was about 18 Yuan. In following two months, similar event happened in many Chinese provinces. China's Health Ministration highly suspected that San Lu infant formula was contaminated by melamine. Melamine is a kind of chemical raw material which can improve test value of protein. If people take in it for a long time, it would cause vesical calculus and kidney stone, and it may induce bladder cancer.

On September 19, 2008, Chinese quality inspection administration announced that three branches of milk, Yili, Mengniu and Guangming, contain melamine. Among the 81 batches of Yili milk in casual inspection, 7 batches were found contain melamine. The check value was 0.8-7 mg/kg. Among the 93 batches products of Mengniu, 6 batches were detected to contain melamine. The value was 0.6-8.6 mg/kg. But there is no melamine in the products of Sanyuan and Nestle. On September 17, 2008, China's the general administration of quality supervision, inspection and quarantine announced to cancel the regulation of industry inspection-free system and the inspection-free sign printing on the products and pack were no longer effective.

(10) Fukushima nuclear leak

On May 11, 2011, 13: 46, Beijng time, an earthquake with 9.0 magnitude happened in Japan and triggered a strong tsunami which up to 10 meter, resulting in No. 1, No. 2 and No. 3 unite operation's shutdown in Fukushima nuclear station of Tokyo Electric Power Company, the reactor control rod insertion and unit's subcritical shutdown. In the subsequent development of the accident, the earthquake leads to the loss of otc ac power. Then, because of the tsunami, ac power (diesel generators) failure, leading to the cooling system of reactor lose function and caused accident.

On July 22, 2013, after two years and several month of the accident, the Tokyo Electric power suggested that water of radioactive contamination from the nuclear station was flowing into Pacific Ocean. Local fish man and experts in nucleon supervision organization had supposed this problem would rise. But the Tokyo Electric power denied firmly. Therefore, Japanese Prime Minister, Shinzo Abe, command government to solve the emergent issue. On August 20, another accident happened in nuclear station. 300 tons of high radiation and concentration of wastewater leaked from wastewater storage tank. The wastewater is enough to do harm to the health of nearby workers. The accident of leaking wastewater is classified into third stage of international nuclear accident.

Section 3 The Process of People's Awareness of Environmental Issues

1. The main environmental problems

Environmental problem refers to the phenomenon of all kinds of degradation and its adverse effect on biosphere, including acid rain, greenhouse effect, ozone hole, the lack of fresh water resources, shortage of resources and ecological degradation.

2. People's awareness of environmental problem

In the 1960s, people regarded environmental problem as pollution problem. They did not link it to natural ecology and social factors, which underestimated the perniciousness and complexity of environmental pollution. The root of the problem was ignored.

In 1972, the United Nation had a conference on human environment in Stockholm, Sweden. It was the first time to link environmental problem with social factor, which was a mile stone. However, it did not work out the basic way to prevent and control environmental problem. The responsibility of the problem did not be identified. It did not emphasize the importance of working together.

In the 1980s, there was a breakthrough in people's understanding of environment. It put forward a strategy of sustainable development and worked out the basic way to solve environmental problem.

In 1992, the United Nation had a conference on environment and development Rio de Janeiro, Brazil, which was a significant turn point of the century. It set a new milestone in history of the relation between human environment and the development.

In 2002, the United Nation held a conference on sustainable development in Johannesburg, South Africa.

In 1973, 1983, 1989 and 1996, China held national environment conference respectively, which pushes the development of environment protection and the understanding of environment forward.

As the environmental problem goes worse, people's understanding of the problem is developing. In the word today, people begin to realize that environmental problem is not the business which only belongs to one country or local area. Acid rain, green house effect, and ozone depletion are beyond the limits of country and regions, becoming global problem. These problems seriously threaten human's existence and development, and become major concern of all countries in the world and a focus of the international attention.

3. The solution to the key problems of the prevention and control of environmental pollution in China

(1) Resource utilization and economic development

From the per capital view, the environment can hardly afford the large population which made China go short of resources. China is one of the resources lacking country, which has a small amount of per-capita resources. From the total view, China has a vast territory and abundant resources, which make China become one of a handful of resources-rich countries. According to comprehensive rank of total natural resources, China is the fourth in the world. The potential value of proven mineral resources reserve is third in the world. Hydro energy, solar power and coal resources are first, second and third respectively. China has the world's eighth largest oil resources. The area of cultivated land and grass are fourth and the area of forest is sixth in the world.

China has various geography and full range of resource. But the quality of the resources is not high and natural productivity is low. The resources distribution is unbalanced and the resources are not matched. Besides, the resource distribution and productivity layout is discordant. Because of the large population, the pressure on limited resources will be increasingly high. The poor supply of reserve resources intensifies the resources crisis. If we not strengthen investment and achieve a breakthrough in geological exploration, petroleum, natural gas and China's key mineral resources will be a serious exhaustion.

The continuing increasing population, with decreasing of cultivated land, the shortage of water supply, scarcity of resources, poor mineral resources, weak foundation of reserve resources, rapidly expending demand of resource, the shortage of all kinds of resources in long term, is the basic pattern of the conflict among population, economic development, and resources supply. The shortage of the resources will be the bottom neck that hinders China's economic development for a long time.

(2) The issue of environmental destruction and resource exploration

With the high-speed development of national economy, rapidly increasing of population and the improvement of people's living standard, the consumption of energy and raw material will multiply increasing. The coal resources have the largest proportion of resources, which is about 75%. And it will be the main energy in China in 21st century. The mineral exploration, especially the process of coal resources exploration, provides country a large amount of energy, raw material and creates huge material wealth, which supports nation's rapidly economic development. At the same time, however, it caused severe damage and pollution to the environment of mining area.

In the region where mineral resources development is put at the first place, mineral mining, washing, smelting process need a lot of water. Therefore, a large amount of sewage and wastewater are emitted, which causes serious water pollution. Mining activity do great damage to water resource. With the channel excavation, more mine water will be discharged, which drains off the water in aquifer above the mining area or around the pit formation. Besides, funnel-shaped zone of groundwater forms and underground water level will fall. The area within 40-50 kilometer can be effect and cause the regional hydrological environmental damage.

(3) The problems of economic development and environment

With the further economic development, our country's environmental pollution becomes severe and threatens healthy living of citizen. What is worse, it will be the main constraint on economic development in future. Industrial emission is the main reason that brings our country serious environmental pollution. With the development of nation's economic and improvement of citizen's living standard, there will be huge amount of resources consumption. At the same time, waste gas, waste water and solid wastes is increasing. The assignment to control pollution is significant and requires a long time. When preventing the pollution incident happening and protect the people's living environment, we should obey the law and regulation. Along with the development of social economy, urbanization and improvement of people's living standard, our country will confront more huge challenge of protecting environment and controlling pollution in following 10-20 years. ①The rate of 7% economic development increases the pressure on environmental control. ② Living standard of well-off society requires higher quality of environment. ③Developed countries' progress on technology raises the international environment standard.

As the old saying goes, whoever started the trouble should end it. The environmental problems caused by chemical substance, for most part, still have to rely on chemical way and method to solve. Environmental chemistry is such a subject. On the one hand, it studies the cause of environmental pollution. On the other hand, it studies the method to control the pollution, playing a significant role on solving various environmental problems. At the same time, because of the complexity of the environmental factor, environmental chemistry is closely related to physics, meteorology, biology, geography, medicine, engineering, and statistics. It is a comprehensive and intersectional subject. To control pollution, western developed counties fight the pollution after polluted. But we will not be same as them. Whether we can control the pollution depends on ourselves. Although it is difficult, we will try our best.

Questions

1. What is the environmental chemistry? How to classify? What are its characteristics?
2. What is environmental science? What is its study field and tasks?
3. Which part composite environmental chemistry?
4. What is the environmental pollution?
5. What are the major hazards in the history of the event?
6. In recent years, which environmental pollution events that has attracted the attention of the world?
7. How does people's understanding of environment develop?
8. What is the solution to the main problems we are facing to prevent and control environmental pollution in China?

Chapter 2
Atmosphere Environmental Chemistry

Key Point: The vertical structure and distribution structure of atmosphere; The characteristics of stratosphere and troposphere; Radiation balance of earth and global warming; Free radical reaction and the formation mechanism of photochemical smog; Acid deposition chemical mechanism; Urban air quality evaluation method.

Section 1 Evolution of Earth Atmosphere

The earth we living on is surrounded by thick atmosphere. There still remains attenuated atmosphere at a height of 1000km above surface of the earth, but 99% of the atmosphere is near surface with the height in the range of 30km. The components of atmosphere are nitrogen and oxygen, which is one of the most principal conditions that higher organism appear and exist on this planet. According to space exploration, the components of atmosphere on earth are very special in the nine planets in the solar system. The components of Venus and Mars are completely different from Earth's. Venus is 90 times the atmospheric density of Earth and 90% of it is carbon dioxide nearly without any oxygen. The surface temperature of Venus is extremely high, even up to 750K, due to the strong greenhouse effect caused by high-concentration of CO_2, while the atmosphere of Mars is very thin and its density is no more than 5% of Earth's. Its main component with 80% proportion is CO_2. As the atmosphere of Mars is very thin, its surface temperature is lower than Earth's and has a bigger diurnal variation.

If only considering spatial parameter like the location, track and factors like its own size and quality, we can calculate the components of Earth's atmosphere theoretically from Venus's and Mars's by interpolation method. It has huge difference from what it actually is. Its actual density of CO_2 is 1000 times lower than its theoretical one, while the density of oxygen is 700 times higher and the density of N_2 is 26 times higher. How does this unique atmosphere form and how does it maintain itself ? It was a question confusing scientists for a long time. Currently, a well-acknowledged view says the appearing of biosphere on Earth and its development changed the primitive elements of atmosphere. And this biosphere is still in control of the basic components of Earth's atmosphere.

Earth is a very unique planet, which can bear lives completely relying on her own special conditions. First, her quality is modest. If it is too small, it can't keep the atmosphere. If it is too large, it might cause plenty impacts of comets. Second, Earth has its autorotation and produces magnetic field. The existence of magnetic field can dispel high-energy ray (solar wind), which is beneficial to protecting lives. Third, Moon is the satellite of

Earth and a spinning braking system that will lessen the rotation of Earth and furthermore, reduce the frequency of volcano eruption and earthquake on Earth. Fourth, there is water on Earth and it is the origin of life and creates existing condition for the birth of any lives. Fifth, there is also steam and carbon dioxide which lead greenhouse effect. It decreases the diurnal amplitude and helpful for the surviving of lives. Sixth, the transformation of atmosphere leads the increase of the proportion of oxygen that, with the help of ultraviolet irradiation, comes into the formation of ozone layer and also absorbs left ultraviolet which is harmful. It protects animals and advances the process of transformation of atmosphere.

Today, the transformation of atmosphere has been through a long process. When the earth begins to take shape, it is still the state of molten mass and its main elements are H_2, He, $H_2O(g)$, N_2 and CO_2. With its gradual cooling solidification, the surface volcanic activity becomes very frequent and the atmosphere also changes aggressively. At this point, its main elements are H_2, CH_4, NH_3, $H_2O(g)$, N_2, Ar and H_2S. It is still a reducing gas. Due to the proper sun earth distance, Earth maintains modest land surface temperature after forming solid core. With proper surface scale, there appears liquid water in a certain state.

Before the water appearing, the life can't survive because of the strong ultraviolet irradiation. After the formation of hydrosphere, the shortwave of ultraviolet is absorbed 10m under water. The solar irradiation ($\lambda > 290$nm) of on which life survives still keep strong and helps the water absorbs some reducing gas and produce simple organic matter with the assistance of light energy. This is the beginning of life.

In this stage, only the photolysis of water can produce a little oxygen.

$$H_2O(g) + h\nu \longrightarrow H_2 + 1/2O_2 \tag{2-1}$$

The density of oxygen in the atmosphere, at the moment, is very low—only 1% of the amount today. Thus, these primitive lives can only gain energy from the fermentation process of organic matter after being through chemical and photochemical reaction. Then, they can have sufficient energy through carbohydrate-manufactured photosynthesis and release O_2.

$$CO_2 + H_2O + h\nu \longrightarrow \{CH_2O\} + O_2(g) \tag{2-2}$$

Therefore, the density of oxygen begins to increase till 600 million years from now, and reach the current amount of 1%. People see this percentage as the first principal density of the biological course. Under this density, ozone concentration of atmosphere will obviously increase based on the following reaction.

$$O_2 + h\nu(\lambda < 240\text{nm}) \longrightarrow O + O \tag{2-3}$$

$$O_2 + O + M \longrightarrow O_3 + M \tag{2-4}$$

The reaction of (2-4) is named *three-body reaction*. M in the chemical equation is the third part. Its function is to absorb excess energy in the reaction in order to stabilize the product. M can be any massive matter in the system and usually it is N_2 or O_2. Meanwhile, there is also the removing reaction of O_3.

$$O_3 + h\nu(\lambda < 290\text{nm}) \longrightarrow O_2 + O \tag{2-5}$$

$$O_3 + O \longrightarrow 2O_2 \tag{2-6}$$

Although the generated ozone is not adequate, the ultraviolet from sunshine has reduced largely when getting the water surface; thereby life can survive on the water surface and release a large amount of O_2 through photosynthesis. By 400 million years ago from now, the density of oxygen has been a tenth of the one today. It is the second key point of the biological course and ozone density in the atmosphere also significantly increases and gradually forms a relatively stable ozone layer with the height of 15 to 30km above the ground.

The shielding effect of ozone layer on ultraviolet irradiation brings a huge change on the appearance of earth. Life develops from ocean to land and green plants begin to grow in a fast speed. O_2 produced by photosynthesis enormously increases as well and becomes one of the main components of atmosphere now. Atmosphere then transfers into an oxidative gas.

To conclude the whole process of atmosphere evolution, we can summarize three features:

① The whole evolution direction is that lighter gas like H_2 and He will slowly escape from the atmosphere. The content of CO_2 will dramatically decrease and the content of O_2 will be on the opposite. N_2 is very stable and the essential element as well as the main nutrient element of life. It accumulates gradually and eventually tends to be stable. Modern atmosphere is a relatively stable system.

② The law of evolution is to change the original reducing atmosphere into the modern oxidizing atmosphere.

③ The specific conditions of the earth put a good basis for the birth of life. The emergence of life and the development of the biosphere have played an important role in the evolution of the earth atmosphere, and they are still controlling and maintaining the "special" stability of the atmosphere.

Section 2 The Vertical Structure and Distribution Characteristics of the Atmosphere

The total mass of the earth atmosphere is 5.2×10^{15} tons, which is equal to a millionth of the earth's mass itself. The density of the atmosphere decreased with the increase of the height since the effect of gravity.

It was found that the temperature, chemical constitution and other properties of the atmosphere showed different variations in different height. Hence, people divided the atmosphere into several levels of which the thermal stratification method is the most common, according to vertical distribution characteristics of temperature. The thermal stratification method normally divided the atmosphere into five layers: Troposphere, Stratosphere, Mesosphere, Thermosphere, Exosphere (as shown in Figure 2-1).

1. Troposphere

The troposphere is the closest layer to the surface of the earth, whose upper bound will change with the latitude and season. The depth of the troposphere is about 16-18km near the equator, and 10-12km or so in mid-latitude, while merely 8-9km near the poles. It is usually thicker in summer and thinner in winter, and these are the most distinguishing features of the troposphere:

(1) The temperature decreases with the increase of height, about 0.6℃ every 100 meters

The temperature of the air, which is closed to the surface, is higher due to absorbing the terrestrial surface radiation heat. The higher the air is, the lesser the absorption is as well as the lower temperature is. But the lower part will swell and go up when heated, while the higher part will go down when cooled, thus, the strong convection movement in the vertical direction formed, named the Troposphere.

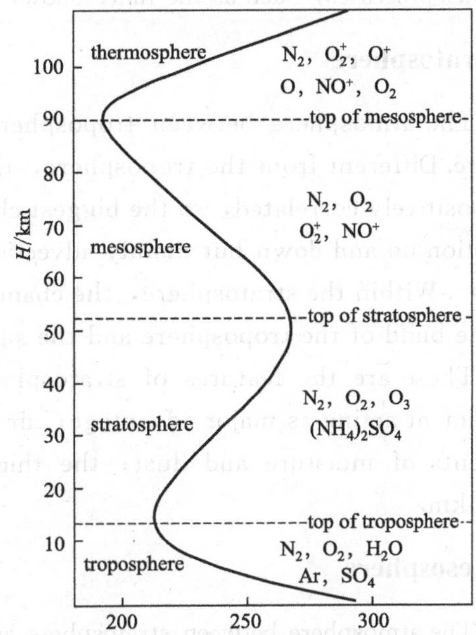

Figure 2-1 The thermal stratification structure

This kind of convection of the atmosphere is very beneficial to the dilution and diffusion of the discharging pollutants from ground. But in the night, the temperature of the ground drops, and it easily leads to the cold under hot, which is called "temperature inversion". This phenomenon usually happens in different degrees in the early morning, yet the temperature of the bottom atmosphere will increase after the sunrise, and the "temperature inversion" will disappear soon. However, when it comes to cold, cloudy or calm windy day, the situation will last a longer time.

Pollution incidents are likely to happen under the condition of inversion because the pollutants emitted into the air can hardly disperse.

(2) Large density, rich water vapor content and various weather phenomena

Troposphere is just a small part of the atmosphere tell from the thickness, but its mass takes up 75% of the whole atmosphere's, and the water vapor is accounted for more than 90%. As for the temperature of the top of the troposphere has dropped to 50℃ below, the water vapor from the ground had already condensed into ice before they arrived here. Thus, the moisture in the atmosphere almost gathered in the troposphere.

The troposphere is often separated into two parts in the practical research due to the different impact of the surface condition.

① The mechanical and thermal acts between the atmosphere and the objects are aggressive under 1-2km from the surface, which is usually called the friction layer or boundary layer. Most of the pollutants in the atmosphere move in this layer, and the buildings, forests, mountains, etc. in the surface have a big effect to the diffusion and transport of the pollutants in atmospheric.

② The free atmosphere which is more than 1-2km above the surface is less influenced. Main weather processes, such as the rain, snow, hail and so on, are formed in this layer.

2. Stratosphere

The atmosphere between troposphere and about 50km from it is called the stratosphere. Different from the troposphere, the temperature of the troposphere and the height are positively correlated, so the biggest characteristic is that the air movement is little in the direction up and down but mainly advection. One of the most key chemical species is ozone (O_3). Within the stratosphere, the change of the ozone layer will make a significant impact on the build of the troposphere and the surface of the biosphere.

These are the features of stratosphere: no air convection movement, the advection movement acting as major advantage; air is much thinner than the troposphere; it has low contents of moisture and dust; the thickness of an ozone layer is 20km in a scale of 15-35km.

3. Mesosphere

The atmosphere between stratosphere and 80km upper from it is called mesosphere. The air here is much thinner, and the temperature decreases with the increase of height since seldom does there appear any capable material absorbing radiation.

The features of the mesosphere include the temperature-decreased along with the increase of height, and the temperature of the top layer can reach approximately −92℃. The vertical temperature distribution is similar to the troposphere.

4. Thermosphere

The thermosphere refers to the atmosphere 80-500km from the ground. The oxygen in this layer can strongly absorb the energy of the ultraviolet irradiation of the wavelength less than 180nm, and then make the temperature increase sharply as the height increases. The temperature within the layer is very high and the change between day and night is large.

5. Exosphere

Exosphere is the atmosphere above the thermosphere, the air is extremely thin and the density is close to the space density, so it is also called outer atmosphere.

In each layer above, the troposphere and the stratosphere have the closest relations with human, namely the parts less than 50 kilometers away from the height, which is also known as the lower atmosphere.

Section 3 Radiation Balance of the Earth

1. The relations between the maximum of radiation wavelength and the temperature

The sunlight, which is incident to the surface of the earth, can be seen as parallel beam. The total intensity of the sunlight incident outside the atmosphere can be named as the solar constant S_0. Solar constant is defined as the amount of light received per unit area that in a plane perpendicular to the direction of light propagation, whose average value is 1368 W/m^2.

The temperature of the sun's surface is about 6000K, and the heated gas emits energy in the formation of electromagnetic radiation. The solar spectrum includes almost the whole electromagnetic spectrum, within which the infrared part accounts for 50%, and the visible light accounts for about 41%, while the rest 9% are X-ray, γ-ray, and ultraviolet.

The sun provides the earth a huge amount of energy input every day, at the same time, the earth also gets rid of this kind of energy at the same speed, therefore, it can keep a constant temperature range on the earth. The earth loses energy in the form of light radiation. Of course, the earth cannot shine like the sun. Hot objects release radiation at a certain wavelength range. With the increasing temperature, the radiation wavelengths would drift in the direction of short wave. This is the reason why an iron is red in the oven while turns to white when the temperature increases. The wavelength released by the earth is so long that we cannot see with naked eyes.

The distribution of the solar and the terrestrial radiation spectrum is as shown in Figure 2-2. The two normative curves are expected to be the "blackbody radiation" curves change with the variation of the temperature in the sun and the earth. The black body spectrum is smooth, but the actual spectrum from the sun and the earth is rugged, since the particular atomic and the molecular jump make contribution to the emission.

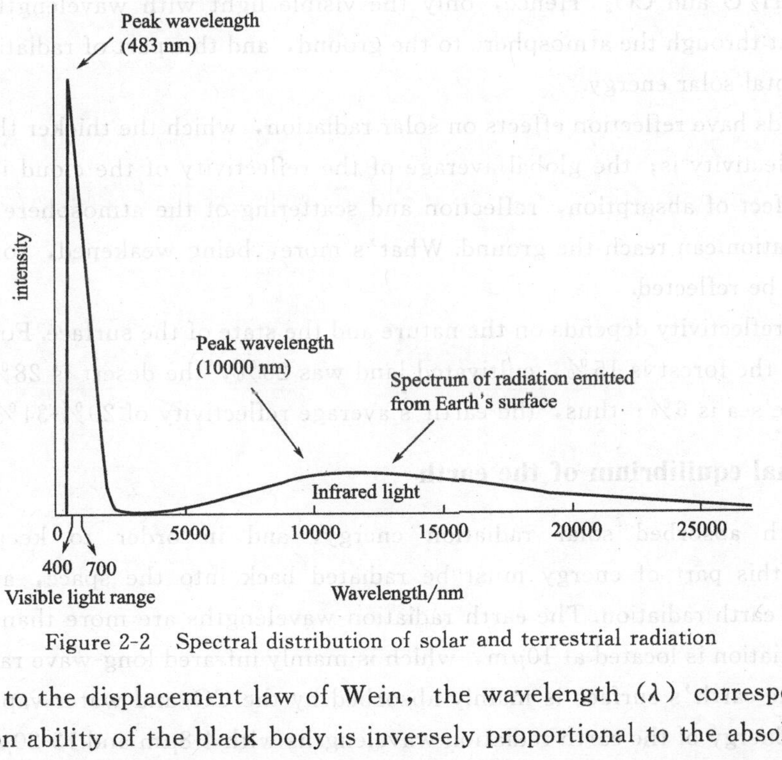

Figure 2-2 Spectral distribution of solar and terrestrial radiation

According to the displacement law of Wein, the wavelength (λ) corresponding to the biggest radiation ability of the black body is inversely proportional to the absolute temperature (T), its mathematical expression is:

$$\lambda_{peak}(nm) = 2.9 \times 10^6 (nm \times K)/T(K) \quad (2\text{-}7)$$

The sun is an extremely hot object whose peak wavelength is 483nm, and corresponding temperature is 6000K. Most of its radiation fell to the wavelength region between 400nm and 700nm, which is visible light region. The reason why it's visible is that our eyes have evolved to adapt to the sun. The earth radiate outward, and its peak wavelength are 10000nm around and the corresponding temperature is 288K. Therefore, when the earth mainly absorbed the visible light radiation, which is the characteristic of the sun's surface temperature, it radiates energy in the infrared zone, which is the feature of the earth's cooling surface.

2. Absorption of solar radiation by atmospheric composition

Due to the absorption and scatter of the atmosphere, the sunlight will weaken when they get through it, which make the radiation projected into the upper atmosphere not all reach the ground. During the course that the solar radiation get to the ground through the atmosphere, the compositions such as N_2, O_2, O_3, H_2O, CO_2 and so forth can absorb the solar radiation of certain wavelength. The radiation with wavelength less than 290nm were absorbed by N_2, O_2, O_3 and then dissociated.

$$N_2 + h\nu(\lambda < 120\text{nm}) \longrightarrow N + N \quad (2\text{-}8)$$

$$O_2 + h\nu(\lambda < 240\text{nm}) \longrightarrow O + O \quad (2\text{-}9)$$

$$O_3 + h\nu(220\text{nm} < \lambda < 290\text{nm}) \longrightarrow O + O_2 \quad (2\text{-}10)$$

Therefore, the solar radiation with wavelength less than 290nm cannot get to the ground, while the long-wave radiations with wavelength between 800-2000nm were almost absorbed by H_2O and CO_2. Hence, only the visible light with wavelength between 300-800nm can get through the atmosphere to the ground, and this part of radiations account for 41% of the total solar energy.

The clouds have reflection effects on solar radiation, which the thicker the cloud is, the larger the reflectivity is; the global average of the reflectivity of the cloud is 20% around. Due to the effect of absorption, reflection and scattering of the atmosphere, only some of the solar radiation can reach the ground. What's more, being weakened, some of the solar radiation will be reflected.

And the reflectivity depends on the nature and the state of the surface. For example, the reflectivity of the forest is 15%, cultivated land was 20%, the desert is 28%, the snow is 80%, and the sea is 6%; thus, the earth's average reflectivity of 29%-34%.

3. The thermal equilibrium of the earth

The earth absorbed solar radiation energy, and in order to keep the thermal equilibrium, this part of energy must be radiated back into the space, and the process known as the earth radiation. The earth radiation wavelengths are more than 4μm, and the maximum radiation is located at 10μm, which is mainly infrared long-wave radiation. Energy radiated by the earth's surface is mainly absorbed by the CO_2 and water vapor in the lower atmosphere. Energy of the earth radiation wavelengths with 4-8μm and 13-20μm is easy to be absorbed by the CO_2 and water vapor, while the radiation of wavelength with 8-13μm was hardly absorbed, which is called "atmospheric window", and this part of long-wave

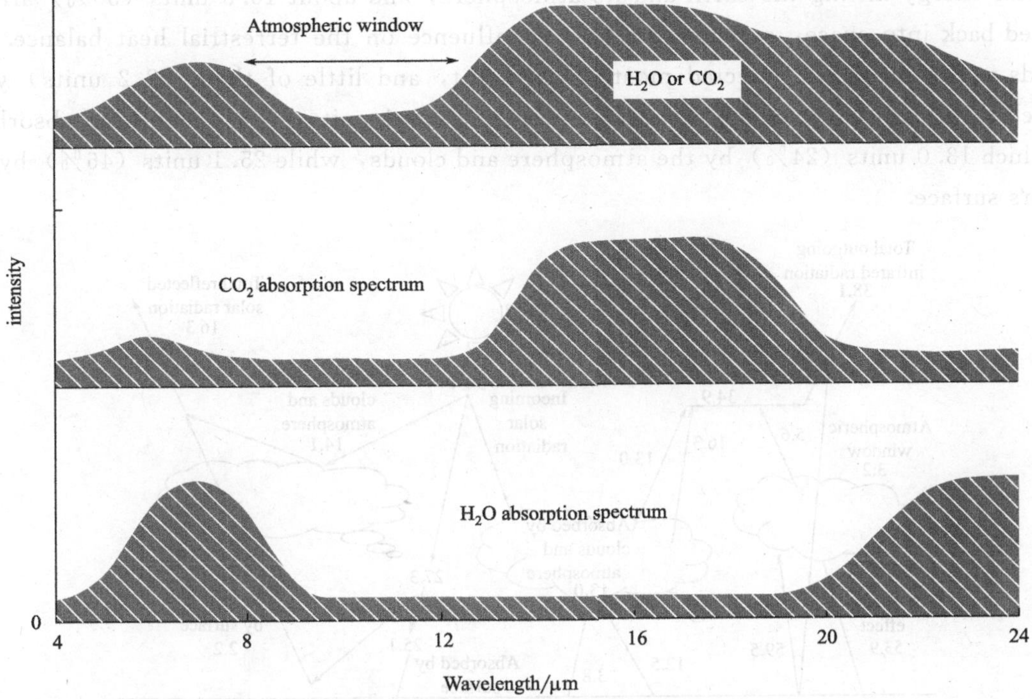

Figure 2-3 Absorption of infrared radiation by H_2O and CO_2

radiation can reach the space through the atmosphere (as shown in Figure 2-3).

After CO_2 and H_2O absorbed the energy of ground radiation, they release the energy in the form of long-wave radiation. This radiation is in all directions, while in the vertical direction there are two parts, up and down, and due to the radiation direction with the ground, the downward part, on the other hand, is called the Atmospheric Counter Radiation. Due to the atmospheric counter radiation effect, part of the terrestrial radiation will return to the ground, and cause fewer loss of calories actually than long-wave radiation. Therefore, the atmosphere of the terrestrial surface played an important role in keeping a suitable temperature range.

The average temperature of the ground remain unchanged for a long time, so we can take it reasonable that incident solar radiation and the terrestrial long-wave radiation keeps balance between income and expenditure.

Thus, the process of energy gain and loss between the earth and atmosphere is closely related to photochemistry and optical absorption of chemical species, especially the O_3, water vapor and CO_2, etc. Thus, the change of these compounds in the atmosphere will be a lot of influence on the earth's energy balance. For instance, when the concentration of the water vapor and CO_2 near-surface increased, they will absorb the ground long-wave radiation, and then form a thermal insulation blanket, which keeps the surface heat and leads to global temperature rise, affect human life and safety directly. That's so-called "greenhouse effect".

The actual energy flow get through the atmosphere is quite complex. As shown in Figure 2-4, the earth's energy input and output can reach 10^{20} kJ per year. About 54.5 units of solar energy hitting the earth and its atmosphere, and about 16.3 units (30%) are reflected back into space, so this part has no influence on the terrestrial heat balance. The clouds and atmosphere reflected most of the light, and little of them (2.2 units) were launched by the surface of the earth. The remaining 38.1 units (70%) had been absorbed, of which 13.0 units (24%) by the atmosphere and clouds, while 25.1 units (46%) by the earth's surface.

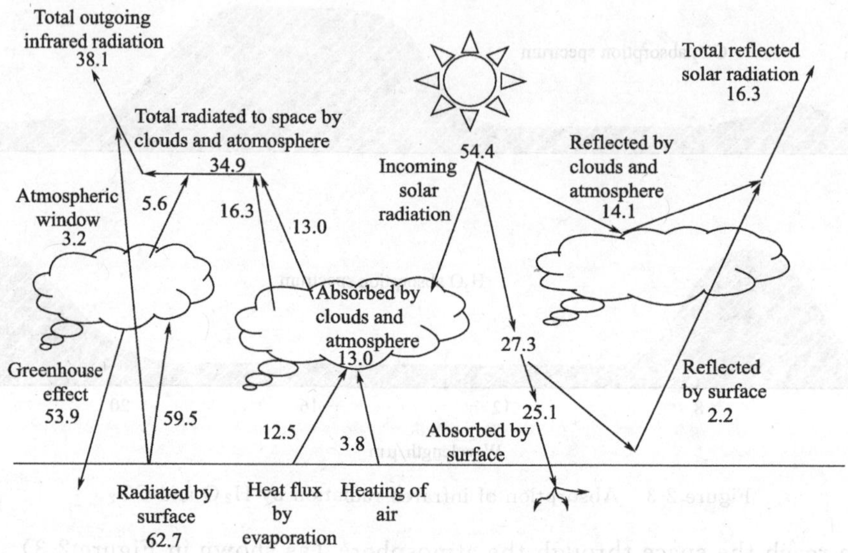

Figure 2-4 Heat balance of the earth (in units of 10^{20} kJ/year)

4. Greenhouse Effect

(1) Infrared absorption and molecular vibration

In CO_2 and H_2O molecules, there can be symmetric stretching vibration, asymmetric stretching vibration and bending vibration, shown in Figure 2-5.

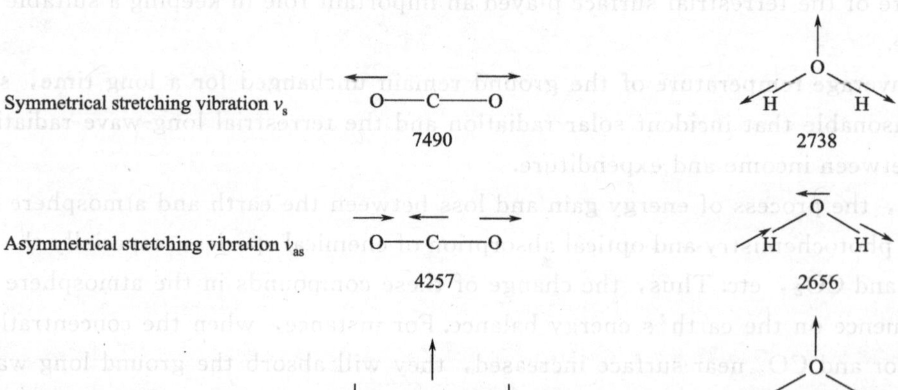

Figure 2-5 Molecular vibrations of CO_2 and H_2O

As mentioned above, the greenhouse effect is to capture low-level heat in the atmosphere. Rays from the sun through the earth's atmosphere permits, but captured from the infrared radiation emitted by the earth's surface.

The main gases in the atmosphere are monatomic molecules, and they can absorb infrared light, because they can't meet the basic conditions of electromagnetic radiation. In contrast, polyatomic molecules have a variety of kinds of vibration; some vibration can change the dipole moment and the infrared activity.

The absorption spectra of water and carbon dioxide are shown in Figure 2-3. As can be seen, water and carbon dioxide can absorb most of the light within the wavelength range 4000nm to 25000nm, but 8000 to 12000nm light cannot be absorbed then escape.

(2) The main greenhouse gas

The substances that cause the greenhouse effect are mainly CO_2, CH_4, N_2O, CFCs, shown in Table 2-1. And the range of absorption wavelength about infrared radiation of greenhouse gas compounds CFCs are shown in Figure 2-6. The sources of main greenhouse gases are as follow.

Table 2-1 Greenhouse gases in the atmosphere

Gases	Concentration/($\mu L/m^3$)	Average annual growth rate/%
CO_2	340000	0.4
CH_4	1650	1.0
CO	304	0.25
$C_2H_2Cl_2$	0.13	7.0
O_3	—	—
CFC-11	0.23	5.0
CFC-12	0.4	5.0
CCl_4	0.125	1.0

① The CH_4 in the atmosphere mainly comes from the wetland, cattle, rice paddies and so on. The concentration of CH_4 is only after CO_2, and the greenhouse effect caused by CH_4 is 20-30 times bigger than CO_2.

② For the soil of China's farmland, the emission of CO_2 is 260×10^6 t, 17.5×10^6 t in CH_4, 9.6×10^6 t in N_2O, accounted for 8%, 50% and 10% of total emissions.

③ CFC is also a type of greenhouse gas, accounted for 25% of the greenhouse effect. CFC-11 and CFC-12 mainly absorb the radiation between 800 cm^{-1} and 2000cm^{-1}; the greenhouse effect produced by every CFC-12 molecules is equivalent to 15000 CO_2 molecules.

④ The strongest material cause the greenhouse effect is CF_3SF_5, which is a new discovery. The greenhouse effect produced by a CF_3SF_5 molecule is equivalent to 10^5 CO_2 molecules do.

(3) The Advantages and Disadvantages of the greenhouse effect

With the climate becomes warmer, greenhouse gases have such features: let the sun through, but strongly absorb the long wave radiation which emitted from the ground, so the temperature rose, and play a role of greenhouse. In 20th century, the average global temperature increased by 0.5 ℃, If greenhouse gases increase at the current rate, By 2030, the global average temperature will increase by 2-3 ℃, and severe weather and extreme weather will become more frequent. Now, global warming is one of the main topics of global environmental research. According to more than 100 analysis of global change information systems analysis, it has found that the global average temperature has risen 0.3 to 0.6 degrees Celsius. Where the 11 warmest years were in the mid-1980s, therefore, there is no doubt that global warming is a fact. Global warming will bring very serious consequences, such as the retreat of glaciers, rising sea levels, desertification, and bring severely effect on ecosystems and agricultural production.

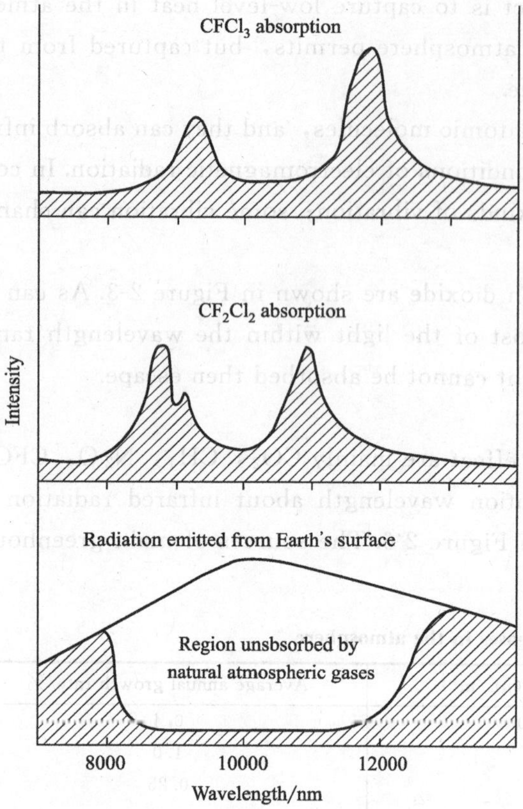

Figure 2-6 The range of absorption wavelength greenhouse gas compounds CFCs

Therefore, exploring the causes of global warming has become an important research topic. Analysis shows that although the history of the evolution of the Earth has happened warmer-colder climate fluctuations on many occasions, the growth of atmospheric greenhouse effect caused by human activities may be the main factor.

Section 4 Tropospheric Atmospheric Chemistry

1. The classification of atmospheric composition

Atmosphere mainly composed of nitrogen, oxygen and several of inert gas, and they accounted for 99.9% of the total atmosphere. In addition to the gas, the atmosphere also suspends a large number of solid and liquid particles. According to the length of residence time, atmospheric composition can be divided into three categories: ① Quasi-permanent gas: N_2, Ar, Ne, Kr, Xe. ② The variable component: CO_2, CH_4, H_2, N_2O, O_3, O_2. ③ Strong variable component: H_2O, CO, NO, NH_3, SO_2, Hydrocarbons (HC), Particulate Matter, H_2S. Troposphere clean air composition is shown in Table 2-2.

Strong variable components of the atmosphere, mainly from anthropogenic sources, are followed by natural sources. The variable component and Strong variable component stays in a short time the atmosphere, and they may be involved in the stratosphere or troposphere chemical changes. Ground Source affects their distribution of time and space in the atmosphere, in different height or regions, which often distributed very differently. For example, In the atmosphere over smelters, power plants contains more soot, SO_2, NO_x and other strong variable component; In the atmosphere surrounding area contains more organic or inorganic chemical substances; When these substances reach a certain concentration in the atmosphere, it is possible to produce local air pollution.

Table 2-2 Composition and the cycle of atmospheric gases

Gases	Average concentration /(mL/m³)	Approximate residence time/a	Cycle
N_2	780840	10^6	Biology and microbial
O_2	209460	10	Biology and microbial
Ar	9340	—	No cycle
Ne	18	—	No cycle
Kr	1.1	—	No cycle
Xe	0.09	—	No cycle
CH_4	1.65	7	Biological activities and chemical process
CO_2	332	15	Human activity and biological activity
CO	0.05~0.2	65/d	Human activity and biological activity
H_2	0.58	10	Biological activities and chemical process
N_2O	0.33	10	Biological activities and chemical process
SO_2	10^{-5}~10^{-4}	40/d	Human activity and chemical process
NH_3	10^{-4}~10^{-3}	20/d	Biological activities and chemical process, below cloud scavenging
$NO+NO_2$	10^{-6}~10^{-2}	1/d	Human activity, chemical process, lightning
O_3	10^{-2}~10^{-1}	—	Chemical process
HNO_2	10^{-5}~10^{-3}	1/d	Chemical process, below cloud scavenging
H_2O	Alterable	10/d	Physical chemistry process
He	5.2	10	Physical chemistry process

2. The sources, sinks of atmospheric composition and the atmosphere cycle

Quasi-permanent gas concentrations in the atmosphere is unchanged, but the whole atmosphere is still a dynamic system; Atmospheric constituents can occur physical, chemical or biochemical processes between the atmosphere and the other by three circles, the ongoing exchange or conversion of the substance that constitutes the so-called "gas circulation".

Gas generation process is called source gas, which comprises, as the combustion of the fuel in the atmosphere of chemical processes, biological activity, volcanic eruptions, and human activities. Removing gases from the atmosphere of the process such as chemical and biological activities, and other physical processes is to be removed sinks gas. For example, the water vapor in the atmosphere, mainly from the evaporation of sea water, a small amount from the rivers, lakes and water evaporation and biosphere, soil and plant transpiration; atmospheric water vapor has condensed form may encounter rain, snow and other precipitation back to the surface, This constitutes the atmospheric water cycle. Due to transpiration is constantly, and the precipitation is intermittent, it resulted in the concentration of water vapor in the atmosphere varies with location.

The atmosphere is rich in nitrogen, but cannot be used by plants directly. Nitrogen-fixing bacteria, blue-green algae, lightning effects can be converted to nitrate nitrogen in the air, and plants absorb nitrates and ammonium from the soil into various amino acids in the body, Then synthesis various proteins. By eating plants, animals obtain nitrogen and protein into the food chain. After the death of animals and plants, of which the protein is broken down into ammonium return to soil organisms, part is directly absorbed by plants; Another portion is gradually transformed into nitrite bacteria and nitrate, which can be absorbed by plants. Denitrified bacteria can also be converted to N_2O, or produced by denitrified N_2; and nitrite to nitrogen effects can also be converted into N_2 or NO by chemical; Soil ammonium salts may also be converted into NH_3 and into the atmosphere. Thus, the nitrogen is circulated through inorganic, organic and microbial action. Other components of atmospheric oxygen also have its own cycle.

3. The important light absorption materials in the atmosphere

The materials of atmosphere can absorb in the 290-700nm wavelength of light that basically has NO_2, O_3, SO_2, HNO_2, RONO, RCOR, $RONO_2$, HNO_3, H_2O_2, HCHO, CH_3CHO, CH_3COCH_3, etc., and in the troposphere, in addition to the SO_2, The rest of the material can produces photolysis reactions.

(1) The Optical absorption properties of NO_2

NO_2 is the most important optical absorption molecular of urban atmosphere, in the lower atmosphere; it can absorb the full range of visible and ultraviolet irradiation from the sun. NO_2 photolysis is the most important photolysis reactions in the troposphere.

$$NO_2 + h\nu(\lambda \leqslant 420nm) \longrightarrow NO + O(^3P) \tag{2-11}$$

$$O(^3P) + O_2 + M \longrightarrow O_3 \tag{2-12}$$

In the troposphere atmospheric, NO_2 photolysis is the only man-made source of ozone.

$\lambda < 380nm$ $\varphi_{NO} \approx 1$

$\lambda = 405nm$ $\varphi_{NO} \approx 0.36$

$\lambda > 420nm$ $\varphi_{NO} \approx 0$

In more than 420 nm wavelength light, it can only transfer to NO_2 excited states NO_2^*, it can not continue to make photo-dissociation.

(2) The photolysis reactions of O_3, HNO_2 and H_2O_2

$$O_3 + h\nu(\lambda \leqslant 320nm) \longrightarrow O(^1D) + O_2 \tag{2-13}$$

$$O(^1D) + H_2O \longrightarrow 2 \cdot OH \tag{2-14}$$

On the type, $O(^1D)$ as the excited oxygen, The above reaction is the most important formation way of ·OH radical that is one of six clean gases.

The photolysis reactions of HNO_2 and H_2O_2 are also an important source of ·OH radicals in the atmosphere, but both are relatively more important in the atmospheric pollution.

$$HONO + h\nu(\lambda < 400nm) \longrightarrow \cdot OH + NO \quad (2\text{-}15)$$

$$H_2O_2 + h\nu(\lambda < 360nm) \longrightarrow \cdot OH + OH \quad (2\text{-}16)$$

(3) The photolysis reaction of HCHO and CH_3CHO

It is the main source of $HO_2\cdot$ radicals in troposphere atmosphere.

(4) The optional absorption of SO_2

$$SO_2 + h\nu(290nm < \lambda < 360nm) \longrightarrow {}^1SO_2\text{(Single excited state)} \quad (2\text{-}17)$$

strong absorption

$$SO_2 + h\nu(340nm < \lambda < 400nm) \longrightarrow {}^3SO_2\text{(Triplet excited State)} \quad (2\text{-}18)$$

weak absorption

An electronic in the SO_2 is excited to a higher energy level and there have two unpaired electrons to spin in the same direction, thus the SO_2 molecule stays in triplet state.

Because of the SO—O bond energy is 5648kJ/mol, It corresponds to the wavelength of 218nm, so that only the $\lambda <$ 218nm light quantum can make SO_2 molecules to dissociate.

$$SO_2 + h\nu(\lambda < 218nm) \longrightarrow {}^3SO + O({}^3P) \quad (2\text{-}19)$$

But the reaction in the troposphere won't happen, tropospheric photochemical reaction of SO_2 is not photolysis but free radical reaction.

It shows that there are two kinds of optical absorption material in the atmosphere. One way in the aftermath of the absorption of optical is dissociation. The free radicals generated by the reaction make active atmospheric chemical processes. Another is that the stimulation does not occur after dissociation reaction; instead it depends on free radical reaction and then the transformation of primary pollutants to secondary pollutants, such as SO_2.

4. Atmospheric radical

Thc atmosphere is an oxidizing atmosphere, the basic characteristics of atmospheric chemistry is oxidized, People even chemical processes occurring in the atmosphere is considered as low-temperature combustion flame, the flame radicals are accelerant, which is to promote the power of atmospheric chemical reactions. Important common atmospheric radicals: •OH (hydroxyl radical), $HO_2\cdot$(hydroperoxy radical), R•(alkyl radical) and RO• (alkoxy radicals). The most important is •OH radical.

(1) Sources of •OH radicals

•OH radicals in a clean atmosphere to form the most important ways from photolysis of O_3

$$O_3 + h\nu(\lambda \leqslant 320nm) \longrightarrow O({}^1D) + O_2 \quad (2\text{-}20)$$

$$O({}^1D) + H_2O \longrightarrow 2\cdot OH \quad (2\text{-}21)$$

In contaminated atmosphere, the Photolysis reaction of HNO and H_2O is relatively more important.

$$HONO + h\nu(\lambda < 400nm) \longrightarrow \cdot OH + NO \quad (2\text{-}22)$$

$$H_2O_2 + h\nu(\lambda < 360nm) \longrightarrow \cdot OH + \cdot OH \quad (2\text{-}23)$$

From •OH sources we can see that it generates directly related to solar irradiation. Therefore, it is characterized by the distribution of atmospheric concentrations: daytime than at night, in summer more than winter, tropical over the highest in the world. Global average density is 7×10^5

Pieces/ cm^3.

(2) The main source of $HO_2\cdot$ radicals

Atmospheric photolysis of formaldehyde:

$$HCHO + h\nu \longrightarrow H\cdot + HCO\cdot \quad (2\text{-}24)$$

$$H\cdot + O_2 + M \longrightarrow HO_2\cdot \quad (2\text{-}25)$$

$$HCO\cdot + O_2 \longrightarrow HO_2\cdot + CO \quad (2\text{-}26)$$

As can be seen from the above reaction, the reaction, which is generating any H atom or a carboxylic acid group, is the source of tropospheric $HO_2\cdot$. Therefore, acetaldehyde also is one of the sources of $HO_2\cdot$, but because of lower concentrations in the atmosphere, as not formaldehyde important.

In addition, the study also found that the source of $HO_2\cdot$ at night, and is directly related to the presence of $NO_3\cdot$ radicals.

$$HCHO + NO_3\cdot \longrightarrow HNO_3 + HCO\cdot \quad (2\text{-}27)$$

$$HCO + O_2 \longrightarrow HO_2\cdot + CO \quad (2\text{-}28)$$

From the sources and distribution of these free radicals, we can see their relative importance at both day and night.

(3) Sources of $R\cdot$, $RO\cdot$, $RO_2\cdot$

The highest level of alkyl radical content is in the methyl ($\cdot CH_3$) . It is mainly from the photolysis of ethanol and acetone.

$$CH_3CHO + h\nu \longrightarrow \cdot CH_3 + HCO\cdot \quad (2\text{-}29)$$

$$CH_3COCH_3 + h\nu \longrightarrow \cdot CH_3 + CH_3CO\cdot \quad (2\text{-}30)$$

Furthermore, O_2 and $HO\cdot$ removal may have extraction with HCs in the atmosphere and generate $R\cdot$.

$$RH + O_2 \longrightarrow R\cdot + HO_2\cdot \quad (2\text{-}31)$$

$$RH + HO\cdot \longrightarrow R\cdot + H_2O \quad (2\text{-}32)$$

CH_3O in the atmosphere is mainly form methyl nitrite and its photolysis reaction.

$$CH_3ONO + h\nu \longrightarrow CH_3O\cdot + NO\cdot \quad (2\text{-}33)$$

$$CH_3NO_3 + h\nu \longrightarrow CH_3O\cdot + NO_2\cdot \quad (2\text{-}34)$$

The peroxy alkyl in the atmosphere is a combination of peroxy alkyl and O_2 in the air.

$$R\cdot + O_2 \longrightarrow RO_2\cdot \quad (2\text{-}35)$$

(4) Collection of radicals

① Mutual conversion between radicals

$$CO + \cdot OH \longrightarrow CO_2 + H\cdot \quad (2\text{-}36)$$

$$\cdot H + O_2 \longrightarrow HO_2\cdot \quad (2\text{-}37)$$

$$CH_4\ O_2\cdot \longrightarrow CH_3\cdot + H_2O \quad (2\text{-}38)$$

$$CH_3\cdot + O_2 \longrightarrow CH_3O_2\cdot \quad (2\text{-}39)$$

$$HO_2\cdot + NO\cdot \longrightarrow NO_2 + \cdot OH \quad (2\text{-}40)$$

$$HO_2\cdot + O_3 \longrightarrow 2O_2 + \cdot OH \quad (2\text{-}41)$$

② The combined effects

$$HO_2\cdot + \cdot OH \longrightarrow H_2O + O_2 \quad (2\text{-}42)$$

$$\cdot OH + \cdot OH \longrightarrow H_2O_2 \quad (2\text{-}43)$$

In addition, there is a radical transformation and removal of many other ways; here are just a few representative examples.

5. Tropospheric photochemical processes Clean Air

(1) Basic photochemical cycle of NO-NO_2-O_3

Assuming no other atmosphere organic components, only nitrogen oxides (NO, NO_2) exists between the NO-NO_2-O_3 will be under the sun, constitutes a steady-state optical system, first of NO_2 photolysis generate atomic oxygen (O), atomic oxygen can rapidly combine with oxygen molecules to generate ozone, and ozone is generated once, then quickly react with NO to generate NO_2.

$$NO_2 + h\nu(\lambda = 300 \sim 400nm) \longrightarrow O + NO \tag{2-44}$$

$$O + O_2 + M \longrightarrow O_3 + M \tag{2-45}$$

$$O_3 + NO \longrightarrow O_2 + NO_2 \tag{2-46}$$

Generally believed atmospheric O_2 concentration is constant, then in the above reaction, involving a total of four species, i. e. , NO_2, NO, O and O_3. Kinetic equation:

$$d[O]/dt = k_1[NO_2] - k_2[O][O_2][M] \tag{2-47}$$

Since the O atom is very unstable and reacts immediately upon generating the disappearance occurred with atmospheric O_2, and therefore the steady state approximation method can be used to process, i. e. , $d[O]/dt = 0$, thus,

$$k_1[NO_2] = k_2[O][O_2][M] \tag{2-48}$$

Thus O_3 steady-state concentration is obtained:

$$[O_3] = k_1[NO_2]/k_3[NO] \tag{2-49}$$

Equation (2-55) illustrate the absence of other reactions involved in the atmosphere, the O_3 concentration depends $[NO_2]/[NO]$. But thus calculated O_3 concentrations in the atmosphere than the actual measured results is much smaller, indicating that in the atmosphere, there are still other reactions must make O_3 concentration, and can react with the reactive (2-52) competition. The study found that these reactions mainly from carbon containing compounds.

(2) Clean basic chemical processes in the atmosphere

Anthropogenic sources in areas are away from the relatively clean air, composed of relatively simple and low concentrations. However, the findings show that these clean natural gases are still a series of chemical and photochemical reactions, where the key species are CH_4, CO, O_3 and NO_x.

Trigger natural tropospheric chemistry is photolysis of O_3. O_3 light into $O(^1D)$, and then generates two OH and reacts with H_2O, triggering a series of other reactions.

$$CH_4 + \cdot OH \longrightarrow CH_3\cdot + H_2O \tag{2-50}$$

$$CO + \cdot OH \longrightarrow CO_2 + H\cdot \tag{2-51}$$

H, CH_3 and O_2 in the air combine to produce peroxy radicals.

$$H\cdot + O_2 + M \longrightarrow HO_2\cdot + M \tag{2-52}$$

$$CH_3\cdot + O_2 + M \longrightarrow CH_3O_2\cdot + M \tag{2-53}$$

Peroxy radicals will clean NO in the atmosphere and convert it to NO_2, while generating $\cdot OH$ and $HO_2\cdot$ radical groups, which play the role of chain transfer.

$$CH_3O_2\cdot + NO \longrightarrow NO_2 + CH_3O\cdot \quad (2\text{-}54)$$

$$HO_2\cdot + NO \longrightarrow NO_2 + \cdot OH \quad (2\text{-}55)$$

$$CH_3O\cdot + O_2 \longrightarrow HCHO + HO_2\cdot \quad (2\text{-}56)$$

The main chain termination reaction is

$$NO_2 + \cdot OH \longrightarrow HNO_3 \quad (2\text{-}57)$$

$$HO_2\cdot + HO_2\cdot \longrightarrow H_2O_2 + O_2 \quad (2\text{-}58)$$

As seen above, through a chain reaction mechanism, free radicals in the atmosphere oxide compounds containing C and N from a low-value state to a high state. The high state of matter will eventually be absorbed by precipitation or removed by surface.

Section 5 Photochemical Smog

1. The emergence of photochemical smog

Atmosphere containing NO_x and VOCs in the strong ultraviolet irradiation will occur a series of complex chemical reactions. It will produce some highly oxidizing substances, such as ozone (O_3)、aldehydes (RCHO)、peroxyacetyl nitrate (PAN)、nitrous acid (HNO_2) and the combination is known as Photochemical Smog.

Later in 1936, first Photochemical Smog appeared in the US city of Los Angeles smog. It is characterized by pale blue smoke. With strong oxidizing, human eyes and respiratory tract have a strong stimulating effect-harming plant leaves (white withered), resulting in the rubber cracking, and reducing atmospheric visibility.

Due to strong sunlight trigger chemical reactions in the atmosphere exists between hydrocarbons and nitrogen oxides caused. Urban atmosphere, the two main sources of precursors is automobile exhaust. Following the Los Angeles photochemical smog pollution around the world continue to emerge. 1974 Lanzhouxigu in our region also observed photochemical smog.

2. Daily changing law of photochemical smog

For the modern metropolis, the maximum concentration of CO, NMHC (non methane hydrocarbon) and NO was 8:00 in the morning and afternoon 18:00. The two peaks corresponding to the time is the busiest traffic phase during people to and from work. As the sun gradually enhancing, the concentration of NO and NMHC gradually decline and the concentration of NO_2 is rising gradually and reach the peak. At the same time, O_3 concentration increased rapidly to the maximum.

3. Forming mechanism of photochemical smog

From the above daily variation characteristics of pollutants, it can be preliminarily estimated that the NO_2 and O_3 is produced by photochemical reaction under the sunlight through the atmosphere, rather than directly by a pollutant source emission.

In order to make clear the mechanism of each species concentration change over time, smoke box experiment is designed. In a large container, atmospheric chemical reactions are simulated under the irradiation of artificial light (xenon lamp). The results showed that NO change into NO_2 under the effect of ultraviolet light when NMHC and NO_x coexistence. NMHC is consumed through oxidation. At the same time, O_3, PAN and other secondary pollutants are produced.

This simplified mechanism can be used to explain the daily changing law of photochemical smog. At sunrise, NO_2 photolysis of O· radicals generated, and then with H_2O molecule collision ·OH radicals. ·OH radicals generated after the start of the oxidation VOCs, and produce HO_2·, RO_2· and other radicals, these two radicals in the atmosphere can be effectively oxidized from NO to NO_2, while generating ·OH, to oxidation HCs. So on ad infinitum, until the atmosphere HCs and NO is consumed. In this process, NO_2 gradually increased, and continue to produce photolysis O_3, the O· radical to accumulate, while NO_2 can also generate PAN, HNO_2 and other stable oxides react with free radicals. When NO_2 decreased to a certain extent, the amount of O_2 affected, and O_2 in the atmosphere when it is accumulated through rate and consumption rate to achieve a balance, O_3 concentration of the first solution and the reaction of NO with HCs and are constantly consumed, maximum. In the afternoon, with the weakening of daylight, NO_2 first solution is suppressed, so the reaction tends to be slow, the concentration of the product has been declining.

4. The influencing factors and the peugeot of formation of photochemical smog

The formation of Photochemical smog requires three conditions. The first condition is the presence of contaminants which such as HCs, NO_x, CO, VOCs, etc. The second condition is a temperature inversion formed when the ground temperature is below the phenomenon of high temperature. Pollutants on the ground cannot diffuse upward and strand the surface' s prone to photochemical reactions when inversion phenomenon occurs. If there is wind at the moment, the inversion phenomenon disappears. The third condition is sunlight. Meet the above three conditions, photochemical smog will happen.

Ozone which is one of the oxidant of photochemical smog formation is the most important antioxidant, accounting for more than 90%, followed by PAN (i. e., peracetic acid), both of which can be seen as a sign of the product of photochemical smog.

The background concentration of tropospheric ozone in the atmosphere is usually between 0.02-0.06μg/mL, with an average of about 0.03μg/mL. Ozone concentrations in urban atmospheric is much higher, generally in the 0.1μg/mL or more. When the photochemical smog occurs, ozone concentration in the atmosphere is up to 0.2-0.5μg/mL, which is the risk of harm to human health and survival to the concentration of biological.

Ozone major harms to human trachea and lungs, but also has an impact on the heart and brain. Small doses of contact for a long time can cause the disease such as the lung disease, lung cancer and aging. Ozone can damage the plant leaves, affect their growth and lower yield, but also crisp the fabric, paper, etc. and reduce the strength of the rubber by aging.

5. PAN

PAN contains ethyl peroxy radicals with NO_2 reaction. When R · is CH_3 ·, When the product was PAN; when R · is ethyl, propyl or phenyl when, will form the corresponding PAN series homologues.

PAN can strongly stimulate the eyes, causing tearing and inflammation; can harm plants, so that the back of a variety of plant leaves was bronze or vitrification. Studies have indicated that it is likely to cause skin cancer agents.

As can be seen from the reaction, PAN or NO molecules in a reservoir in the troposphere, you can transfer and release of NO. Molecules, not only expanded the scope of pollution, but also the potential to become another initiator of photochemical smog. In addition, PAN NO dissociation energy in the rain. And organic, acidification participate precipitation.

6. Photochemical smog pollution control

Many research results have shown that a region can be generated between the maximum of ozone concentration and its precursor is not a linear relationship. That is to say, not double atmospheric concentrations of nitrogen oxides or volatile organic compounds can generate double ozone concentration accordingly.

If the concentration of nitrogen oxides remain unchanged, generating little impact of changes in ozone concentration of hydrocarbons, but to reduce the concentration of the hydroxide results in a significant reduction in the concentration of ozone.

Far away from the city's suburbs and rural atmosphere atmospheric conditions typically have such characteristics. Due to the high $NMHC/NO_x$, that NO_x is relatively small, the generation of O_3 is mainly limited by NO_x. In this case, the photochemical smog can be controlled by use of controlling NO_x.

It should be noted that, in order to control effectively photochemical smog pollution in a city or region, people must find out the variation law of pollutants and formulate the corresponding pollution control measures according to the atmospheric conditions and characteristics of this region, including distribution and emission characteristics and weather conditions.

Section 6 Acid Deposition Chemistry

Acid deposition chemistry research began in acid rain. 1950s Europe found acidic precipitation gradually increasing trend, acid rain issue has received wide attention. As the harm of acid rain becoming bigger, the formation process is complex and the influence is widespread as well as a fairly long-term effect. It can also be transmitted over a long distance. Acid rain receives worldwide attention. States have been vigorously for the research of acid rain, acid rain monitoring sites have been established to develop a long-term research programs, international co-operation. In recent years, the dry deposition of acid found in acid rain research settlement cannot be underestimated; the environmental effects are often caused by comprehensive results of dry and wet deposition. Therefore, "acid rain" widely cited in the past has been gradually replaced by "acid deposition" .

1. The definition of acid rain and precipitation acidification process

(1) The definition of acid rain

Rainwater is generally considered "natural" acidity of pH 5. 6. This value comes from the following considerations: a factor affecting the natural pH of precipitation is only the present of CO_2 in the atmosphere, according to the global atmospheric CO_2 concentrations of 330 mL/m^3 and water balance of:

$$CO_2(g) + H_2O \xrightleftharpoons{H} CO_2 \cdot H_2O$$

$$CO_2 \cdot H_2O \xrightleftharpoons{K_1} H^+ + HCO_3^-$$

$$HCO_3^- \xrightleftharpoons{K_2} H^+ + CO_3^{2-}$$

Type: H for Henry constants of CO_2; According to the neutral principle:

$$[H^+] = [OH^-] + [HCO_3^-] + 2[CO_3^{2-}]$$

$$= K_W/[H^+] + K_1 H p_{CO_2}/[H^+] + H K_1 K_2 p_{CO_2}/[H^+]^2$$

Where: K_W ion product of water, p_{CO_2} partial pressure of CO_2 in the atmosphere, K_1, K_2, respectively $CO_2 \cdot H_2O$ primary and secondary ionization constants; clean rainwater can be calculated pH of 5. 6, so the pH is less than 5. 6 rainwater is considered to be acid rain.

Through many years of rainfall observation, in recent years has been able to pH 5. 6 as the boundaries of acid precipitation and discriminant boundary dispute of an anthropogenic pollution. Tang Xiaoyan, etc go far in the "atmospheric environment chemistry" summarize its main points are as follows. ①high clean air, in addition to CO_2 outside, there are still a variety of acid and alkaline gaseous and aerosol substances, they go through the process of cloud and precipitation erosion path into the rain, the precipitation acidity is the result of the combined effect of all kinds of acid, alkali substances, which may not be exactly pII 5. 6. ②As a decisive influence on the pH of acid precipitation, especially sulfuric acid and nitric acid, strong acid does not come from human sources. SO_2 emissions such as volcanic eruptions and SO_4^{2-} in sea salt contribute to the rainwater. ③Precipitation pH>5. 6 in the region does not mean that there is no man-made pollution, acidic substances pollution in some areas is severe, but alkaline dust or other alkaline substances such as high NH_3 content, the results of erosion of the pH of precipitation is greater than 5. 6. This involves the ionic balance of precipitation. ④H^+ concentration is not a conserved quantity; it does not indicate the degree of contamination of precipitation. The same acidity of precipitation, the SO_4^{2-} and NO_3^- content can vary widely.

In remote areas, low precipitation pH does not necessarily mean that pollution is serious; precipitation near the city which sometimes is not low pH, has actually been contaminated. Precipitation pH is associated with the balance of including acid, alkali ions.

In summary, pH 5.6 precipitation determine whether or not a subject to reasonable limits acidification and anthropogenic pollution, therefore, put forward the background precipitation pH value and precipitation pollution of discriminant standard question, which has done a lot of research.

Due to different natural conditions in various areas of the world, such as geology, meteorology, hydrology, will cause the precipitation pH of different among regions. pH of precipitation in some regions of the world is 4.79 to 5.00, less than or equal to 5.0 (Table 2-3), which will be 5.0 as the boundaries of the pH of acid rain is more in line with the actual situation.

Table 2-3 pH of precipitation in the world some background points

Place	Sample number	The average pH
Lijiang, China	280	5.00
Amsterdam	26	4.92
Workflot	16	4.94
Katherine	40	4.78
Sancarlos	14	4.81
St. Georges	67	4.79

Some people think that pH greater than 5.6 precipitation that was not affected by the human disturbance of acid, because even if someone is interfering but not very strong, because of the rain had enough buffer capacity, won't make acidic rain; And pH between 5.0-5.6 rain water may be influenced by human activities, but no more than the influence of natural background sulfur, or even the existence of human impact, beyond natural cushion adjustment ability, because of the rain and natural sulfur balance the pH value is 5.0. If the rain pH is less than 5.0, it is believed that the presence of human impact. So put forward to 5.0 as the boundaries of the pH value of acid rain is more exact.

Since the actual atmosphere is a very complex system, so the precipitation acidity is a result of the combined action of various substances into the water, it depends on the acid and alkaline ion concentration of contrast. Many facts show that as fossil fuel consumption increased year by year, SO_2, NO_x emissions also increased, resulting in increased atmospheric acidification. According to some background samples of continuous observation, the precipitation pH has a downward trend year by year. Such as Japan Iwate ryori precipitation pH was from 5.20 to 4.85 between 1976 and 1992. This shows that the global atmospheric climate is gradually changing, which is no doubt that the global acidic pollutants emissions increases year by year; and acidification buffering capacity of aerosol decreased are closely related. CO_2 concentrations in the atmosphere is increased at an annual rate of 2 mL/m^3, currently has nearly 360 mL/m^3. Thus, the global precipitation acidity of background values is not stable and unchanging, in this sense, the background value drops to a certain value of little practical significance. Hence, now still at pH 5.60 as a general acid precipitation judgment standard.

(2) Acidification of precipitation

Atmospheric precipitation acidity is concerned with the nature and ratio of the acid, alkali. Following is a brief introduction that these substances go into the precipitation and cause acidification of the precipitation process.

The formation of acid rain process includes the rain and washed out. In the free atmosphere, because of the presence of 0.1-10μm range of condensation nuclei caused by condensation of water vapor, and then touch and coalescence process further grow to form cloud droplets and raindrops. In the clouds, cloud droplets and touch each other or with the touch of aerosol particles and, at the same time absorb atmospheric gaseous pollutants, chemical reactions inside the cloud, this process is called in-cloud scavenging or rain out of pollutants. In the process of raindrops falling, rain washed out by gases and aerosols in the air, rain internal chemical reaction will occur, this process is called under the cloud of pollutants removal or flushing (below-cloud scavenging or washout). The process is the precipitation of atmospheric particulate matter and gaseous material removal process, acidification is occurring in these clearance process, shown in Figure 2-7.

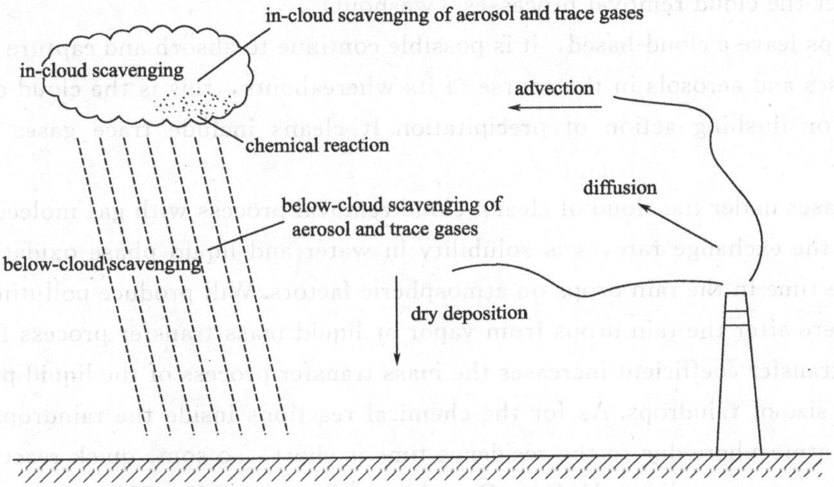

Figure 2-7 Precipitation acidification process

① Within the cloud removal process (rainout).

Such as sulfate and nitrate aerosols in the atmosphere can be used as active condensation nuclei involved in cloud process; In addition, the water vapor can also generate nucleation when super saturation. Because of the water vapor condensation on cloud droplets and touch between the cloud droplets, the cloud droplets grow continuously, at the same time, various pollution gases dissolved in cloud droplets and various chemical reactions; when cloud droplets become mature after rain falling from the cloud base. Atmospheric pollutants within the cloud cleared (except) the rain process including the rain of aerosol particles and trace gases in addition to the rain.

Aerosol particles can pass through the cloud droplets by three mechanisms: Ⅰ. The activity of aerosol particles as condensation nuclei for water vapor to enter the cloud droplets. Ⅱ. Touch between aerosol particles and cloud droplets. Aerosol particles and cloud droplets has a touch by Brownian motion and turbulence movement. The aerosol particles whose sizes are less than 0.01 microns will all go into the cloud droplets by this mechanism. Ⅲ. Aerosol particle movement by force, and move in the direction of the vapor pressure gradient into the cloud droplets. In the troposphere, if the aerosol concentration is less than 200-300$\mu g / m^3$, almost all particles are cleared in a cloud process. As an active condensation nuclei into the cloud droplets, participate rainout process is the primary mechanism of aerosol particles rainout.

In addition to rain pollution gases and aerosols affect rain cloud of water, in addition to equally important. Rain trace gases in addition to depending on the mass transfer process and reactive gas molecules in solution, and also with the cloud type and cloud droplet spectrum related. In the contaminated gas cloud removal in the process, some of them be oxidized substances, such as S (Ⅳ) is oxidized to S (Ⅵ); chemical oxidation rate is the fast step of the process, liquid phase oxidation reaction rate depends on the types and concentrations of oxidizing agent, and pollution gas solubility in the cloud droplets depends on the concentration of the gas phase and pH value of cloud droplets.

② Under the cloud removal processes (washout).

Raindrops leave a cloud-based, it is possible continue to absorb and capture atmospheric pollution gases and aerosols in the course of its whereabouts, this is the cloud of pollutants under clear or flushing action of precipitation. It cleans include trace gases and aerosol clouds.

Trace gases under the cloud of clear: cloud removal process with gas molecules in liquid phase under the exchange rate, gas solubility in water and liquid phase oxidation rate and the residence time in the rain drops on atmospheric factors. Will produce polluting gases into the atmosphere after the rain drops from vapor or liquid mass transfer process from the gas phase mass transfer coefficient increases the mass transfer process of the liquid phase decreases with the size of raindrops. As for the chemical reactions inside the raindrops, the raindrops in the atmosphere due to the residence time is short, so some quick reactions such as ion reaction, strong oxidants H_2O_2, O_3, free radical ·OH, HO_2· and metal ions Mn^{2+}, Fe^{3+}, etc. of S (Ⅳ) oxidation reaction, only the chemical composition of raindrop impact, while less affected most of the slow response of raindrops. Under the cloud of pollutant removal in the process, rate of mass transfer between gas and liquid and the liquid phase Rate should be jointly determine pollution gas in the liquid phase reaction rate. Liquid mass transfer rate control large raindrops in the liquid phase reaction rate, the chemical reaction rate is controlled liquid phase reaction rate is small raindrops. Maximum solubility in water or participate in the solution only for fast ionic reaction increased solubility of trace gases and their removal under the cloud is irreversible, its removal rate and independent of the concentration of the liquid phase has been entered, only the gas phase concentration. In irreversible removal of gas in the liquid phase to the total concentration increases linearly with raindrops landing distance. Gas irreversible clear clearance has nothing to do with the other gas, only related to its own features and raindrop spectrum and rainfall, Clearance rate increased with the increase of rainfall. Research shows that the rain cleared coefficient of SO_2 gas and rainfall intensity, gas concentration of SO_2, NH_3 concentration, oxidant concentration and the initial pH of rainwater concerned, there are complex interactions between them.

Aerosol clouds clear: Aerosol is captured, the soluble part such as SO_4^{2-}, NO_3^-, NH_4^+, Ca^{2+}, Mg^{2+}, Mn^{2+}, Fe^{2+}, H^+ and OH^- will be released, thus affect the chemical composition and acidity of the rain. Shuairen Liu studied the role of atmospheric aerosols rain acidification process under the cloud. The results showed that: (a) aerosol impacts on the acid rain, if the pH value is lower than the rain of aerosol of pH value, it plays a role of acidification, aerosol. On the contrary, basification role; acidification of aerosols is stronger than basification. (b) In general concentration (each containing 10^3 cm^3), the acidic aerosols are an important source of H^+ rain, alkaline aerosol can consume rainwater H^+; the contribution of aerosols on rainfall SO_4^{2-} is small. (c) Acid aerosols acidification of rain weakened with increasing SO_2 concentration, and alkaline alkalizing effect of aerosols on rainfall and enhanced with increasing concentration of SO_2; cloud clearance process is one of the important sources of SO_4^{2-} rain cloud aerosol removal process under small contribution to SO_4^{2-}. (d) HNO_3 contribution to rain H^+ is several times than with the concentration of SO_2, aerosol's contribution to the rainwater NO_3^- is the equivalent of 1 ng/mL the contribution of HNO_3, with the increase of concentration of HNO_3, aerosol the relative contribution of rapidly reduced. (e) Aerosol is an important source of NH_4^+ in the rain, the equivalent of 5-8 NH_3 concentration ng/mL's contribution to the rain NH_4^+.

Cloud cleared and removal processes is affected the degree of air pollution and many environmental parameters. Cloud cleared about the relative importance of acid rain formation is different in different geographical regions and different source emission and different meteorological conditions, etc. Observation results show that some heavy pollution area in our country, under the cloud removal process is very important. Under the cloud removal process numerical simulation, such as Chongqing and Beijing area, results show that chongqing water source of H^+ is given priority to with SO_2 oxidation under the cloud, plays a role of basification aerosol; Beijing source of H^+ rain is given priority to with cloud clearance process, alkaline gas NH_3 under the cloud and aerosol. Visible, high concentrations of NH_3 in the Beijing area is the leading cause of no acid rain, at the same time also make high SO_4^{2-} concentration in the rain.

Thus, the formation process of the acid rain is very complex. The precursors of acid rain of SO_2, NO_x, in addition to the local source emission, also can be transmitted over a long distance. Such as acid precipitation in south China has local scour and dual mechanism of long distance transmission. The SO_2 concentration of south atmospheric is higher in some mountainous cities, the main source of precipitation acidification is for local scour. The southern spring often appears large area of acid precipitation; its main source is long distance transport of pollutants. Pollutants in a certain weather conditions can transmit hundreds of kilometers or more and make the knot lee area and precipitation acidification. The southern most of the acid precipitation is for local scour and dual source of transmission. When the cloud water acidity keeps the same, cities of the concentration of acid gases and aerosols buffering capacity determines the local scour and transfer to the relative contribution of precipitation in acid.

As described above, the main cause precipitation of acidic substance is sulfuric acid, followed by nitric acid, as well as organic acids and other acids. Now cause precipitation acidification of SO_2, NO_x, for example, an overview of the formation process of acid rain.

(a) By source emissions of gaseous SO_2, NO_x should be generated by H_2SO_4, HNO_3 or sulfate, nitrate aerosols.

(b) Cloud formed, aerosol particles containing SO_4^{2-} and NO_3^- go into precipitation with the form of condensation nuclei.

(c) Cloud droplets absorb SO_2, NO_x gas phase oxidation in water to form SO_4^{2-}, NO_3^-.

(d) Absorption of SO_2, NO_x, raindrops fall again into SO_4^{2-} and NO_3^- in the water phase.

(e) When raindrops fall down, it absorbs SO_2, NO_x, and then converts them into the aqueous phase SO_4^{2-}, NO_3^-.

Pathway (b), (c) for rain, (d), (e) as the scour (cloud under clearance process); engaged in SO_2, NO_x absorption and liquid-phase oxidation and erosion while the rainout process; H_2O_2, O_3 and $\cdot OH$, $HO_2\cdot$ for liquid phase oxidation of SO_2 and NO_x plays an important role.

Other gaseous substances in the atmosphere and NH_3, H_2O_2 and O_3 and hydrocarbons, could also be cleared into precipitation, some of the material (such as hydrocarbons) can occur oxidation conversion, thus to precipitation acidification. Therefore, the formation of acid rain is acidification process of chemical process and clear physical intertwined.

2. The harm caused by acid rain pollution

(1) Effects on the soil

Acidic wet settlement is not only by rainfall, but also through dry soil subsidence. On the one hand the soil calcium, magnesium, potassium and other nutrients are leached, resulting in increasing soil acidification, impoverishment, affect plant growth; on the other hand acidification of soil microbial activity.

(2) Effects on plant

Leaves by acid rain, which chlorophyll content decreased, due to photosynthesis blocked, so that lower crop yields, but also can reduce the growth rate of the forest.

(3) The influence for surface water

Acid rain can make the surface water acidification, such as lakes, rivers, polluting drinking water. When the water pH <5, the growth and reproduction of fish will be severely affected; watershed soil and lake, river sediments of toxic metals, such as aluminum is dissolved in water, poison fish. Acidic water will cause changes in the structure of aquatic ecological; acidified lakes and rivers, the fish would be reduced or even extinct.

(4) The impact on the material

Acid rain accelerates the corrosion for many building structures, bridges, dams, industrial equipment, water supply networks and telecommunications cables and other materials, but also serious damage to monuments, historic buildings and other important cultural facilities.

(5) Effects on human health

Acid rain cannot only cause great economic losses, but also harmful to human health, the hazard may be indirect or straightforward. Scientists of the Chinese academy of sciences ecological environmental research center and other units were first in the west of Beijing acid rain monitoring in 1974; and then was monitoring again running in the city of Chongqing, Guiyang, Wuhan, Nanjing and Beijing, and in the south of the Yangtze River in the late 1970s. Cause people attention.

3. Present situation and the control countermeasures of acid rain pollution in China

(1) Pollutions

Our research on acid rain began in the late 1970s. At that time local studies were launched in Beijing, Shanghai, Nanjing, Chongqing and Guiyang and other cities. They found that these areas existed in varying degrees of acid rain problem; the Southwest is very serious. From 1982 to 1984 in our country has carried out the investigation of the acid rain, in order to explore the space-time distribution of precipitation acidity and chemical composition, from 1985 to 1986, deployed 189 monitoring stations and 523 precipitation sampling points across the country, analysis of the precipitation data conducted a comprehensive system. The results showed that the main region of average annual precipitation pH less than 5. 6 is south of the QinLing Mountains and Huaihe River, just north of the individual areas; The main regions of annual average precipitation pH less than 5. 0 is in southwest, east China and the southeast coastal area. The acid rain in our country gradually from north to south is aggravating, south of the Yangtze River is a common problem of acid rain. Acid rain is the most serious in southwest China, such as Chongqing and Guiyang; rainwater acidity on average is almost under pH 5.

Matter is the main cause of acid sulfate acid rain in China; the precipitation is generally having high content of SO_4^{2-}.

① The present situation of acid rain in Zhejiang province.

Zhejiang Province, covering basic acid rain, acid rain rate has reached 83. 7%; precipitation pH (acidity) annual average is 5. 6 or less, the 96. 9% of the area precipitation pH (acidity)$<$ 5. 0, especially Linan, civilized, Anji, is less than 4. 5; Precipitation mainly acidic pollutants are SO_4^{2-}, but the effects of NO_3^- is increased year by year. If the rain samples collected from a place is acid, it can not be regarded as acid rain region, because there are dozens of rain a year, some rain may be acid, some may not, so whether the region is acid rain region depends on the annual average.

Nowadays scientific standard definition of acid rain region is still under discussion, but it is generally believed that: first, the non-acid rain region is the place where the average annual of pH of rain is above 5. 65 and proportion of acid rain is 0-20%, second, the mild acid rain region is the place where the average annual of pH of rain is between 5. 30 to 5. 60 and proportion of acid rain is between 10% to 40%, third, the moderate acid rain region is the place where the average annual of pH of rain is between 5. 00 to 5. 30 and proportion of acid rain is between 30% to 60%, forth, the heavy acid rain region is the place where the average annual of pH of rain is between 4.70 to 5.00 and proportion of acid rain is between 50% to 80%, and last, the heavy acid rain region is the place where the average annual of pH of rain is less than 4.70 and proportion of acid rain is between 70% to 100%. All above is the so-called "five stars criteria". In fact, acid rain can also be collected from Beijing, Xining, Lanzhou and Urumqi, but the average annual pH value of acid rain and the proportion of acid rain belong to non-acid. Thus, these places are regarded as non-acid region.

② The acid rain in Yunnan, Guizhou, Sichuan.

In the early 1980s, some township enterprises just started in the area of Yunnan and these enterprises produce sulfur by indigenous methods just to pursuit the profits. The technology is out of date and the recovery rate is just 30% and the rest of the sulfur turn into SO_2 by combusting. The SO_2 is clouded in the atmosphere, coupled with the wet climate in Yunnan and the SO_2 is turn into acid rain in the end. It has been detected that the rain are all acid rain whose pH value is between 3.2-3.8 in these regions. The consequence is that there are no insects in the trees, no ants on the underground, and no birds in the sky and no plants in the arable land. The problem has been alleviated since a series of measurements is established.

Producing sulfur by indigenous methods also happened in the northern and northwestern of Guizhou. They use pyrite as raw material to produce sulfur. The recovery rate of sulfur is just 45%, more than half of the sulfur flow to the atmosphere in the form of SO_x, which will cause pollution. It is so surprising that within a radius of 400 meters of a sulfur plant with an annual output of tons of sulfur, there are no plants growing and the concentration of SO_2 in the air is more than 6 mg/m^3. Besides, within two square kilometers, the concentration of SO_2 is 0.5 mg/m^3.

The situation in some counties of Southern and East Sichuan is more serious. According to statistics, there are two thousand factories whose capacity is 8-100 tons. About 1.8-2.2 tons of SO_2 and H_2S gases were discharging through producing one ton of sulfur. After monitoring concentration of SO_2 is 2.2-1087 mg /m^3 near the refining sulfur factories, average concentration of is 78-185 mg /m^3 which is exceeding a hundred times over. The concentration of SO_2 is 1.1-27.5 mg /m^3 in the living area and the average concentration is 8.8 mg /m^3 which is actually exceeding 27 times over. People cannot stand with that.

The rate of acid is 100% near the production areas. Annual precipitation pH value is 3.0-5.2 and the average value is 3.75, which ranks in the heavy acid rain area. There are 350-400 square kilometers of earth is polluted by the indigenous methods, which result in 5500 kilograms of grain losses. In the heavy acid rain area, the yield of tomato is 3600 kilogram and the yield of reinwardtia 1850 kilogram. But outside the heavy acid rain area the yield of tomato is 5675 kilogram and the yield of reinwardtia is 2425 kilogram. The Emeishan in Sichuan is so beautiful in the past, which is on account of the deep mountain and lots of trees. But over the past decade, the acid rain makes the fir forest into a piece of dead forest. Jinding is 3077 meters above sea level, the average pH value of rain is 4.34, the rate of acid rain is 85.7%, and the percentage of sulfate is 60 percent of total cations. Other scenic such as terra top, Prince Ping, Seven Mile slope and Lei Dongping are also severely polluted due to acid rain. The fir forest is the major scenic in the Emei, 7.00 square kilometers forest are moderately damaged which the mortality rate of the forest is more than 30%, 2.28 square kilometers forest are seriously damaged which the mortality rate of the forest is more than 50%. 4% of the trees are dead near Qilipo, 600 acres forest are all-dead near the Jinling and nothing is beautiful. In these places monkeys are gone.

③ The acid rain in Chongqing.

Chongqing and its suburbs had black rains, which is black and acid. People were panic about that and asked the authority what happened. According to analyzing the black substance is carbon particles that the fossil fuel cannot combustion completely and turn into carbon particles and vent directly to the atmosphere. So they made a conclusion: black rain is a strong acid rain and it can be said that it is the development of acid rain in some extreme cases. Coincidentally, the Himalayas in China had some "black snow" in 1991. It came from the Middle East war. When Iraqi troops retreating from Kuwait, they put a fire and made the main oil and gas wells ablaze which was mixed up with SO_x and NO_x. The SO_x and NO_x went Straightly into the sky and also drifted to east with the wind until it met the Himalayan mountains, which was too high to cross, and it fall with snow and became acidic black snow. Few people reached to this place but this place fails to escape from this disaster.

④ The acid rain in Fujian.

Sanming City is an emerging industrial city in Fujian. It is located in the central mountainous inland whose area is about 1000 square kilometers. Mountains surround Sanming city. The distance from the east to the west is long and that from the north to the south is short. It is surprising that for the recent years rains are acid in the east of this city and even the pH value reached 4.0. But in the west of the city which is only four kilometers far from the east of the city, the rains was significantly alkaline and even the pH value reached 9.89. It never happened that in the same rain, the east of the city is acid but the west of the city is alkaline. Scientists have made a reasonable explanation through constant monitoring and investigating of the pollution. The city is a bason surrounded by mountains; pH of rain is mainly caused by local environmental pollution. The east of the city locates last direction of wind and the precipitation is mainly affected by cumulonimbus layer in the high altitude atmosphere. This layer contains SO_x and NO_x, which are emissions from coal-fired. The SO_x and NO_x are precursor of acid and when it falls down on the ground, it becomes to acid rain. But the west of the city located last direction of wind. Especially there is Sanming Iron and Steel Plant in the west of the city. The plant discharges alkaline dust and particulate matter, which piled up in the west at low altitude. In the end the rain can wash it and makes the precipitation was significantly alkaline. Such alkali rain is different from the alkali rain in Xinjiang. The former contains only basic particles, while in the latter, the major composition is alkaline dust and particulate matter interact with acid rain, so the content of anion and cation is very high in the latter acid rain and the content of ions is very high in the latter contains, so the pollution is too serious to solve.

(2) Control Countermeasure

The country has began to organize a large-scale study of acid rain since during the period of "65", at the same time, state and local governments have taken a number of measures to control and reduce the development of acid rain. NPC Standing Committee adopted Atmospheric Pollution Prevention Act which was newly revised. The new law stipulated that the acid rain control area and sulfur dioxide control area should be classified. The two kinds of areas are so-called "two control zones".

In the "two control zones", the annual average concentration of SO_2 exceeded the secondary standard of national environmental quality and the daily average concentration of SO_2 exceeded the third standard of national environmental quality. These two standards are the lowest concentrations which protect people and environment from damaging and acute hazards. The Acid Rain Control Zone refers to the place whose pH value of precipitation is less than 4.5, and the sulfur settlement exceeds the critical loads of acid settlement while considering the economic development. The critical loads of acid settlement refer to the maximum amount of acid deposition, which the ecosystem can withstand. According to such classification, the total area of the "Two control zones" is 1.09 million square kilometers. The area of the Acid rain control areas, south of the Yangtze River, Sichuan, the east of Yunnan region, is approximately 800000 square kilometers. And the area of the sulfur dioxide control is about 29 square kilometers.

There is a close interrelationship between pollution of acid rain and the pollution of sulfur dioxide, so the control countermeasure should coordinated. To reduce the concentration of sulfur dioxide and to improve local air quality, it is important to control the pollution resources, which can produce most sulfur dioxide. However to control acid rain, it is not just to control the local resource but also focus on the emissions sources of sulfur dioxide and nitrogen oxide which make contribution to the acid rain and the formation of supercritical load zone from a regional perspective. Especially with the continued development of the national fuels consuming and coal-fired power plants, the average height of emission source of sulfur dioxide emissions will continue to increase and the interaction between regions will be growing, so it is more and more important to control coordinated.

(3) The main measures

① Technical measures for the use of coal.

More than 50 percent of China's primary energy is from coal, therefore, 50 percent of China's sulfur dioxide emissions are from coal using. So sulfur dioxide emission control is a crucial work in the process the coal consumption. So it is very important to control from the source by limiting the high flow of coal mining. Two control zone will stop building mines whose concentration of sulfur is more 3%; For building, rebuilding and expanding mines whose concentration of sulfur is more 1.5%, for the built mines whose concentration of sulfur is more 2%, it is very important to build up coal washing facilities. At the same time, the use of low-sulfur coal and desulfurization before using the coal by filtering, washing and other technical desulfurization are promoted. Useful coal, coal vaporization and CWS are vigorously promoted; central heating and other new technologies will help improve the efficiency of coal. Desulfurization by all effective and economically viable technologies in the end the burning of coal is also an effective way to reduce dioxide emissions to the atmosphere.

② Energy measure based on adjusting the energy structure and improving the energy efficiency.

Energy restructuring plan should be considered in the acid rain region such as to replace the disposable energy which we rely too much on with clean gas, oil or hydropower and nuclear energy, especially to reduce the amount of domestic coal combustion. At the same time, it is to improve the energy efficiency by the technological transformation and technological innovation.

③ Implementation of an effective management system and economic policy.

(a) To control the amount of sulfur dioxide emissions in the region and implement of permit system to the key sources of pollution emission.

The authority issued a total control targets to the provinces, municipalities and autonomous regions based on the amount of sulfur dioxide emissions at different stages of the two control zones and basic control unit. Environmental protection departments at all levels confirmed the permitting amount of sulfur dioxide emissions according to pollution control targets of the two control zones. And to discharge total amount of sulfur dioxide to other units by using the emission permits system to supervise and manage the sources of sulfur dioxide emissions.

(b) Try out trading policy about sulfur dioxide emissions.

On the basis of the permit system, Firms can try out for the sale of government allocation which is allowed the sulfur dioxide emissions targets under condition that it does not exceed the total amount of emissions control targets premise. On this occasion, firms can choose to implement total pollutant discharge permit limits control targets and minimum cost method to reduce the sulfur dioxide emissions.

(c) Implementation of the fee policy of the sulfur dioxide emission in favor of pollution control.

Improve the fee of sulfur dioxide emission to be equal to or higher than the cost of governance. Only in that way can make the pollution control costs become part of the total cost and form the mechanism that no matter who makes pollution will be punished in the economy. This measure can make enterprises to increase investment in pollution control initiative.

In addition, we should vigorously strengthen environmental monitoring in the two control zones while implement of control programs. Environmental monitoring is an important means of acid rain and sulfur dioxide pollution condition and the scientific assessments of implementation in the two-control zone. Besides, the monitoring results also provide further guidance for controlling.

Section 7 Urban Air Quality Problems

1. Status quo and development trend of China's urban air quality

Our country is in the period of social and economic rapid development, in the process of our development is facing the challenge of all kinds of environmental problems. With the process of urbanization and industrialization, pollution caused by transport has become increasingly prominent environmental problems in our country. In the process of development of national economy, transportation infrastructure is the main condition and the automotive industry is the one, which supports the infrastructure.

(1) City cars development

The development level of automotive industry is still relatively low in China currently. Meanwhile, car ownership in China is extremely uneven regional distribution. The density of the population and vehicles is very high in highly populated cities and rapid economic development region due to rapid urbanization. Motor vehicles are mainly distributed in Beijing, Shanghai, Guangzhou and other major cities. For example, the car social ownership accounted for about one tenth of the country in Beijing.

In addition, the development of motor vehicles in China is accompanied by two different characteristics from the developed countries and the two characteristics are the poor road conditions and poor car conditions. The level of the pollutants discharged by the automobile is higher than the other countries, car emissions is 10-20 times than Japan's and the truck's emissions is 1-8 times than the United States. While the atmospheric environment problem caused by the auto industry has not taken place in China, but from the strategic perspective of sustainable development and the lessons learned in developed countries, the seriousness of this problem should be fully recognized.

(2) The pollutants discharged by motor vehicle

Pollutants caused by a motor vehicle include primary pollutants and secondary pollutants substances formed by atmospheric chemical processes such as fine particles and O_3. The main pollutants include: carbon monoxide, nitrogen oxides, volatile organic compounds, lead, particulate matter, carbon dioxide.

These primary pollutants will in the process of various atmospheric chemical reactions to produce secondary pollutions such as fine particles and atmospheric oxidants which are harmful to human health and ecosystems. Globally, the motor vehicle the main factor which causing the deterioration of visibility and oxidizing atmosphere to enhance. It should be pointed out that at present the fuel that China's motor vehicle using has a big difference between past and abroad, in order to improve the antiknock performance of unleaded gasoline, the content of olefins and aromatics in gasoline has increased greatly, these substances can get into the atmosphere by evaporation, combustion and other ways and are more likely to cause photochemical pollution due to their stronger photochemical activity.

(3) The impact of vehicle emissions on the environment

The rapid development of motor vehicles, have a profound essential impact on the atmospheric environmental quality in China and is mainly reflected that the types and features of atmospheric pollution in our country has changed and the area affected by the pollution is expanding.

China is a traditional country of coal-burning pollution; SO_2 and atmospheric aerosols have been the main pollutants in our atmosphere.

In recent years, with the rapid development of China's economy and the structure of the regional energy conversion between parts of the city and the region, pollution emissions and fuel gas vehicle emission pollution was rapidly increasing in some cities. The concentration of nitrogen oxides is higher than the sulfur oxides in Guangzhou and Beijing respectively in 1985 and 1993, which explained that the character and nature of the urban air pollution have changed. For example, photochemical pollution has formed complex atmospheric pollution and how to control the NO_x pollution and light chemical pollution while control the SO_2 and atmospheric aerosol pollution must be considered and carefully handling.

The photochemical has been found in petrochemical area of Xigu, Lanzhou in the late 1970s. It has been researched that photochemical smog is caused by emissions of NO_x and VOC emissions from the oil industry plants and its impact is limited to the local area. Photochemical smog has been found in Beijing in the summer of 1986. The photochemical smog in Beijing is mainly caused by vehicle exhaust and emissions of coal-fired. In the past 30 years, with the rapid growth of economic development and the number of vehicles, photochemical smog has been frequently found in Shanghai, Guangzhou, Shenzhen and other cities. Vehicle emissions have become a major source of air pollution and NO_x pollution has become the main source.

Photochemical smog will be more and more serious in the coming period in China due to the atmospheric fine particle pollution and also due to the improving atmospheric oxidation capacity, which results from the pollutants discharged from motor vehicle, improve in the processes of atmospheric chemical. Fine particles and the transportation of atmospheric oxidants spread the pollution from the city to the suburban areas which forms the regional atmospheric oxidant pollution.

2. Urban air quality daily

In recent years, more and more cities start to provide air quality daily service. Such as the Beijing Environmental Protection Monitoring Center provides daily pollution index of sulfur dioxide, carbon monoxide, ozone, nitrogen dioxide and respirable particulate matter and indicate the primary pollutants and air pollution. Table 2-4 and Table 2-5 respectively give the pollution index corresponding to the pollutant concentration limits and air quality conditions and the data can be for reference.

The calculation method of air pollution index:

When the sub index of a pollutant meet the following condition

$$C_n < C_i < C_n + 1$$

① The sub index of I_i is:

$$I_i=\frac{(C_i-C_n)}{C_{n+1}-C_n}\cdot(I_{n+1}-I_n)+I_n$$

② After the pollution sub index of all the pollutants have been calculated, the maximum sub index is called API which stands for The air pollution index of the region or city. That is:

$$API=\max(I_1,I_2,\cdots,I_n)$$

Example. If the concentration of the Sulfur dioxide in air is 1. 8 mg/m^3, please calculate the pollution index.

The answer is:

$$I_i=\frac{(C_i-C_n)}{C_{n+1}-C_n}\cdot(I_{n+1}-I_n)+I_n$$

The answer is:

$$I_i=\frac{(1.8-1.6)}{2.1-1.6}\times(400-300)+300=340$$

Table 2-4 Pollution index and limited concentrations of pollutants

Pollution index API	Limited concentrations of pollutant/(mg/m^3)				
	Total suspended particles	Sulfur dioxide	Nitrogen oxides	Carbon monoxide	Ozone
500	1. 000	2. 620	0. 940		
400	0. 875	2. 100	0. 750	>20	1000
300	0. 625	1. 600	0. 565	10	800
200	0. 500	0. 250	0. 150	6	200
100	0. 300	0. 150	0. 100	40	160
50	0. 120	0. 050	0. 050	2	120

Table 2-5 Air pollution and air quality level

API	Air quality level	Air quality	Health level	Usable range of the air quality
0-50	Ⅰ	Superior	Normal activity	Reserves, scenic spots and some other places needing special protection
51-100	Ⅱ	Fine	Normal activity	Determined residence zones on the town planning, shopping-traffic-mixed residences, cultural centres, general industrial zones, and rural areas
1-200	Ⅲ	Mild pollution	With a long-term exposure, symptoms of vulnerable groups mildly increased, and healthy people generated stimulate symptoms	—
201-300	Ⅳ	Moderate pollution	Be exposed after a period of time, heart disease and pulmonary disease patients' symptoms significantly increased, with exercise tolerance decreased, while the heathy developed symptoms commonly	—

(continue)

API	Air quality level	Air quality	Health level	Usable range of the air quality
>300	V	Serious pollution	Healthy people generated strong symptoms obviously, and exercise tolerance decreased, meanwhile some diseases appear in advance	—

Questions

1. How many layers the atmosphere can be divided into, according to the temperature variation of the earth's surface from the vertical direction?

2. How much temperature can reduce as 100 meters higher in the convection atmosphere?

3. Please describe the formation and function of "atmospheric window".

4. Please describe the formation mechanism of acid rain and its harm.

5. If the concentration of sulfur dioxide in the air is 2.43, please calculate its pollution index.

Chapter 3
Water Environmental Chemistry

Key Point: Water resources and utilization; Water distribution on the earth; Water cycle; The conformations of natural water; Water body pollution and self-purification; Pollutants in water; The movement procedure of pollutants after discharging into the water; Water pollution effect on human body; Parameter of organic pollution; Degradation of some organic pollutants in water; First treatment for wastewater; Secondary treatment for waste water; Tertiary treatment for wastewater; Typical process of municipal wastewater treatment.

Section 1 General Situation of Natural Water

1. Water resources and utilization

The amount of water is fixed on the earth, about 1.386 billion cubic kilometers, but the percentage of available fresh water is only 0.3%. Water resources are limited, but if the water resources have a suitable spatial and temporal distribution and keep proper water quality, it still can meet the need of fresh water in the current and future. It's a pity that the spatial and temporal of fresh water in the earth is extremely asymmetry, with the addition of increasing water pollution and more and more water for industrial, agricultural and domestic, critical shortage of water resources appeared in many countries and regions. The shortage and pollution of water not only affected survival of organisms, but also threatened to human's life directly or indirectly and even caused significant financial losses.

At the moment, the mainly form of water shortage are: resource-oriented water shortage, engineering water shortage and quality-induced water shortage. Because of water shortage, the crops grow, industrial manufacture has been affected, and the human's health, ecological security and national security have been threatened.

2. Water distribution on the earth

More than 70% of the earth has been covered by Water, so the earth is also called "water planet". Water on the earth distribute in seas, lakes, marshes, rivers, glaciers, snowfield, atmosphere, living body, soil and stratum. The amount of water is about 1.386 billion cubic kilometers (as shown in Table 3-1, Table 3-2), among them, seawater is nearly 96.5% and fresh water is 0.35 billion cubic kilometers, 2.35% of the total water. Because of exploit difficulties and technique, economic limited, seawater, deep underground fresh water, ice-snow solid-state fresh water, salt lake water is few used directly now. The fresh water which is easy to development and utilization and relate to the human's life and production closely reserves about 4 million cubic kilometers, only 11% of total fresh water, 0.3% of total water.

Table 3-1 The distribution of water in natural environment

Water type	Water yield/km^3	Proportion/%
Seawater	1338000000	96.5
Groundwater	23716000	1.71
Meltwater	24064100	1.74
Lacustrine water	176400	0.013
Bog water	11470	0.0008
River water	2120	0.0002
Atmospheric water	12900	0.001
Biowater	1120	0.0001
Total	1385984610	

Table 3-2 The reserves of fresh water in the world

Water type	Water yield/km^3	Proportion/%
Groundwater	10846500	30.96
Meltwater	24064100	68.70
Lacustrine water	91000	0.260
Bog water	11470	0.033
River water	2210	0.006
Atmospheric water	12900	0.037
Biowater	1120	0.0032
Total	35029210	

Surface water runoff volume is about 2.8 trillion cubic meters in our country, underground water resource is about 800 billion cubic meters, the average of melted water from glacier is about 50 billion cubic meters, and offshore seawater is about 5 million cubic meters. Available water resource in our country is about 1100 billion cubic meters per year at present. Every person occupies surface water resource about 2700 cubic meters, the 88th in the world, the average per capita water availability in China is lower than one fourth of that of the world average. Surface water per acre was 1755 cubic meters, only half of the world average. All in all, fresh water resource is not rich in our country, our country is in water shortage condition, and water resources is unsuitable in spatial and temporal distribution, more in southeast and less in northwest, agricultural acreage occupied only 33% of Yangtze river basin and south of Yangtze river, water resource is about 70% in our country. Water is in severe shortage while water pollution is very serious. On the other hand, because of population explosion, industrial and agricultural production are rapid development, like many western countries, our country is facing water quality decline and shortage of water resource, so control water pollution, protect water resources is an urgent task for us.

3. Water cycle

The water in nature can not only flow along the slant direction under the influence of earth's gravitation, but also change its form because of solar energy and heat on the earth's surface. The vaporing water transfers with the movement of airflow, and condenses to cloud or falls on land as rain when cold. The water on the earth's surface can evaporate, condense, and fall again. This is a process in cycles, which is called water cycle. The cycle of water in part of area, such as land and ocean, is called minor-cycle. Part of the moisture evaporated from the ocean moves to high up in the air, falls on land as rain, and then comes back to the ocean through river, lakes and seas or seeping into underground. This repeating process between ocean and land is called major cycle of water. Water cycle in environment is made up with minor-cycle and major cycle, and always work on everywhere all over the world. There are three reasons influencing the water cycle.

① The physical quality of water decides the possibility of water cycle.

② Solar Irradiation provides power for water cycle.

③ The structure and property of the cycle route, especially cycle on earth's surface, such as geology, landform, and the type and quality of living beings, can not only affect the distribution and transportation of rainfall, but also influence infiltration and the quality of water duct.

The direction and strength of the cycle of natural water are influenced by these reasons, which cause a complex hydrology phenomenon in nature. Evaporating, rainfall and runoff are basic factors of water cycle. The runoff of river and stream are basic sources for human's need of water, it's an active part of water cycle on earth, even the water yield of riverbed in the world can change every 16 days. So it has a very strong ability to clean itself.

4. The conformations of natural water

(1) Definition of the water

In the term of water pollution chemistry, the water refers to river, lake, swamp, glacier, groundwater, sea and so on. The water consists of not only water, but also suspended substances, colloidal substances, dissolved substances, sediment and aquatic organisms in water, So the water is a complete ecosystem or a natural complex covered by water. It's important to distinguish water and water body. For example, It's easy for heavy metal (such as Pb, Cu, Cr, Zn, Cd, etc.) pollution to transfer to sediment from water by some ways like precipitation, adsorption and chelation. As a general result, there are not too much heavy metal in water. Although it seems that water is not polluted, maybe the water has been polluted seriously.

(2) The compositions of natural water

In natural circulation of water, although the flow of surface runoff is not large, it has great influence in nature. In the process of weathering (including physical, chemical and biological weathering) and water migration, the minerals and organic matters in soil are transported, which makes natural water owns different components. Besides water, the natural water contains many substances. According to their different statuses in water, these substances can be divided into three kinds—suspended substance, colloidal substance and dissolved substance. We only discuss the dissolved substance's types and existing form here.

The substances dissolved in natural water can be divided to main ions, dissolved gas, nutrients, trace elements and organism. The content of dissolved substance in water does not only relate to matter and nature, but also relate to climate conditions, hydrological characteristics, forms of rock and soil. Table 3-4 and Table 3-5 show us the content of parts of elements and probable existing forms in theory respectively.

Table 3-3 The compositions of natural water

Sort	Main substance
Suspended substances	Bacteria, viruses, algae and protobiont, silt, clay and such particulate matters
Colloidal substances	Hydrated oxide colloid material of Si, Al and Fe, clay mineral gel material, humus and such organic polymer compounds
Dissolved substances	O_2, CO_2, H_2S, N_2 and such dissolved gases, halide of Ca, Mg, Na, Fe, Mn and such ions, salts such as carbonate and sulfate, and other soluble organic matter

Table 3-4 Some elements in seawater and their possible existing forms

Element	Average content/(μg/L)	Possible speciation
Na	10.77×10^6	Na^+
K	3.99×10^5	K^+
Mg	1.29×10^6	Mg^{2+}
Ca	4.12×10^5	Ca^+
Sr	7.9×10^3	Sr^{2+}
Cl	19.35×10^6	Cl^-
S	9.05×10^5	SO_4^{2-}, HSO_4^-
C	2.8×10^4	HCO_3^-, CO_3^{2-}
Br	6.73×10^4	Br^-
F	1.3×10^3	F^-
B	4.4×10^3	$B(OH)_3$
N	6.40×10^2	NH_4^+, NO_3^-
P	60	HPO_4^{2-}, $MgPO_4^-$
Si	2.20×10^3	$Si(OH)_4$, colloid
Fe	2	Colloid, $Fe(OH)_2^+$
Mn	0.2	Mn^{2+}, $MnCl^+$
Cu	0.25	$Cu(OH)^+$, $CuCO_3$
Cd	0.11	Colloid, $CdCl_2$
Hg	0.03	$HgCl_4^{2-}$, organomercury
Ni	0.56	Ni^{2+}, $NiCO_3$
Pd	0.03	$PdCO_3$, colloid
As	3.7	$HAsO_4^{2-}$, organoarsenic
Sb	0.24	$Sb(OH)_6^-$
Sn	0.004	$H_3SnO_4^-$, organic tin
Zn	4.9	Zn^{2+}, $Zn(OH)^+$
U	3.2	$UO_2(CO_3)_3^{4-}$

① Main ion.

The main positive ions in the natural water are Ca^{2+}, Mg^{2+}, Na^+, K^+, etc. The main negative ions are Cl^-, $SO_4{}^{2-}$, HCO_3^-, etc. And the 8 kinds of ions are able to occupy more than 95 percent to 99 percent quantum of the total dissolved solids in the water. The Na^+ and Cl^- in the seawater content are dominant, while Ca^{2+} and HCO_3^- ions' content are high in the river. $HCO_3^- > SO_4^{2-} > Cl^-$, $Ca^{2+} > Na^+ > Mg^{2+}$ are the general order in land water. $Cl^- > SO_4^{2-} > HCO_3^-$, $Na^+ > Mg^{2+} > Ca^{2+}$ are the general order in seawater. Groundwater is limited by local environmental geological conditions, its dominant ions has significant changes.

② Trace Element.

Trace element refers to the content of the elements in ng/mL level. There are many kinds of trace elements in the natural water, such as As, Cd, Hg, Ni, Pb, Sb, Sn, Zn, Mn, Cu, etc. Most of trace elements belong to heavy metal element. Their content and existence form have been showed in the Table 3-4 and Table 3-5.

Table 3-5 The contents and possible existing forms of sectional elements in river water

Element	The median/(μg/L)	Content range/(μg/L)	Possible speciation
As	0.5	0.2～230	As(Ⅲ, Ⅴ), arsenic methide
B	15	7～500	$B(OH)_3$
Br	14	0.05～55	Br^-
C	1.1×10^4	6×10^3～1.9×10^4	HCO_3^-
Ca	1.5×10^4	2×10^3～1.2×10^5	Ca^{2+}
Cd	0.1	0.01～3	Organic complexes, chelate compound
Cl	7.0×10^3	1×10^3～3.5×10^4	Cl^-
Cu	3	0.2～30	Organic complexes, chelate compound
F	100	50～2.7×10^3	F^-
Fe	500	10～1.4×10^3	Colloid
Hg	0.1	1.0×10^{-4}～2.8	Organic complexes, chelate compound
K	2.2×10^3	500～1.0×10^4	K^+
Mg	4.0×10^3	400～6.0×10^3	Mg^{2+}
Mn	8	0.02～130	
N	50	2～1.8×10^3	NO_3^-
Na	6.0×10^3	700～2.5×10^4	Na^+
Ni	0.5	0.02～27	
P	20	1～300	$H_2PO_4^-$
Pb	3	0.06～120	
S	3.7×10^3	200～4.0×10^4	SO_4^{2-}
Sb	0.2	0.01～5	Sb(Ⅴ)
Si	7.0×10^3	500～1.2×10^4	
Sn	9.0×10^{-3}	0.004～0.09	Sn(Ⅳ), stannum methide
Sr	70	3～1.0×10^3	Sr^{2+}
Zn	15	0.2～100	Zn^{2+} organic complexes and chelate compound

③ The Nutrients.

The nutrients refer to ions which are concerned with biological growth, such as N, P, Si, etc., non-metallic elements and some trace elements (Such as Mn, Fe, Cu, etc.). The contents of these elements are generally in mg/mL-ng/mL. Their existing form is concerned with the acid-base and the oxidation-reduction quality of the water.

Nitrogen and phosphorus are nutrient elements, which are essentially used to ensure the growth and breeding of aquatic creatures. But if the content of nitrogen and phosphorus in the water is too high, "eutrophication" phenomena will happen. Iron is a kind of familiar mineral element, its main existing form in the underground water is Fe (Ⅱ), and content can reach to mg. As the dissolved oxygen is enough, iron usually exists by Fe (Ⅲ) in the surface water. The silicon of natural water is mostly from silicate and hydrolysis of the aluminosilicate.

④ Organic substance.

There are many kinds of organisms in natural water. They are products of photosynthesis of aquatic plants and mixtures of resolving products of aquatic animals during different stages. Organisms in water generally are divided into corrosive mediums and non-corrosive mediums. Non-corrosive mediums include carbohydrate, fat, protein, vitamins, organisms of low molecular weight and so on. Most of organisms in water are brown or black amorphous corrosive mediums; the range of their molecular weights is hundreds to tens of thousands. Generally according to the conditions of dissolving of corrosive mediums in acidic and alkaline environment, they are divided into humin and humic acid (HA). Humic acid is divided into fulvic acid (FA) and humin, as shown in the Figure 3-1. In addition to a large number of benzene rings, the corrosive medium contains active groups such as hydroxyl (—OH)、carboxyl (—COOH)、carbonyl ($>O=O$) and so on, so it has many chemical activities including low-acid, the ion exchange ability, coordination compound, redox and so on. It can combine with metal ions in water and form stable water-soluble and water-insoluble compound, it also can interact with organisms. The corrosive medium can make a big difference on migration and transformation of pollutants in water such as heavy metals and so on.

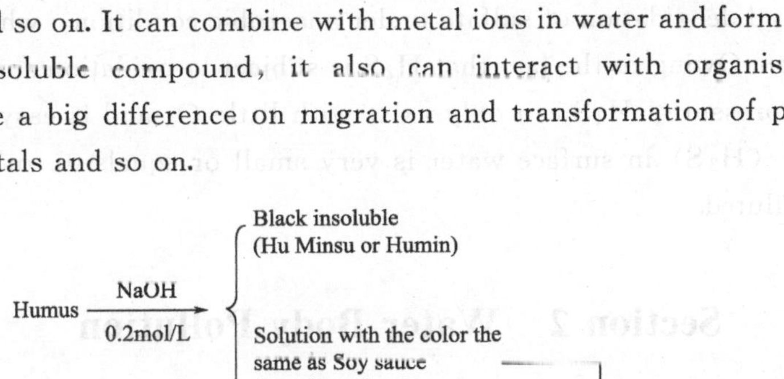

Figure 3-1 The classify of corrosive mediums

⑤ Dissolved gas.

Generally, gases in natural water include oxygen, carbon dioxide, sulphureted hydrogen, nitrogen, methane and so on. These gases come from dissolution of all kinds of gases in atmosphere, activities of aquatic plants and animals, chemical reactions and so on. Gases in seawater maybe also come from seabed volcanic eruption. There are content ranges of the main dissolved gases in seawater as shown in the Table 3-6.

Table 3-6 Main gases content dissolved in seawater

Gases	O_2	N_2	CO_2	H_2S	Ar
Content range /(mg/L)	0~8.5	8.4~14.5	34~56	0~22	0.2~0.4

Oxygen (O_2) and carbon-dioxide (CO_2) are prime gases dissolved in water, they can affect the survival and multiplying as well as the chemical and biochemical reaction such as combination reaction of the substances in water. O_2 in seawater springs from two sources, one is gathering from O_2 in air, which occurs when $c(O_2)$ in seawater is below its saturation concentration. The other is O_2 released by photosynthesis of aquatic plant. The oxidation of organics and respiration of organism as well as ferment and rot of skeletal remains of sea creatures consume O_2 in water. The subject $c(O_2)$ in natural water generally 0-14 mg/L. CO_2 in water is from oxygenolysis of organism and metalbolism of marine flora and fauna and solution of CO_2 in the air, and goes to dissolution of carbonate as well as the photosynthesis of marine plants. Besides from O_2 and CO_2, there may exist hydrogen sulfide (H_2S) in natural water when ill ventilated. H_2S in water is marginally from decomposition of sulfur organic compounds and deoxidation of sulfate under anaerobic condition, while is mostly from paroxysmal eruption. Owing to the fact that H_2S is subject to oxidation and make up a small quantity of partial pressure, H_2S can only exist with little O_2 and is easy be released from water. So usually $c(H_2S)$ in surface water is very small or equals 0, otherwise the water must have been polluted.

Section 2 Water Body Pollution

1. Water body pollution and self-purification

Water body pollution: is primarily caused by flowing of contaminant emitted by human into water bodes like rivers, lakes, oceans or underground water, which changes physical and chemical properties or biocenosis composition of water and sediment, thus cut down the use value of water body. This is called water pollution.

Self-purification of water body: broadly speaking, it means contaminated water lessenings concentration of contaminates gradually and at last reaches its previous unpolluted condition after a period of time by means of physical, chemical, biological and other actions. In a limited sense, it means the effect of purifying water body by method of microorganism's decomposing organic pollutant.

There are many factors influence water purification, such as the terrain and hydrological conditions of rivers, lakes, oceans, varieties and quantities of microbes in water, water temperature and oxygen status and properties and concentration of pollutants, etc. Principles of self-purification of water include precipitation, dilution, mixture and something like these physical processes. Oxidation-reduction reaction and decomposition-combination reaction and adsorption-cohesion and something like these chemical processes, physical chemical processes and biochemical processes are also included. Each process can occur in the same time and influences each other.

There are three types of self-purification of water.

① Physical self-purification. Insoluble solid substances sink to the bottom and generate mud after pollutants get into water. Because of the mixed dilution, the concentration of suspended substances, colloid and soluble pollutants will reduce.

② Chemical self-purification. When pollutants get into the water, they'll be purified by complexation, oxygen-reduction reaction and precipitation reaction, etc. For example, insoluble sulfide can be oxidized to soluble sulfate in water under some certain conditions.

③ Biological self-purification. The quantity, concentration and toxicity of pollutants will reduce or even disappear under creatures. For example, organic pollutants suspending and dissolving in water need creatures to be oxidized into simple and stable inorganic substances, such as H_2O, CO_2, nitrates and phosphates in order to purify the water. In general, physical and chemical processes dominant in the whole self-purification of water. In terms of organics, however, biological self-purification is most significant. Self-purification of water can be not only in the same mediator but also in different mediator. For example, the process of self-purification of rivers as follows. When polluted water enters the river, mixed dilution and diffusion will occur. Precipitations formed by reaction and solids with pollutants on them will sink to the bottom of the water to make the concentration of pollutants down. Then, the most important step of self-purification of water relies on the microorganism. They regard pollutants as resources of nutrition and change complex compounds to simple through biochemical processes. Final products after these processes are H_2O, CO_2 and some other inorganic substances. In addition, all kinds of aquatic creatures absorbing large solids and other livings including bacteria and plants also contribute a lot to the self-purification of the river. Moreover, photosynthesis of green plants and algae are beneficial to the purification.

The effect of self-purification in water is limited. If human beings put a large amount of pollutants into the water directly or indirectly, waters are going to be contaminated. In principle, pollutants in water can be purified sooner or later. However, speed and difficulty of purification are different because of difference of environment, difference of properties of pollutants and extend of contamination. Knowing the properties, contents, forms of existing and chemical behaviors of pollutants is critical for studying the ability of self-purification in water and taking action to prevent and eliminate the harms caused by contaminating.

2. Pollutants in water

Water pollution is mainly caused by man-made pollutants. There are many kinds of pollutants, including inorganic and organic toxic substances, oxygen-consumption organic matter, petroleum substances, radioactive substances, thermal pollution, pathogenic microorganism and so on. Here is the main water pollutants will be explained respectively.

(1) Pathogen pollutants

Domestic sewage, waste water from livestock and poultry farms, waste water discharged from leather making factories, wool washing factories, slaughtering industry and hospitals usually contain various kinds of pathogens, such as viruses, bacteria and parasites. Water contaminated by pathogens would spread diseases, such as schistosomiasis, cholera, typhoid, dysentery, viral hepatitis etc. Some of the plagues, which were popular in history, were caused by water-borne infectious disease. For example, British cholera epidemics in 1848 and 1854 caused over ten thousand of deaths, and in the cholera epidemics in Germany in 1892, over 750 people were killed. All these were caused by water pollution.

After contaminated by pathogens, the amount of microorganism in water increases sharply, most of which is bacteria, disease eggs and viruses. Usually they coexist with other germs and escherichia coli, thus the total amount of germs, the escherichia coli index and bacteria number of values are usually regarded as direct indicators of pathogen pollution. The features of pathogen pollution include: ①large amount, ②wide distribution, ③long survival, ④fast breeding, ⑤prone to resistance and hard to extinct, ⑥after traditional secondary biochemical sewage treatment and chlorine disinfection, some pathogenic microorganisms and viruses can still survive. Common ways of treatment such as coagulation, sedimentation, filtration, and disinfection can remove more than 99% of the viruses in water. For example, when water turbidity is more than 0. 5 degrees, the penetration of viruses still exists. Pathogens pollutants can enter the water through a variety of ways, as long as the conditions are suitable, human diseases would be caused.

(2) Oxygen contamination

Organic substances such as carbohydrates, protein, fat, lignin etc. exist in domestic sewage and waste water discharged from food-processing and papermaking industries. These substances exist in suspended or dissolved state in wastewater and can be decomposed by microbial biochemical effects. During the decomposition process, they need to consume oxygen, thus this process is called oxygen-consumption pollution. These pollutants can cause the decrease of water-dissolved oxygen, affecting the growth of fish and other aquatic organisms. After the water dissolved oxygen depleted, organic matter goes into anaerobic decomposition, then produces smelly gases, such as hydrogen sulfide, ammonia and mercaptan etc. causing water quality further worsened. Composition of organic matter in water is very complex. The concentration of oxygen-consumption organisms per unit volume of water is usually expressed by the amount of oxygen consumed in the process of biological decomposition. That is to say, it is expressed by biochemical oxygen demand (BOD), generally by biochemical oxygen demand (BOD_5) in five days when it is 20℃.

(3) Plant nutrients

Plant nutrients such as nitrogen and phosphorus mainly refer to stimulate the growth of algae and aquatic plants, water purification, enable higher BOD_5 substance. "Eutrophication", caused by the excess nutrients of the water, was harm to lakes and low flow velocity water body and this phenomenon has become a serious problem for water conservation.

Eutrophication refers to a phenomenon that under the influence of human activity, nitrogen, phosphorus and other nutrients which creatures needed enter into the slow flow water bodies such as lakes, estuaries, baies, which cause rapid reproduction of algae and other plankton, water dissolved oxygen decrease, water quality deterioration, death of fishes and other creatures. Under the natural conditions, lakes maybe transform oligotrophic state to eutrophic state, sediment grow rapidly, become swamp first, and then turn into land. This natural process is very slow, and it often needs thousands or even tens of thousands of years. And anthropogenic emissions of industrial wastewater and domestic sewage, which contain a lot of nutrients, led to eutrophication phenomenon, which can appear in a short time.

Plant nutrients sources are abundant, such as domestic sewage (organic matter, detergent), agricultural (chemical fertilizer, farmyard manure), industrial wastewater, waste and so on. About 50g nitrogen per person each day discharge into wastewater. The phosphorus of domestic sewage mainly comes from washing wastewater, and 50%-80% of chemical fertilizer which were used in farm was discharged into rivers, lakes and underground water. The content phosphorus and nitrogen (particularly phosphorus) of natural water can control the growth of plankton in a certain extent. When a large number of nitrogen and phosphorus of plant nutrients discharged into water body, they will prompt some creatures like algae growing sharply, shorten the growth cycle. The dead algae and other plankton were decomposed by aerobic organism, constant consumption of dissolved oxygen in the water, or decomposed by anaerobic organism. They continuously produced the gas such as hydrogen sulfide and made water quality deterioration, killing a large number of fish and other aquatic organisms. In the process of decay of algae and other plankton residue, nitrogen, phosphorus and other nutrients creatures needed were released into water, which algae and other creatures can reuse. Therefore, after the eutrophication of water body, even though the outside sources of nutrients were cut off, the water body can still hard to self-purify and return to normal levels. Eutrophication is serious, the growth of algal blooms and its debris can silt up some lakes, and they will become swamps even dry lands. Local sea area can be turned into "Dead Sea", or led to "red tide" phenomenon.

Commonly the contents of nitrogen, phosphorus, productivity (O_2) and chlorophyll-α were used as indicators of eutrophic state. Prevention of eutrophication has to control the high content nitrogen and phosphorus of fertilizer discharge into the water body.

(4) Toxic pollutants

The toxic pollutants refer to a certain amount of substances that accumulate into the organism, which can make changes in biochemical and physiological functions of the body fluids and tissue, cause temporary or permanent pathological condition and even be life-threatening such as heavy metals and organic pollutants that are difficult to break down. The intake body and the number of toxic contaminants are closely related to. Toxic pollutants are also related to its existing form closely. Valence or different forms of them can be very different toxicity. For instance, Cr(Ⅵ) is more toxic than Cr(Ⅲ) and As(Ⅲ) is more toxic than As (Ⅴ) as well as the toxicity of methyl mercury is much greater than inorganic mercury. In addition, toxic pollutants are also related with the several combined effects closely. In the view of traditional toxicological, the combined effect of toxic pollutants on organisms is in three ways.

① additive effect is that when two or more poison coexists, the total effect is substantially the sum of the effects of each component.

② The synergistic effect is that if there exist the coexistence of two or more poison, one ingredient promotes to increase dramatically another ingredient toxicity. Such as when copper and zinc coexist, their toxicity is 8 times than when they are separated.

③ Antagonism, when two or more poison coexistence, its toxicity can offset part or parts. Such as zinc can inhibit cadmium toxicity; and under certain conditions selenium can produce antagonism on mercury. In short, in addition to considering the content of toxic pollutants, we must also consider its existing form and combined effect, so as to in-depth understanding of the impact of pollutants on water quality and human health.

Toxic pollutants are the following categories.

① Heavy metals, such as mercury, cadmium, chromium, lead, vanadium, cobalt, barium, among them, mercury, cadmium and lead are harmful; arsenic, selenium and beryllium are toxic as well. Heavy metals in nature generally difficult to disappear, they have been enriched through the food chain; these substances in addition to direct effects on the human body caused by disease, certain metals may also promote the development of chronic diseases.

② Inorganic anions, mainly NO_2^-, F^-, CN^- ions. NO_2^- is carcinogenic. Highly toxic cyanide mainly comes from industrial wastewater discharge.

③ Organic pesticides, polychlorinated biphenyls. There are about 6000 kinds of organic pesticides in the world; about 200 kinds of them are commonly used. Pesticides are sprayed on farmland, and so into the water through leaching, resulting in the pollution effect. Organic pesticides can be divided into organophosphorus pesticides and organochlorine pesticides. Organophosphate pesticides are greatly toxic, but generally susceptible to degradation, accumulation ability is not strong, so the impact on the ecosystem is not obvious; the vast majority of organochlorine pesticides are toxic and hardly to degradate, their accumulation ability is very high, so there is a significant impact on the ecosystem. Polychlorinated biphenyls (PCB) are a general term for a part of the hydrogen molecule biphenyl or all of hydrogen substituted by chlorine after various isomers of the resulting mixture of. PCBs is toxic, fat-soluble large, easily absorbed by biological, and its chemical properties are very stable and difficult to have effects with acids, alkalis, oxidizing agents and so on, and it fully decomposes under a high degree of heat resistance at 1000-1400 ℃ high temperature, so it is difficult to degradate in water and biological.

④ Carcinogens. Carcinogens are generally divided into three categories: polycyclic aromatic hydrocarbon (PAHs), such as 3,4-benzopyrene, etc; heterocyclic compounds, such as aflatoxin, etc; aromatic amines, such as A, B aniline, benzidine, etc.

(5) Oil pollutants

Oil pollution is one of the important types of water pollution, especially more prominent in estuaries, coastal waters. The oil that discharges into the sea estimated at up to thousands of millions of tons of tons per year, accounting for about five thousandths of world oil production. Petroleum pollutants mainly from industrial emissions, cabin cleaning oil transport vessels, mechanical and accidents, offshore oil, etc. can cause oil pollution. The tanker accidents are concentrated sources of explosive, which is devastating harmful.

Oil is alkanes, alkenes and aromatics mixture, which is multifaceted hazard after being discharged into the water. Such as to form a film on the water, the water can hinder the role of reoxygenation. Oil adheres to the gills, fish can suffocate; Oil adheres to the algae, plankton, leads them to death. Oils inhibit waterfowl spawning and hatching, what is worse, it even makes a lot of birds die. Oil pollution also makes aquatic product lower quality.

(6) Radioactive contaminants

Radioactive contamination causes after radioactive substances are discharged into the water. Radioactive contamination mainly comes from nuclear power plant cooling discharged water, jettisoning of radioactive waste, nuclear explosion fallout landed a body of water, nuclear-powered ship nuclear fuel spills; When mining, refining and making use of radioactive materials, if handling improperly, it can also cause radioactive contamination. Radioactive contaminants in water can be attached to the surface of the organism and enter the organism to accumulate, they can also produce internal exposure through the food chain to people. Main natural radioactive elements in water include ^{40}K, ^{238}U, ^{286}Ra, ^{210}Po, ^{14}C, etc. Currently, ^{90}Sr and ^{137}Cs are measured almost in any sea in the world.

(7) Acids, bases, salts of inorganic contaminants

Various acids, bases, salts and other minerals (acid, alkali and salt formation, their interaction with certain mineral waters certain salts) increase salinity of freshwater resources and effect various water qualities when they are into the water. Salt contamination mainly comes from sewage, mining waste and some industrial waste. In addition, due to the growing scale of acid rain, the acidification and the salinity of groundwater increase. That Inorganic salts in water increase can improve the osmotic pressure of the water, which makes an adverse effect on the growth of freshwater organisms and plants. In salinization area, surface water and groundwater will have a greater impact on soil quality salt.

(8) Thermal pollution

Thermal pollution is an energy pollution, it is caused by wastewater of high temperatures discharged to water by the industrial and mining factory. If the cooling water of some thermal power plants and various industrial process is failure to take measures to be directly discharged into the water, this can make the water temperature increase and chemical reactions happen in water, in addition, it will accelerate the rate of biochemical reactions so that the toxicity of some toxic substances (such as cyanide, heavy metals, etc.) increase, dissolved oxygen decrease, and that affects fish survival and reproduction, accelerates certain bacteria multiply and makes fueled plants cluster, anaerobic fermentation.

There is an optimum growth temperature range for fish. If water temperature is too high or too low, it is not suitable for the growth of fish, even leads them to death. The adaptability of water temperature for different fish is different. For example, tropical fish is adapted to 15-32 ℃, warm octopus fishes adapted to 10-22 ℃ and cold octopus fish is adapted to 2-10 ℃. Another example although the temperature of water for the life of the trout is 24 ℃, their breeding temperature is below 14 ℃. Temperature limit for life of generally aquatic organisms is 33-35 ℃ range. Except the eight classes of pollutants, detergents and other surfactants on the water environment of the water main harm is that they produce foam to prevent air and water contact and reduce dissolved oxygen, and because the biochemical degradation of organic matter leads to the consumption of dissolved oxygen in water, so that the oxygen in water decreases. High concentrations of surfactants on microorganisms have significant toxicity. There are many examples of water pollution, for example, there are many factories on both sides of the Beijing-Hangzhou Grand Canal (Hangzhou section), they discharge a large number of wastewater into the canal every day, so that water, suspended solids, organics, heavy metals (Zn, Cd, Pb, Cu, etc.) and phenol cyanide and other content much higher than surface water standards, some more than a few times, so that the water body is reduced state anaerobic, black smelly and fish disappear, even the water cannot be used for agriculture and life. The self-purification capacity of water is poor, if this is out of control, water pollution will be further expanded.

Water contamination in the environment on the whole can be divided into inorganic and organic pollutants. Heavy metal and organic pollutants are more important and studied more in the water environment chemistry. Chemical Water Pollution study in China began in the 1970s with research for pesticide pollution from heavy metals, oxygen organics, DDT, BHC, at present, the research focus has shifted to organic pollutants, especially refractory organics, because of its persisting in the environment for their long period, and easily passing along the food chain (net) accumulation (enrichment), the threat of biological growth and human health, so people pay more attention to it. This chapter focuses on heavy metals and organic pollutants in the water environment, chemical conversion migration behavior.

3. The movement procedure of pollutants after discharging into the water

The various sports of pollutants immediately occur after into water bodies. The following case example for a brief introduction to the ocean, other water bodies can be by analogy. Movement of pollutants after discharged into water bodies are as shown in Figure 3-2.

Process 1 and 2 in Figure by diluting contaminants in seawater, diffusion, biological activity caused by movement of pollutants spatial location. The main factors affecting the process is sea hydrological conditions, including multiple variables tides, currents, water advection and eddy current and so on.

Process 3 is a contaminant in the process of transformation in the ocean, including the role of chemistry, physics, chemistry, water chemistry and biochemistry. Through the transformation process, the concentration of pollutants, in particular form will change,

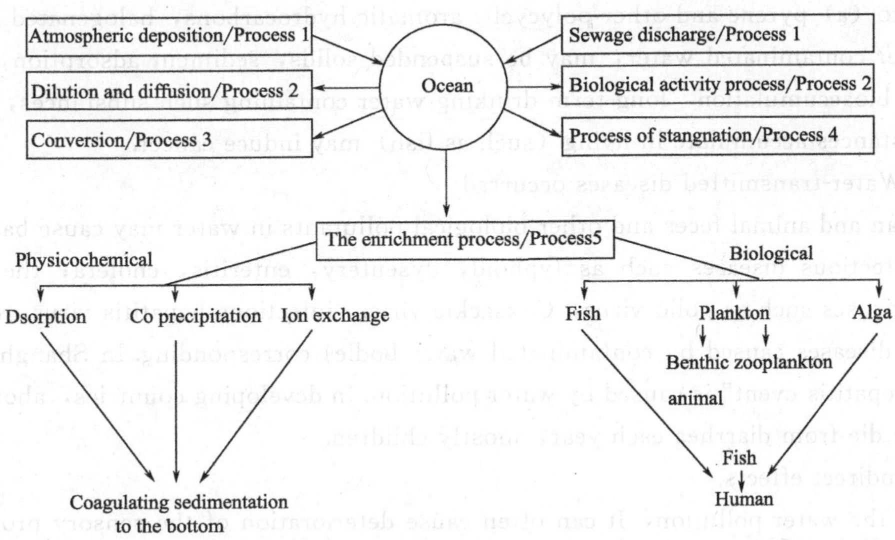

Figure 3-2 Movement of pollutants after discharged into water bodies

thus affecting the distribution of contaminants in the ocean, migration, stay, enrichment and their toxicity, sediment of secondary pollution. This is an important chemical pollution research.

Process 4 is a process to stay the course of pollutants in seawater. This process involves the question of the duration of pollutants endanger duration; it depends on the exchange rate updates, and water and sediment colloids, suspended solids seawater. Contaminants stay time τ in seawater can be calculated:

$$\tau_i = A_i/(dA_i/dt)$$

Where A_i is the amount of water discharged pollutant i, dA_i/dt is the rate of deposition of pollutant i in the ocean. Under normal circumstances, the activity of pollutants in the sea, the more short residence time.

Process 5 is a contaminant in the process of enrichment in the ocean, it depends on the selective absorption of other physical and chemical adsorption and sedimentation enrichment food chain, as a result of pollutants from the water, purify the latter, while the in varying degrees harmful organisms, and increase the accumulation of sediment contaminants may cause secondary pollution of seawater.

4. Water pollution effect on human body

Water pollution hazard performed in many ways, briefly outline the impact of water pollution on human health.

(1) Cause acute and chronic poisoning

After water contaminated toxic and hazardous chemicals, it may cause poisoning through drinking water or food chain. The famous Minamata disease, Itai-itai disease is caused by water pollution.

(2) Carcinogenic effects

Some are carcinogenic chemicals such as arsenic, chromium, nickel, beryllium, aniline, benzo (a) pyrene and other polycyclic aromatic hydrocarbons, halogenated hydrocarbons, after contaminated water, may be suspended solids, sediment adsorption, but also in aquatic bioaccumulation, long-term drinking water containing such substances, or eating these substances accumulate in living (such as fish) may induce cancer.

(3) Water-transmitted diseases occurred

Human and animal feces and other biological pollutants in water may cause bacterial intestinal infectious diseases such as typhoid, dysentery, enteritis, cholera; the intestine common viruses such as polio virus, Coxsackie virus, infectious hepatitis virus, etc. , can infectious diseases caused by contaminated water bodies corresponding. In Shanghai, when 1989, "Hepatitis event" is caused by water pollution. In developing countries, about 6000×10^4 people die from diarrhea each year, mostly children.

(4) Indirect effects

After the water pollution, It can often cause deterioration of the sensory properties of water, such as at a certain concentration of certain pollutants on human health Although there is no direct harm, but it can happen smelly water, color, showing the foam and film, etc. , hinder normal use of water bodies. Copper, zinc, nickel and other substances in certain concentrations can inhibit microbial growth and reproduction, thus affecting the biological decomposition and oxidation of organic matter in water, the water purification capacity decline, affecting the health status of water bodies. Water pollution can seriously harm the ecosystem, but also cause serious economic losses.

Section 3 Water Environment Chemistry of Organic Pollution

1. Parameter of organic pollution

There are so many types of complex organic pollutant in water that it is difficult for the modern analysis technology to determine the content of them respectively. Therefore, we can summarize their common characteristics and make use of the parameters, which indicate the pollution degree of organic matter in water indirectly. Common parameters: dissolved oxygen, biochemical oxygen demand, chemical oxygen demand, total organic carbon and total oxygen demand.

(1) Dissolved oxygen (DO)

Dissolved oxygen is the content of dissolved oxygen in water under prescribed temperature and pressure. It is one of the important parameters of water quality. Dissolved oxygen in water is affected by two roles, one is the oxygen consumption, including oxygen consumption of organic matter degradation, biological breathing, etc. , which reduces the DO; The other is are-oxygenation mainly including dissolving oxygen in air, photosynthesis of aquatic plant, which increases the DO. The interplay of the two roles makes DO presents the change of time and space. In addition, DO decreases with temperature rises, and also decrease with the increase of the depth of the water. Under normal temperature, Do in water is 8-14 mg/

L; In the water full of algae, the DO is probably in saturated state; In the water full of organic pollution, oxygen consumption works greater than the re-oxygenation, as a result the DO reduces; If the Organic pollution is too serious, the DO is zero. In the water lack of oxygen, the growth of aquatic animals and plants will be suppressed which even lead to death. For example, when Do is less than 4 mg/L, the fish will die. So through the determination of DO in water, we can evaluate the pollution degree and self-purification capacity of water. The methods of DO determination in water are iodine quantity method, sodium aside correction method, the $KMnO_4$ correction method and membrane electrode method, in which the iodine quantity method the most commonly used.

(2) Biochemical oxygen demand (BOD)

Biochemical oxygen demand (BOD) is the consumption quantity of dissolved oxygen during the process of microbial decomposition of organic matter in water, whose unit is mg (O_2)/L. BOD reflects the total quantity of organic matter which can be decomposed by microbes in water. Microbial oxidation decomposition of the organic matter processes in two stages. The first stage is mainly the conversion of organic matter to inorganic matter such as CO_2, H_2O and ammonia; the second stage is the conversion of ammonia to NO_2^- and NO_3^-. For the environmental impact of the second stage is small, BOD generally refers to the oxygen demand of microbial decomposition of organic matter in the first stage. The speed and degree of microbial decomposition of organic matter related to the temperature and time. For example, at 20℃, organic matter in domestic sewage can basically completed the first phase of biochemical oxidation in about 20d, but it can also be completed around 70% of the first phase transformation after 5d. To shorten the measurement time and make the BOD values comparable, adopt the determination of biochemical oxygen demand after the cultivation of 5d under the condition of 20℃ as standard method, which is expressed as BOD_5. BOD can basically reflect the oxygen consumption during the oxidative decomposition of organic matter under natural situations and manifest the environmental impact of aerobic organic pollutants. However, because measure time is long and microbial activity is controlled in wastewater with high toxicity, it's difficult to measure BOD accurately. But we can get acquire of contamination status of organic matter in water quicker through the determination of chemical oxygen demand.

(3) Chemical oxygen demand (COD)

Chemical oxygen demand (COD) is the consumption quantity of oxidizing agent during the chemical oxidation of the matter that can be oxidized in water under prescribed conditions, whose unit is mg (O_2)/L (the number of milligrams of consumption oxygen per liter of water). The higher the COD of water presents, the more serious the organic matter pollution is. Causing the chemical oxidation reaction of various organic matter in water is in different difficulty level, COD means to be the sum of oxygen demand of the matter that can be oxidized in water under prescribed conditions. The common methods for determination of COD are $KMnO_4$ method and $K_2Cr_2O_7$ method. The former one is suitable for the determination of cleaner water for the relatively weak oxidability. The latter one is usually applied to polluted water and industrial wastewater. Determination result of the same water sample with the above two methods is different. Therefore, it is essential to indicate the determination method when reporting the result of COD.

Compared with the BOD, COD is not restricted by water quality conditions, the measurement time is short, but COD can't express the amount of microbial organic matter that can be oxidized well. Aerobic oxidation can't oxidize some aerobic organic matter, but can oxidize inorganic reducing substances (sulphide, iron, etc.). So, as an evaluation index of aerobic organic pollution, the chemical oxygen demand BOD fit better. But when the conditions are not available to do BOD determination or being limited by water quality, COD is available instead. In addition, under conditions of relatively stable water quality, between COD and BOD have a relatively close correlation. Generally, the order should be potassium dichromate COD, BOD_5 and potassium permanganate method COD.

(4) Total organic carbon (TOC) and Total oxygen demand (TOD)

TOC (total organic carbon, referred to as TOC) is almost entirely carbon content of organic matter in water. The total amount of oxygen required for COD (total oxygen demand, referred to as TOD) substances in the water is almost all that can be oxidized (essentially organic) into a stable oxide. Because BOD measurement is time-consuming, to achieve that reacting organic pollution rapidly, TOC and TOD assay measuring is suggested to use. It needs 3 minutes every time and it operates continuously and automatically. They are all available for the determination of chemical combustion, carbon measurement results indicate the former, and the latter is indicated by aerobic oxygen content of organic matter. They are an indicator of evaluating aerobic organic pollutants in water. However, the determination of total organic carbon and total oxygen demand is by no means the complete oxidation of organic matter in water, oxidizing conditions when measured and nature oxidation very far, for a total impact of inorganic oxygen demand may not be the nature ozone-depleting substances standardization substances, as well as the measuring device is not fully resolved, so they can't be evaluated as a universal water aerobic organic pollutants indicators. Because the process of oxygen consumption determination is different, and there is a variety of different organic components in water, chemical and biological processes are large differences, so between the various water quality, TOC or TOD and BOD_5 there is no fixed correlation. Under the same water conditions, water BOD_5 between TOC and TOD has a certain degree of correlation. Such as Japan Tamagawa river, BOD and TOC correlation coefficient is 0.6, the regression equation is:

$$BOD_5 = 1.72TOC - 1.9 \text{(TOC for } BOD_5 \text{ return)}$$

$$BOD_5 = 1.04TOC + 4.2 \text{(} BOD_5 \text{ return for TOC)}$$

The correlation coefficient between TOD and BOD_5 0.86, the regression equation is:

$$TOD = 1.34BOD + 4.7 \text{(TOD return for BOD)}$$

$$TOD = 1.15BOD + 10.4 \text{(BOD return on TOD)}$$

DO, COD, BOD_5 is usually used as indicators of water quality assessment, surface water classification criteria in Table 3-7.

Table 3-7 Surface water quality standards (mg / L)

Water class		Ⅰ	Ⅱ	Ⅲ	Ⅳ	Ⅴ
DO	≥	Saturation 90%	6	5	3	2
BOD_5	≤	Under 3	3	4	6	10
COD_{Mn}	≤	2	4	6	8	10

(5) Biological contaminants indicators

Biological contaminants mainly refer to pathogenic microorganisms and other organisms in wastewater. Wastewater, especially domestic wastewater often contains many microorganisms. Most of them are harmless, but which also contains harmful pathogens to humans and other biological pathogens. Such pollutants are mostly from hospitals, slaughterhouses, Institute of Biology, and sewage and so on.

The biological indicators are mainly bacteria, coliform count, pathogens and viruses. The total number of bacteria refers to the total number of all kinds of bacteria in 1 mL of water. Water quality analysis, is to inoculate a certain amount of water in agar medium and incubated at 37 ℃ for 24 hours, the number of colonies of bacterial growth, and then calculate the number of bacteria per milliliter of water contains. Coliform count refers to the number 1 L of water-contained coliform. Because of coliform bacteria in the external environment, living conditions and intestinal infectious parasite eggs similar, and the number of coliform much easier to test, so make the number of E. coli as a biological pollution indicators.

(6) Sensory pollutant index

Wastewater color, turbidity, foam, etc. can cause a bad smell unpleasant sensory contaminant people called sensory pollutants, especially in exploring the waters, the impact of these sensory pollutants are more serious. Common sense said that pollutant indicators color, color, smell and taste like. Color refers to the color of the type of waste, usually text description, such as: dark blue, brown, light green, dark, red. Chroma refers to the degree of color that wastewater presented. Chroma has two-color representation. First, Platinum-cobalt standard colorimetric, stipulate the color that In 1L of water containing chlorine platinum potassium (K_2PtCl_6) 2.491 mg and cobalt chloride ($CoCl_2 \cdot 6H_2O$) 2.00 mg, that is in 1 L water containing 1 mg of platinum and cobalt 0.5 mg, is 1 °. Second, Dilution method, wastewater was diluted, Chroma represent by the dilution of diluting water to near colorless.

Smell and taste is one of the sensory indicators to judge the merits of water quality. Clean water is no odor, will produce a variety of odor after contamination. There common are rotten smell, feces odor, the smell of gasoline, rotten egg smell, chlorine taste and so on. For now, representation of the smell is that use text describing the type of smell and use strong or weak represent smelly strength. More accurate quantitative methods are smelly threshold method, that use dilution of XiuShui-free water dilutes sample to be tested near odorless degree represents the strength of smell, multiples greater indicating the greater odor.

2. Degradation of some organic pollutants in water

Degradation of organic matter in the water environment is the main process of natural purification of environmental pollutants, it is mainly achieved by hydrolysis, oxidation, photolysis, biochemical decomposition . Some substances in water is more easily degradable, such as carbohydrates, fats, proteins, etc. , some is biodegradable, such as organ chlorine pesticides, PCBs, PAHs, etc. Following is a brief introduction of degradation of organic pesticides, synthetic detergents, oil, PAHs and PCBs.

(1) The degradation of organic pesticides

The world has more than 1000 kinds of organic pesticides, about 200 kinds of commonly used. According to the application, Organic pesticide can be divided into insecticides, fungicides, herbicides, seed selection, according to the chemical composition; pesticides can be divided into organ chlorine pesticides, organ phosphorus pesticides, organic mercury pesticides, and carbonate pesticides. Organ chlorine pesticides have a several types. Most are used as an insecticide, such as DDT, BHC, aldrin and so on. Their features are chemically stable, not easy to decompose, slowly toxicity, and residual long time, slightly soluble in water, soluble in fat and easily to be accumulated. Also the enrichment coefficient of aquatic life can be up to tens of thousands of times. Organ chlorine pesticides currently have limited to using, our country has stopped production in 1983. Most organ phosphorus pesticides are used as pesticides, such as parathion, trichlorfon, dichlorvos, its characteristics are toxic, easy to break down, weak accumulation effect, Thus it is not has obvious impact on the ecosystem, has a tendency to replace organ chlorine pesticides. Carbamate pesticides, such as Carbaryl insecticide, herbicides, etc. It is low toxicity pesticides on animals, shortly the residence time, easy to break down. Organic mercury fungicides are mostly pesticides, such as San Saili, West of hygiene, due to mercury contamination, has been to reduce the use.

After the pesticide into the water body, it can be in contact with all kinds of substances, a series of physical, chemical and biochemical reactions. Their behavior can be summarized as follows: the water is absorbed particles, adsorption and bioaccumulation, degradation reaction, decreased pesticide content. Following is a brief discussion of some common organic pesticide DDT, BHC, organ phosphorus and carbamate degradation.

① DDT degradation.

Photochemical degradation of DDT: Under the action of ultraviolet light, DDT may produce free radicals chlorobenzene by cracking carbon-carbon bonds, the latter combined with each other to generate dichloro-diphenyl. Such photochemical reactions may also have associated trichlorobiphenyl, tetrachlorobiphenyl, 3,6-dichlorobenzene and furans, and may generate PCBs by radical course.

Catalytic degradation of DDT: Metal can restore degradation of DDT by catalytic. Dechlorination of DDT may be two ways: draw off one chlorine atom becomes DDD, then dechlorination becoming DDMS and DDEt; or at the same time to get off three chlorine atoms and becoming DDEt.

Toxicity of DDD is same as DDT, and the toxicity of DDT DDEt is less than one-thousandth to DDT, so it is the best product of DDT reduction degradation.

Biochemical degradation of DDT: Studied about the biochemical degradation of DDT is much, and some people have used various microorganisms do culture test under anaerobic and aerobic conditions. Degradation products in different situations was detected, including DDD, DDE, DDMV, DDMS, DDNV, DDA, DDM, DBH, DBP, Kelthane and DDCN like. DDT degradation pathway has not yet clear. Rhead (1975) suggest possible ways to DDT degradation in Figure 3-3.

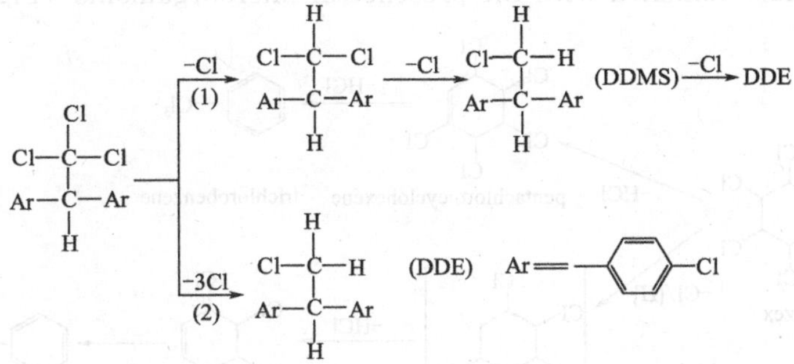

Figure 3-3 Possible ways to DDT degradation

The degradations of different microorganisms on DDT are not identical. DDE and DDD are general degradation products, while DDE is much more stable than DDD. DDCN is a staple product of anaerobic degradation in sewage sludge. It is supposed that DDCN has a larger possibility of being formed via chemical process. The biological actions in sewage sludge probably contribute to the maintenance of appropriate reduction conditions, which indirectly influence the transformation between DDT and DDCN.

② The degradation of benzex.

Compared to DDT, benzex is easier to be degraded by the approaches of photochemical, chemical and biochemical reactions as well.

The photochemical degradation of benzex: Generally, it is not easy for benzex to directly absorb photons to occur photochemical reaction. However, when some substances (aromatic compounds such as anilines and aromatic phenols, inorganic anions such as X^- and $S_2O_3^{2-}$) exist in the water, their molecules or ions can occur photochemical reaction under the excitation of light and produce some reactive intermediates with reducibility, such as hydrated electrons (e_{aq}^-). e_{aq}^- can react with benzex molecule, leading to the degradation of benzex by fracture of chemical bonds, which is similar to directly receive optical energy. Someone suggest that photochemical reaction of benzex belongs to first-order reaction on the basis of dynamic stable-state processing method, and the speed constant of photochemical reaction (K) is proportional to the concentration of electron donors in the water, while inversely proportional to the concentration of electron acceptors (such as Cu^{2+}, Zn^{2+}, DO and H_3O^+) which can decline the photochemical reaction rate of benzex. Thus, the photochemical reaction rate of benzex is very slow with the presence of DO in the water.

The chemical degradation of benzex: Someone suggests that the chemical degradation products of benzex are tetrachloro cyclohexene and pentachloro cyclohexene. HCl needs to be removed during the degradation, indicating that benzex is easy to be degraded in alkaline medium while relatively stable in acid medium.

The biochemical degradation of benzex: Microorganisms are crucial to the degradation of benzex. γ-Benzex can transform to γ-pentachloro cyclohexene by removing hydrogen chloride, or transform to α, β and δ-benzex isomer in anaerobic condition and be rapidly degraded in the water. Experimental results indicated that after soaking for two months, four isomers of benzex almost vanished with the presence of microorganisms (Figure 3-4).

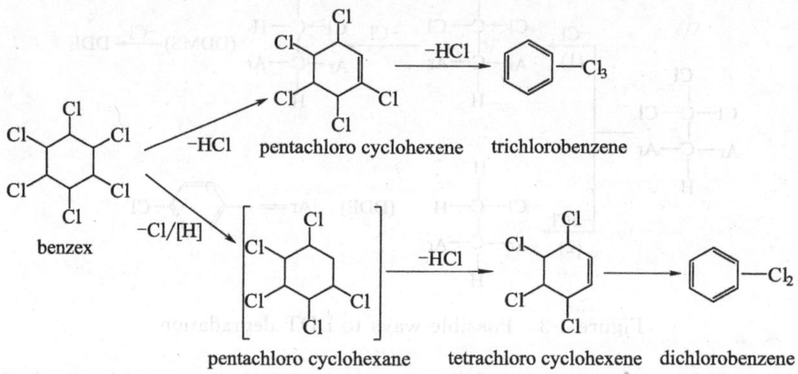

Figure 3-4 The degradation pathway of benzex

③ The degradation of organ phosphorus pesticides.

Organ phosphorus pesticides cannot only be directly hydrolyzed, but also be degraded by microorganisms. For example, under the effect of microorganisms, organic pesticides such as methyl parathion, ethyl parathion and fenitrothion can be degraded via the process of oxidation, reduction, hydrolysis and so on. Figure 3-5 shows the biochemical degradation

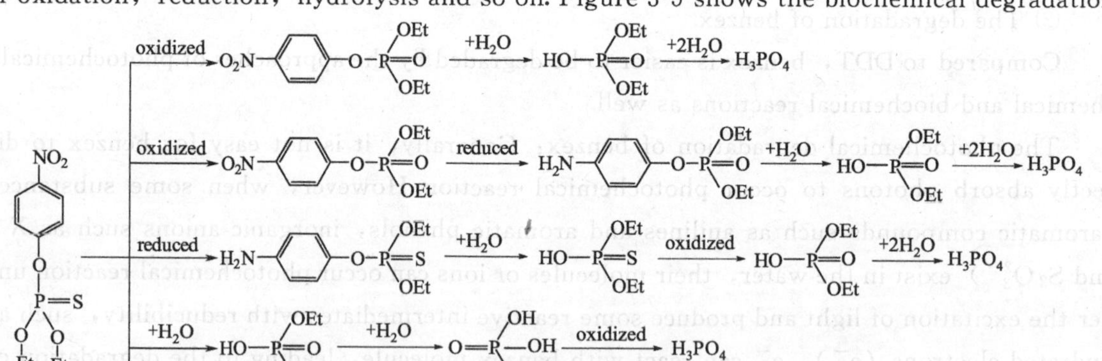

Figure 3-5 The degradation of ethyl parathion

pathway of ethyl parathion. The oxidation process of ethyl parathion refers to the desulfurized oxidation of phosphorothioate, the reduction process refers to the reduction of nitro to amidogen, and the hydrolysis process refers to fracture of ester bonds to form corresponding compounds.

Other organophosphorus pesticides such as malathion, DDVP and dipterex can also be degraded by microorganisms.

It is worth pointing out that hydrolysis is an important degradation way of pesticides in environment. The hydrolysis rates of different pesticides are influenced by temperature and acidity of medium. For example, at the temperature of 25℃, the order for stability of pesticides in distilled water is: malathion>methacrifos>fenitrothion>methidathion, while the order in alkaline solution is: fenitrothion>methidathion>methacrifos>malathion. For another example, DDVP and methyl parathion both belong to compounds with unstable hydrolysis. In the condition of surface water (25℃, pH=7), their $t_{1/2}$ is 43.9 h and 83.5 h respectively, while in the condition of underground water (15℃, pH=7), their $t_{1/2}$ is 133.3 h and 3013 h respectively. These two pesticides also have certain stability in acid medium. Their $t_{1/2}$ is 533 h and 2772 h respectively in the condition of pH 3.04 and 25℃.

Therefore, each kind of pesticides has its own degradation process. Even for the same pesticide, its degradation process will not be identical under different conditions or degradation by different microorganisms.

(2) Polychlorinated biphenyl (PCBs) and polyaromatic hydrocarbon (PAHs)

The pollution of polychlorinated biphenyl (PCBs) and polyaromatic hydrocarbon (PAHs) is widespread around the world, and exists in many samples of natural water and soil. Organisms can easily absorb PCBs and PAHs with high toxicity, large fat solubility and stable chemical properties. Moreover, some PAHs can lead to carcinogenesis, teratogenesis as well as mutagenesis, which cause a threat to human health. People pay more and more attention to the study about environmental chemical actions and biological effects of PAHs in the water.

According to the literatures, the biodegradation product of monochlorinated biphenyl and dichlorinated biphenyl is chlorobenzoic acid and dichlorobenzoic acid respectively. Chlorobenzoic acid can continue to be degraded in a slow rate, of which the stagnant period depends on bacterial inoculum.

In 1980, R. E. Ballay et al. found that organisms in river water can rapidly degrade three isomers of monochlorinated biphenyl. Their degradation rates in low concentration are similar, while the biodegradation is relatively slow in high concentration. However, no degradation phenomenon of 2,2′,4,4′-tetrachlorinated biphenyl is observed after cultivating for 98 d. Minxue Li et al. studied the migration and distribution of PCBs and organochlorine pesticides in Second Songhua River. The results showed that PCBs in river are hardly degraded by light and microorganisms under natural conditions. The vapour pressure of PCBs is very low and the frozen season of Second Songhua River lasts for nearly five months. Thus, the volatilization quantity of PCBs in river is very small. PCBs are easy to be absorbed by suspended particles and then sink into the bottom of river, which ascribes to their larger viscosity and proportion as well as slight solubility. Thus, the content of PCBs in sediment can be one order higher than that in river water. Though the detection ratio of PCBs in the water of Second Songhua River is not high, the enrichment amount in fish bodies is larger. It is considered that PCBs are difficult to volatilize and degrade in water environment. Furthermore, a large fat solubility of PCBs leads to their enrichment in the body of aquatic animals. Enrichment concentration in fish bodies can reach to a high level via the transmission of food chain. The enrichment coefficient reaches to 10^4.

Hongguo Liu et al. investigated the microbial degradation of PCBs. The selective enrichment culture was conducted by utilizing PCBs as sole carbon source. 25 PCBs-degrading strains were isolated from soil polluted by transformer oil in transformer substation. The results of substrate growth test showed that they could utilize various PCBs as sole carbon source. The degradation ability of eight strains on all kinds of PCBs was detected by gas chromatography and chloridion measurement. The degradation rates were generally 20%-50%, while the higher ones could reach to 60%-100%. The degradation ability decreased with the increasing chlorinity in PCBs. And there was a great difference between the degradative complexities of different isomers. The concentration of PAHs in natural water is relatively low (generally 0.001-10μg/L). PAHs have fat solubility and slight solubility. Thus, PAHs entering into water are mainly absorbed on suspended solids, aquatic organisms or sediment and ultimately sink into bottom mud, which is the main existence form of PAHs in the water.

Lingren Kong (1986) investigated the photochemical degradation of benz (a) anthracene (BaA). It showed that when DO existed in the water, oxygen in the ground state could be motivated by ultraviolet radiation. Oxygen in the excited state could easily react with BaA in the ground state or being motivated, generating the oxide of BaA, that was benz (a) anthracene, 7,12-quinone. The oxidation reaction between BaA and DO occurred here, and light exhibited a catalytic action for it. Liansheng Wang et al. (1991) systematically studied the photochemical degradation of over 17 common PAHs in the solution of methanol-water or acetonitrile-water. The results showed that photolysis rate constants (k) of most PAHs were relative to their half-wave potential of polarographic oxidation. The low half-wave potential meant the fast photochemical degradation rate. PAHs could not be degraded under dark condition. Moreover, it was detected that photolysis product of benzene was phthalic acid, and naphthalene could not be photolyzed in solution without DO. Thus, the photolysis of most PAHs belongs to optical oxidation reaction.

It can be concluded that the degradation process of organic pollutants is complicated. In the actual water, one organic pollutant can often be degraded by several degradation pathways simultaneously, or mainly by one degradation pathway due to the change of environmental conditions. The ultimate decomposition of organics is completed by the effect of microorganisms.

Section 4 Method for Wastewater Treatment

There are lots of methods for wastewater treatment. Their classification method is also different. Here two of them are introduced. The first one is classification according to the deletion way of pollutants from wastewater, which classified the methods into three types: ①Separation treatment, which is utilizing various methods to separate pollutants from wastewater and usually not changing the chemical properties of pollutants. ②Transformation treatment, which is utilizing chemical or biochemical methods to transform pollutants in wastewater to harmless substances, or transform to segregative substances, which are separated afterwards. ③Dilution treatment, which can't change the chemical properties of pollutants as well as separate them. This method declines the concentration of pollutants by dilution and mixture, while the total quantity of pollutants is constant. Thus, it is a passive method.

The second one is classification according to the degree of wastewater treatment, or the stage of treatment. Wastewater treatment can be usually classified into three levels according to the degree of treatment: first treatment, secondary treatment and tertiary treatment. First treatment is also called primary treatment, which can only delete suspended solids or floaters with large particles in wastewater, and is hard to reach the discharge standard. Secondary treatment usually can delete tiny or gelatiniform suspended solids and organics. Generally, it can reach the discharge standard. Tertiary treatment is also called advanced treatment, which can further delete colloids or pollutants in dissolved state in wastewater, and usually can achieve the purpose of recycle. In the following, common methods for wastewater treatment are summarized based on this classification method.

1. First treatment for wastewater

(1) Gravitational separation method

Utilize the effect of gravitation to separate pollutants. It can be classified into settlement separation and float up separation. Settlement separation is used for the deletion of pollutants with a larger density than water, while float up separation is used for the deletion of floaters with a smaller density than water.

(2) Resistance interception method

This method utilizes the different geometric dimension between sieves and suspended solids to intercept suspended solids. It includes interception methods by gratings, sieves or granulous medium.

(3) Dilution method

That is utilizing water of no pollutants or low-concentration pollutants to mix with water of high-concentration pollutants, which can reduce the concentration of pollutants.

(4) Neutralization method

Utilize the theory of acid-base neutralization to delete acidic or alkaline pollutants in wastewater.

First treatment for wastewater is relatively simple and can't reach the discharge standard in most cases. Thus, further treatment is required.

2. Secondary treatment for wastewater

(1) Gas flotation method

That is utilizing the hydrophobicity of pollutants in wastewater or adding some kind of medicament, which can make pollutants in wastewater hydrophobic. Then bubbles are passed into wastewater. The hydrophobic pollutants will be adsorbed on bubbles, with which the pollutants float to the surface of water and generate the foam layer. Pollutants can be separated from water when separating the foam layer and water.

(2) Coagulation method

That is adding electrolytes or coagulants into wastewater, or utilizing mechanical agitation to cohere pollutants, which exist in a colloid state at wastewater. Then a large and heavy flocculating constituent is formed and separated by gravitational sedimentation.

(3) Extraction method

According to the theory of distribution law, utilizing a organic solvent, which is immiscible with water and easy to dissolve some pollutants in wastewater, for the separation of pollutants from wastewater.

(4) Oxidation-reduction method

That is adding oxidizing or reducing agents into wastewater to oxidize or reduce the harmful pollutants to harmless or slightly harmful substances.

(5) Electrolytic method

Electrolytic treatment is utilizing the electrolytic fundamental for the transformation of pollutants in wastewater to harmless substances by the oxidation or reduction reaction happened on cathode and anode during the electrolytic process, in order to achieve the purification of wastewater.

(6) Biological method

That is a process during which microorganisms are used for the oxygenolysis of pollutants. It contains aerobic method and anaerobic method. Aerobic method is utilizing aerobic and facultative microorganisms to decompose pollutants when the water contains DO. Anaerobic method is utilizing anaerobic and facultative microorganisms to decompose pollutants when no DO exist in the water. Now biological method is used widely as a secondary treatment method, especially for municipal wastewater which is almost decomposed by biological method.

(7) Blow-off method

This method is used to delete gaseous or volatile pollutants in wastewater. Expose wastewater fully to air, and the gas or volatile pollutants dissolved in wastewater can be deleted via diffusing into air.

(8) Vaporizing extracts method

This method is also used to delete volatile pollutants in wastewater. Steam is utilized to directly heat wastewater to boiling, and volatile pollutants are deleted via escaping with water vapor. When the concentration of pollutants is high, the vapor can be condensed to recycle the pollutants.

3. Tertiary treatment for wastewater

(1) Adsorption method

Expose wastewater to solid adsorbents, which can adsorb pollutants in state of molecule or ion. Then pollutants can be separated from water after separating the water and adsorbents. In general, adsorbents can be recycled after regeneration.

(2) Separation membrane method

According to the difference of action principle, separation membrane method can be divided into ultrafiltration, electrodialysis and antiosmosis.

Ultrafiltration is also called as secondary filter. It is utilizing filtering medium to delete pollutants in wastewater. Unlike general filter, the aperture of medium used in ultrafiltration is very small (generally 1-0.1μm). This method can delete colloidal substances or macromolecular pollutants in water. Electrodialysis is getting wastewater through passageways composed by the alternative arrangement of anion-exchange membrane and cation-exchange membrane. Ions can selectively penetrate different membranes in the effect of direct-current electric field. Thus, pollutants in some passageways are concentrated, while wastewaters in other passageways are purified. Antiosmosis is utilizing semipermeable membrane to separate wastewater and clear water. Pressure is exerted on the surface of wastewater, making water molecules pass through the semipermeable membrane while pollutants can't pass through it. Therefore, pollutants are separated or concentrated.

(3) Magnetic filtration method

Depending on the effect of magnetic field, magnetic pollutants are intercepted by magnetic filter with high gradient. Or add magnetic seeds to adsorb nonmagnetic pollutants, and then separate them.

(4) Ion-exchange method

Expose wastewater to solid ion-exchangers and pollutants in ionic state can exchange with same-charged ions on the solid ion-exchanger. Thus, the harmful ions in wastewater are separated.

4. Typical process of municipal wastewater treatment

There is typical treatment process for municipal wastewater. The typical process diagram of first and secondary treatment is showed in Figure 3-6, and the typical process diagram of tertiary treatment is showed in Figure 3-7.

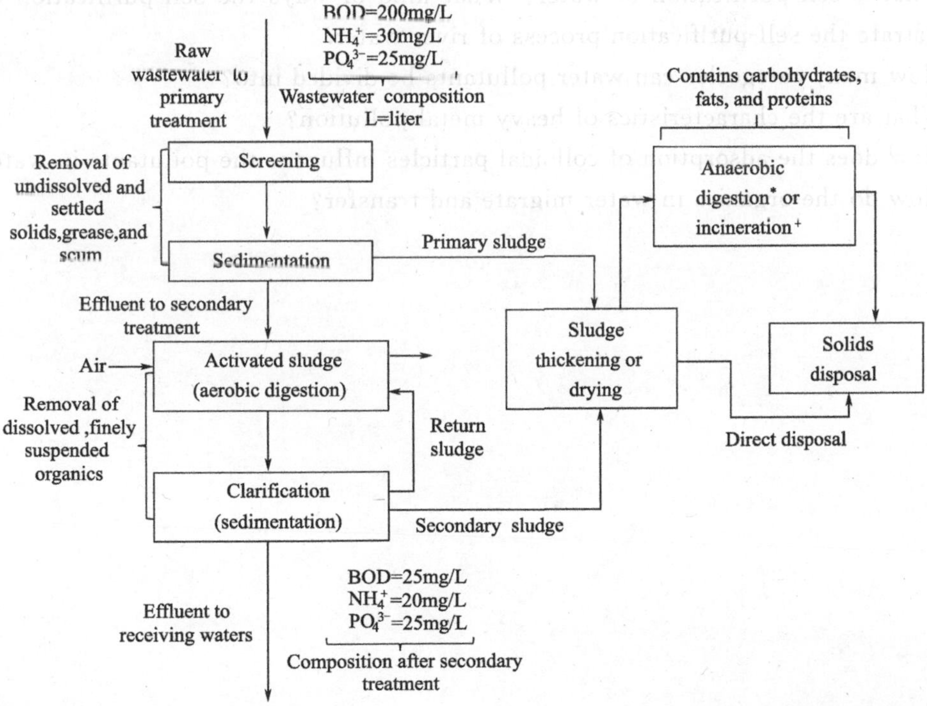

Figure 3-6 Typical process diagram of first and secondary treatment for municipal wastewater

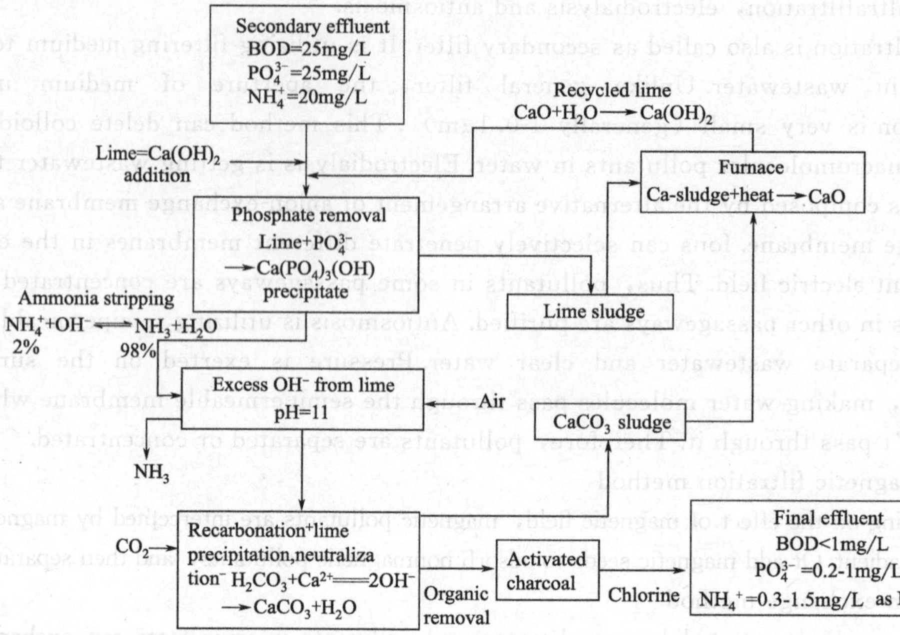

Figure 3-7 Typical process diagram of tertiary treatment for municipal wastewater

Questions

1. Illustrate the fundamental types of substances in natural water.
2. What is self-purification of water? What kind of ways the self-purification of water has? Illustrate the self-purification process of river water.
3. How many categories can water pollutants be divided into?
4. What are the characteristics of heavy metal pollution?
5. How does the adsorption of colloidal particles influence the pollutants in water?
6. How do the organics in water migrate and transfer?

Chapter 4
Soil Environmental Chemistry

Key Point: The essence of soil formation; Natural factors for the formation of soil; Human role in formation of soil; Soil mineral; Soil organic matter; Soil solution; Soil air; The physical property of soil; The colloidal property of soil; The composition and chelate of soil; The redox properties of soil; Acid and alkali of soil; The biological properties of the soil; Self-purification of soil; Soil pollution source; The main pollution of soil; Control and eliminate soil pollution source; Soil improvement and repair; Development of green food and promotion for human health.

Soil is one of the environmental elements. It is between living nature and non-living nature system, which is the basic of the all creature's existence. With the development of modern industrial and agricultural production, chemical fertilizer and pesticide are heavily used, water waste of industrial mining ceaselessly invade farmlands, and urban industrial waste and other artificially synthesized substances constantly enter into the soil. These lead to serious pollution accidents constantly occur. When pollutants enter into soil, they may cause secondary pollution to surface water, underground water and so on. Furthermore, pollutants in soil will influence the growth and development of plant, as well as the change of bio-group inside soil and the transformation of substance. Pollutants can even enter into human body via the soil-plants system and food chain, which threaten human health. Therefore, the prevention of soil pollution is one of the most important research topics of environmental science. And realizing the existence, immigration and transformation of pollutants in soil is an important basis for taking control measure.

Section 1 Formation of Soil

1. The essence of soil formation

The essence of soil formation is the contradiction and unity of the geological large-cycle and biology small cycle.

(1) Geological large-cycle

Geological large-cycle: Plant nutriments flow from mainland to ocean, and then from ocean back to mainland. Afterwards these substances flows to new ocean form new mainland. This cyclic process of nutriments is called as geological large-cycle of plant nutriments.

Under the effect of weathering, chunks of rock are ceaselessly disintegrated, reduced in size and thickness, finally becoming loose and porous scattered-broken bodies. In this process, some mineral which can be dissolved in water are also produced (for example some nutrient elements in silicates, like kalium, sodium, calcium, magnesium, phosphorus and sulfur). These substances are constantly washed by rainfall, flow from higher to lower places with surface runoff, and ultimately flow into ocean via river. After flowing into ocean, the diagenesis makes these substances cemented and hardened with rock and sediments in the ocean. And the sedimentary rocks are formed via the processes of concentration, precipitation and accumulation. Sedimentary rocks rise from the bottom of ocean and generate the mainland during the rising movement of earth crust. When sedimentary rocks are exposed to the surface of earth, weathering and leaching are conducted again. Thus, this kind of action is repeated.

(2) Biological small-cycle

Biological small-cycle: The process of repeating absorption and utilization as well as accumulating nutriments of plants (including all organisms participated in this process) is called as biological small-cycle of plant nutriments.

Parent material is produced due to the result of weathering. The friability, porosity, breathability, water permeability and water retentivity of parent material provide the growth of plants with many conditions, such as moisture, air and nutrients, which means the possibility for plants to grow on parent material. However, what initially lived on the original youth soil are some lower plants which need fewer nutrients. From the original youth soil, they absorb mineral nutrient and water as well as obtain other living materials and conditions in order to build their own organisms. Thus, some soluble nutrients in the geological large-cycle are preserved. When these plants die, the nutrient elements in organic residues are transformed to inorganic substances by the decomposition of microorganisms, and a part of them reenter into the process of geological large-cycle, while another part of them can be absorbed and utilized by plants again.

(3) The contradiction and unity of the geological large-cycle and biological small-cycle

The formation process of soil is established on the basis of geological large-cycle (releasing and leaching process of nutrient elements) and biological small-cycle (absorption, accumulation and releasing process of nutrient elements by organisms).

Biological small-cycle is a part which constitutes the motion of surface substances in geological large-cycle. Geological large-cycle leads to the constantly downward leaching loss of nutrient elements, while biological small-cycle constantly accumulates the nutrient elements that are necessary to organisms from geological large-cycle. According to the development process of biology, the requirement of primitive organisms for nutrient elements is less categories and low quantity. Thus, few nutriments are accumulated in soil. With the evolution of biological species, more and more nutriments are accumulated in soil through the selective absorption of organisms. The impact of biology on soil is gradually enlarged and deepened.

During the process of interaction and continuous development ascribe to the contradiction of the geological large-cycle and biological small-cycle, the soil fertility also gets continuous development.

2. Natural factors for the formation of soil

Natural factors for the formation of soil: Natural factors refer to the factors which influence the formation process of soil without the impact of human economic activities. Natural factors for the formation of soil include biology, climate, parent material (or rock), topography and time.

$$S=B+C+P+Tg+Ti(+M)$$

(1) Parent material

Parent material is the product of rock weathering and the material basis for the formation of natural soil. Its impact on physical and chemical properties of soil is quite obvious. For example, the feldspar and mica in granites are easily weathered and highly contained potassium, while quartz is not easy to be weathered and usually stayed as sand in soil. Thus, the sand ratio of soil developing on granitic parent material is usually moderate. Because shale is formed through the hardening of substances that highly contain clay, the character of soil generated on its weathering product is relatively adhesive. Quartz sandstone is mainly composed of quartz particles. Thus, the soil generated on its weathering product usually has strong grittiness, less nutrients and contains more grit.

(2) Biology

When plants begin to grow on parent material, the formation of soil is set out. The physical and chemical effects produced by plants and their dead bodies unceasingly improve the fertility of soil. Higher green plants that play the main role in improving the fertility of soil can synthesize organic matters through selective absorption for nutrients, and after the death of them the organic matters are accumulated in soil. Microorganisms in soil and tiny animals decompose the organic matters to release nutrients, and simultaneously synthesize steady humus substances. Thus, not only the content of organic matters increases, but also the physical properties of soil are remolded to form various soil structures. The fertility of soil gets continuously development due to the results of biological effects.

(3) Climate

The impact of climate on soil formation is very complicated. Climate directly influences the moisture and heat which play an important role in the formation process of soil. Simultaneously, climate decides the distribution of various vegetation types largely. Thus, the decomposition and synthesis of minerals and organic matters in soil are influenced.

Temperature directly influences the strength and direction in the formation process of soil. In cold zones, the chemical effect of soil is relatively weak, the growth of plants is relatively slow, the formation amount of organic matters is low, the activities of microorganisms in soil are not vigorous, and the transformation of nutrient in soil is also very slow. Contrary, in tropical zones, the minerals in soil are mostly decomposed except for quartz, the growth of plants is fast, the formation amount of organic matters is high, the activities of microorganisms are vigorous, and the biological small-cycle is faster than that in cold zones.

Precipitation also has a remarkable impact on the formation of soil. Under the conditions of arid climate, the salts are constantly accumulated, which leads to the occurrence of salinization in soil. Under the conditions of wet climate, base ions are unceasingly eluted by rainfall, leading to the unsaturated state of soil colloid. These phenomena above are intensively manifested in the formation process of various types of soil in our country.

(4) Topography

The effect of topography on the formation process of soil is extensive. Firstly, topography can influence the reassignment of heat. Slopes with different gradient and orientation accept different amount of heat from the sun. The largest amount is accepted by the southern slopes, of which soil temperature is high. While small amount is accepted by the northern slopes, of which soil temperature is low. Distribution of plants on slopes with different orientation varies due to the difference of temperature and humidity. Therefore, in some regions, distribution of soil types on different slopes is not same.

Topography can also influence the distribution of moisture, nutrients and mechanical composition of soil. At watersheds and slope regions, moisture as well as nutrients (including salts) and tiny particles of soil carried by moisture move to downgrade and low-place in the way of surface runoff or soil runoff, causing the difference between the soil of sloping fields and watersheds. Upper soil of sloping field frequently maintains a well drainage conditions, of which the layer is thin, the character is thick and the nutrients (base as well) are less. The moisture content of soil in lower and plain regions is larger because moisture is concentrated here.

The impact of topography is also showed by the change of absolute height of altitude. Climate becomes colder and wetter with the increase of altitude, causing a corresponding change to the hydrothermal conditions of soil and vegetation. Therefore, the distribution of montanic soil is closely related to the change of altitude.

(5) Time

The formation of soil progresses with time. The impact of any factors for soil formation is deepened over time. The degree for the formation process of soil is transferred by time. The degree for the leaching and accumulation of substances in soil varies with the difference of time remaining. Therefore, the formation process of soil is inevitably influenced by the local geological age. When other conditions of soil formation are same, the soil with different developmental age has different fertility. For example, the curing degree of garden soil varies with the length of cultivation time.

(6) The relationship between five elements

Each of five elements of soil formation can not only directly influence the process of soil formation, but also make a difference by indirectly influencing other elements. For example, on the one hand, the element of climate directly influences the process of soil formation. On the other hand, it also influences the composition of vegetation, the amount of organic matter and microbial decomposition. Vegetation can also make an effect on the near-surface atmosphere. They produce a special microclimate through which affects soil hydrothermal regime.

3. Human role in formation of soil

Human produce activity directly affects the fertilization of soil and makes a difference on the elements of soil formation. From long time ago, a lot of natural soil was reclaimed by human. To improve soil productivity, human actively control natural factors, leading to a direction which does good to production. For example, they improved soil fertility by intensive cultivation, adequate fertilizer, irrigation and drainage, etc.

Section 2 Soil Composition

Soil is the loose layer on the land surface with fertility and plant growth. It is produced by the process of ground rock weathering and of parent material into soil. It's essential feature is fertility. Soil, whose thickness is about 2 m, is a thin layer of the lithosphere. It provides plant with growth substances and energy, which is necessary.

Soil is composed of solid, liquid and gas phase composition. Solid phase includes soil mineral and organic matter, which is 90%-95% of soil weight and is about 50% of its volume. Liquid phase refers to soil moisture and soluble substances. It is also called as the soil solution which is 20%-30% of soil volume. There are large amount of bacteria and microorganism. Therefore, soil is a heterogeneous multiphase system which is solid phase-based. The three phases connect and restrict each other, forming a organic whole as shown in Figure 4-1. The components which closely related to soil pollution behavior in soil mainly are substances, organic matters and microorganism.

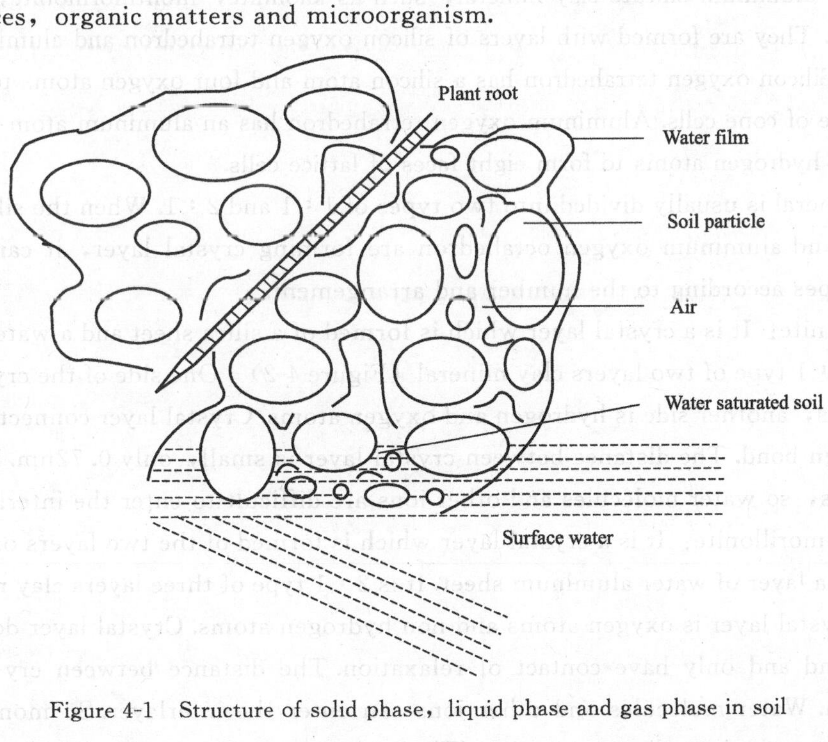

Figure 4-1 Structure of solid phase, liquid phase and gas phase in soil

1. Soil mineral

According to the causes of soil mineral, it can be classified into primary and secondary mineral.

Primary mineral refers to the original rock forming mineral which did not change its chemical composition during weathering. They mainly are quartz, feldspar, amphibole and mica, etc. For example, the particle size of sand is 0.02-2mm and the particle size of silt is 0.002-0.02mm. It has a solid and stable lattice. It is impermeable and non-expansive. It does not have absorption properties of physics and chemistry.

Secondary mineral is the new minerals which were formed by rock chemical weathering. It has the smaller particle size. Most secondary mineral are exist as clay and colloids (particle size is less than 0.002mm) dispersed state. Many secondary minerals have an active lattice, strong adsorption and Ion exchange capacity. They can expand after absorbing water and have the obvious characteristic of colloid. Secondary mineral is the main component of soil. It has a great impact on the behavior and fate of inorganic pollutants in soil.

Secondary minerals can be divided into crystalline and amorphous. Amorphous secondary mineral was mainly adhesive film state. It is wrapped on the surface of the soil particle, such as hydrous ferric oxide, aluminum and silicon, etc. There are very fine soil particles which like grainy gel, such as water aluminum stone, etc. The latter is a non-fixed composed of silicon and aluminum oxides. And it has the higher cationic and anionic exchange capacity. Especially amorphous oxide has a huge surface and the higher chemical activity. Crystalline secondary mineral is mainly aluminum silicate clay mineral, such as kaolinite, montmorillonite, illite, vermiculite, etc. They are formed with layers of silicon oxygen tetrahedron and aluminum oxygen tetrahedron. Silicon oxygen tetrahedron has a silicon atom and four oxygen atoms to form a triangular lattice of cone cells. Aluminum oxygen tetrahedron has an aluminum atom and six oxygen atoms or hydrogen atoms to form eight faces of lattice cells.

Clay mineral is usually divided into two types of 1∶1 and 2∶1. When the silicon oxygen tetrahedron and aluminum oxygen octahedron are forming crystal layer, it can be divided into three types according to the number and arrangement.

① Kaolinite: It is a crystal layer which is formed of a silica sheet and a water aluminum sheet. It is 1∶1 type of two layers clay mineral (Figure 4-2). One side of the crystal layer is oxygen atoms, another side is hydrogen and oxygen atoms. Crystal layer connects each other with hydrogen bond. The distance between crystal layer is small, only 0.72nm. It has Small internal voids, so water molecules and other ions are difficult to enter the interlayer.

② Montmorillonite: It is a crystal layer which is formed of the two layers of silica sheet sandwiching a layer of water aluminum sheet. It is 2∶1 type of three layers clay mineral. The surface of crystal layer is oxygen atoms and non hydrogen atoms. Crystal layer does not have hydrogen bond and only have contact of relaxation. The distance between crystal layer is 0.96-2.14nm. Water molecules and other ions can enter the interlayer. So montmorillonite has the higher cationic exchange capacity (Figure 4-3).

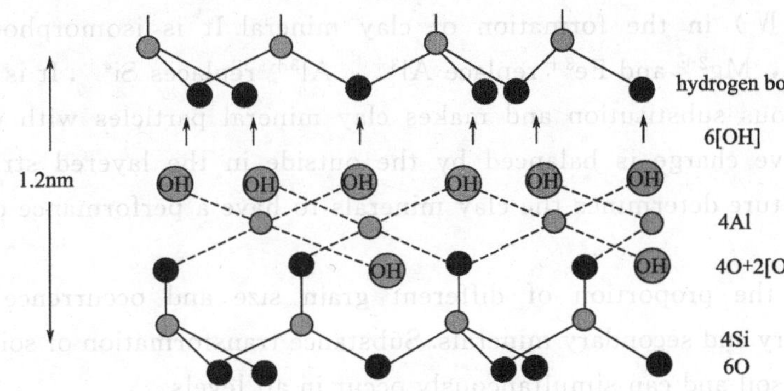

Figure 4-2　1 ∶ 1 type clay mineral（kaolinite）

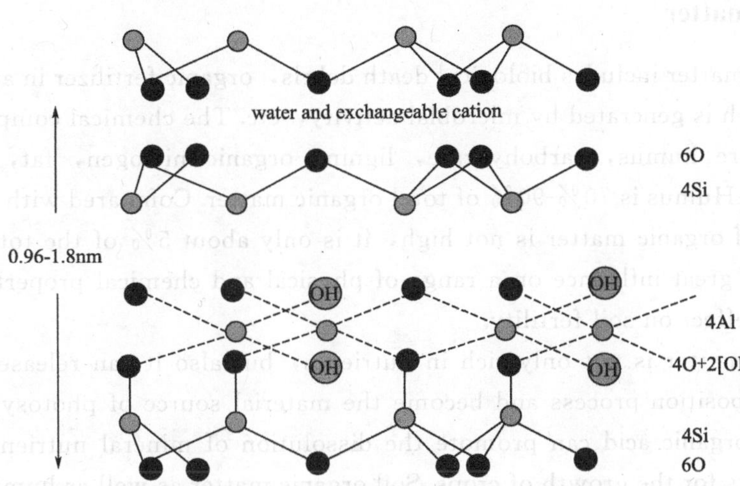

Figure 4-3　2 ∶ 1 type clay mineral（montmorillonite）

③ Illite: It is 2 ∶ 1 type crystal lattice which is formed of the two layers of silica sheet sandwiching a layer of water aluminum sheet. But part of the silicon is replaced by aluminum in illite type crystal lattice. The positive charge is compensated by deficiencies in the two interlayer potassium ions. The potassium ion seems to play the role of a bridge. It links the two adjacent crystal layers（Figure 4-4）. Ionic radius similar often replaces part of aluminum

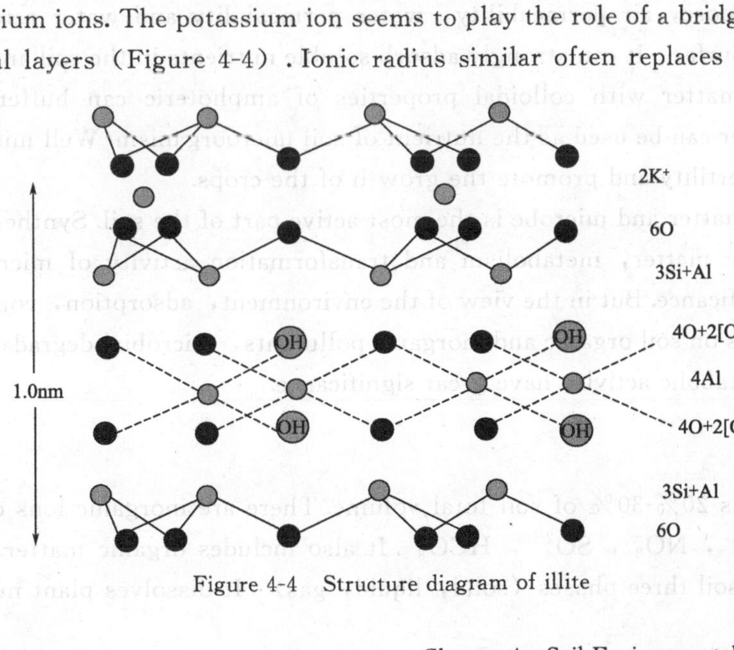

Figure 4-4　Structure diagram of illite

(Ⅲ) or silicon (Ⅳ) in the formation of clay mineral. It is isomorphous substitution effect. For example, Mg^{2+} and Fe^{3+} replace Al^{3+}, Al^{3+} replaces Si^{4+}. It is as same as the result of isomorphous substitution and makes clay mineral particles with excess negative charge. This negative charge is balanced by the outside in the layered structure of K^+, Na^+, etc. This feature determines the clay minerals to have a performance of ion exchange adsorption.

According to the proportion of different grain size and occurrence level, soil is composed of primary and secondary minerals. Substance transformation of soil can be in each component part of soil and can simultaneously occur in all levels.

2. Soil organic matter

Soil organic matter includes biological death debris, organic fertilizer in application, organic matter which is generated by microbial activity, etc. The chemical compositions of soil organic matter are humus, carbohydrate, lignin, organic nitrogen, fat, wax, organic phosphorus, etc. Humus is 70%-90% of total organic matter. Compared with the soil mineral, the content of organic matter is not high, it is only about 5% of the total soil. But organic matter as a great influence on a range of physical and chemical properties of soil and has a significant effect on soil fertility.

Soil organic matter is not only rich in nutrients, but also it can release CO_2 to air in their slow decomposition process and become the material source of photosynthesis. At the same time, the organic acid can promote the dissolution of mineral nutrients and provide abundant nutrients for the growth of crops. Soil organic matter as well as humic acid has aromatic polybasic phenol functional groups. It can enhance plant respiration, increase permeability of cell membrane and promote the growth of root system. Organic vitamins, growth hormone and antibiotics for growth promotion have growth promotion and disease resistant effect on plant. Organic matter can also promote the formation of good soil structure. It can increase the soil loose, air permeability, water permeability and water retention. Humus has huge specific surface. It can strongly adsorb soluble nutrients in the soil and maintain soil fertility. Organic matter with colloidal properties of amphoteric can buffer soil solution pH. Organic matter can be used as the nutrient of soil microorganism. Well microbial activity can increase soil fertility and promote the growth of the crops.

Soil organic matter and microbe is the most active part of the soil. Synthesis and decomposition of organic matter, metabolism and transformation activity of microbes not only have fertility significance. But in the view of the environment, adsorption, complexation and chelation of humus on soil organic and inorganic pollutants, microbial degradation of organic pollutants and metabolic activity have great significance.

3. Soil solution

Soil solution is 20%-30% of soil total volume. There are inorganic ions of Na^+, K^+, Mg^{2+}, Ca^{2+}, Cl^-, NO_3^-, SO_4^{2+}, HCO_3^-. It also includes organic matter. Soil moisture is key element of soil three phases (solid, liquid, gas). It dissolves plant nutrients of soil and atmosphere into nutrient solution, transports to plant root and provides maximum available to plants. Therefore, soil water is the main media of plants to absorb nutrients.

Soil water mainly comes from rain, snow and irrigation. In the case of underground water level is close to the ground (2-3m), the upper groundwater is an important source of soil water. Capillary force adsorption surface soil particles and fine pores can put into the soil retain water. The firm degree of soil solid holding water largely determines soil water movement and water use of plant. Water interacts with other components of substance when it falls into the soil. Some of these soluble substances such as salt and air will dissolve in water. The soil water with salt and air is called soil solution.

4. Soil air

Soil is a porous system. The air filled with pore in the condition of water unsaturated. Soil air mainly comes from atmosphere. Besides, it comes from the biological and chemical processes in the soil.

Soil air is not continuous. It exists in the soil pore which is spaced from soil solid. Its composition is different in different place. There is great difference between soil air and atmospheric composition.

① The content of CO_2 is generally higher than that in atmosphere.

While, the content of O_2 is lower than atmosphere (Table 4-1). The reasons for this difference are the O_2 consumption and the CO_2 release by organic matter degradation and synthesis of plant root respiration in soil and Microbial activity.

Table 4-1 Composition of soil air and atmosphere

Gases	O_2/%	CO_2/%	N_2/%
Soil air	10.00～20.03	0.15～0.65	78.8～80.24
Near-surface atmosphere	20.95	0.03	78.09

② The water content of soil air is generally higher than that in atmosphere.

When soil moisture is appropriate, relative humidity can be close to 100%. In addition, soil air is often saturated by moisture. Therefore, under poor ventilation conditions, small amount of reducing gas such as CH_4, H_2S, H_2 which are generated by anaerobic bacteria activity is accumulated in soil air.

The content and composition of soil air largely depend on the relationship between water and land. The soil air of gas mixture only enters the soil pore which is not occupied by water. A large proportion of fine pores in soil often have poor ventilation conditions. Water is dominant at this type of soil. The content and composition of soil air are not suitable for the growth of plants. The water and air keep in the soil pore, their relative content often changes with the change of natural condition.

Section 3 The Basic Property of Soil

Soil, which is closely related to plant life activities are closely, has storage, conversion of solar energy and biological energy function. Soil is an important hub of agricultural ecological system binding biological and non-biological circle, organic and the inorganic world. The soil is the site where the interactions of various environment parts take place. Soil is a complex material system, where small living organisms survive. And there are a variety of inorganic and organic compounds in it. Various physical, chemical, biological and chemical changes take place in each interface of the phases.

The organic and inorganic contaminant in environment can enter soil-plant system through various channels. Soil itself is a live filter. It can produce physical effects of filter and dilution on pollutants. The adsorption of soil has great influence on the migration of heavy metals and other pollutants transformation. The chemical, biological and chemical reactions which are produced by soil microorganism and life of plant also have a significant role in the purification and metabolism on pollutants. The property of the soil is multifaceted. Following is a brief introduction of the basic properties about migration and transformation of pollutant.

1. The physical property of soil

The soil is multiphase system of solid, liquid, gas phase coexistence. The higher degree of dispersion of soil solids, the smaller the soil particle diameter is, the greater the surface is. Substance with larger surface area has higher surface energy. It shows different chemical and physicochemical properties, like cementation, absorption, etc.

For example, fine particles have a particularly strong cementing property and adsorption capacity. Generally, soil internal space does not be fully filled. It is always in a certain arrangement, in which there is many pores. Soil pore system consists of a large number of intergranular pore with different shapes and sizes. These pores are connected by channels with diameter smaller than it is. Soil pore space can be classified into large pores (non capillary porosity) and small pore (capillary porosity). Pores between soil aggregates mainly are non capillary porosity. While pores in soil aggregates mainly are capillary porosity. Capillary porosity can absorb water, but not easy to ventilate and flood. Non capillary porosity cannot absorb water, but easy to ventilate and flood.

Because the soil is loose and porous, some pollutants volatile or are in gaseous state. Other pollutants dissolve in water or adsorb onto the solid particles. They follow the movement of air and water in the soil pores, then volatilize, spread, dilute and concentrate, so they move out of soil finally. This process is related to soil temperature, water content, and the size, quantity, distribution of soil pore.

2. The colloidal property of soil

Soil colloid can be divided into inorganic colloid, organic colloid and organo-mineral colloidal complex.

(1) Organic colloid

Organic colloid of soil is mainly humus. Humus colloid is amorphous substance. It has huge specific surface and its range is 350-900 m^2/g. Due to the H^+ dissociation of carboxyl or hydroxyl in colloidal surface, humus has negative charge and the average of negative electricity is 200mmol/100g which is higher than the layered silicate colloid. Its cation exchange capacity is 150-300mmol/100g, even up to 400-900mmol/100g.

Humus can be divided into humic acid, fulvic acid, humin, etc. Humin closely link with soil mineral. It is generally believed that Humin has little effect on soil adsorption property. Humic acid and fulvic acid are nitrogen-containing hydroxyl acid. They are the most active dispersion-material in the adsorption process of soil colloid. Their feature is having many functional groups and large negative electricity. Therefore, its cation adsorption capacity is quite high.

The main differences between humic acid and fulvic acid are the latter has strong mobility, high acidity and a large number of oxygen-containing functional group. In addition, fulvic acid has the larger cation exchange capacity. Under the same conditions, the cation exchange capacity of humic acid and fulvic acid is 200-670mmol/100g and 180-500mmol/100g, respectively. Therefore, acid has high chelation and adsorption on cations of heavy metal. Its chelates are generally soluble in water. After adsorption of heavy metal, fulvic acid is in sol state and is easy to move with soil solution. It can be absorbed by plants and flow out of soil to other environmental mediums. Humic acid is easy to form soluble substance with divalent metal ion (such as K^+, Na^+). With other metal ions, it forms insoluble flocculation material. This can maintain soil organic carbon and nutrients, as well as hold toxic heavy metal ions, alleviating the toxicity of plant. Thus, the humus which has high level of humic acid can greatly improve the capacity of heavy metals in soil. In the study of soil environmental capacity, it should consider the relative proportion of the humic and fulvic acid (H/F) in humus.

(2) Inorganic colloid

Inorganic colloid includes secondary clay mineral, iron and aluminum hydrous oxide, hydrous silica amphoteric colloid. Secondary clay minerals are mainly montmorillonite, illite and kaolinite. They are phylloaluminosilicate whose average particle sizes are less than 5nm. They can strongly adsorb molecular and ion pollutants. The reason is that:

① Clay mineral has tiny particles and large surface area. The type of montmorillonite has the largest surface area(600-800m^2/g) among them. It has not only outer surface, but also huge inner surface. Next are illite(100-200m^2/g), and kaolinite minimum(7-30m^2/g). Large surface area associates with a huge surface energy, so it can be adsorbed into the soil gas and liquid pollutants.

② The clay mineral is with negative charge and high cation exchange capacity. It has strong ability of fixed exchange ion of pollutants in soil. The cation exchange capacity of montmorillonite and kaolinite are 80-120mmol/100g, 3-15mmol/100g, respectively. Partial negative charge comes from isomorphous substitution effect of the intergranular layer. Part of it comes from the variable negative charge of H^+ hydroxyl dissociation of lattice surface in the isoelectric point of colloid. According to the research, permanent negative charge of montmorillonite type is 95% of the total charge amount, illite 60%, kaolinite 25%. The metal cations which are adsorbed by kaolinite are located on the surface of the ion exchange of lattice points, and they are easily desorbed. Well the base ions which are adsorbed by montmorillonite and illite are partly located in the lattice, and they are not easily desorbed.

(3) Organo-mineral colloidal complex

It is a kind of colloid which is combined of inorganic and organic colloid. Its property is between the above two kinds of colloids. Soil colloid is mostly organo-mineral colloidal complex.

As mentioned above, the soil colloid has great specific surface and charge, therefore it has the special ability of adsorbing and exchanging ion from the soil solution . Soil colloid adsorption for metal ions are mainly surface molecular adsorption and cation exchange. The adsorption is related to the surface area and charge quantity of colloid. The adsorption of organic pollutants on soil colloids are mainly surface adsorption and partition. Therefore, pollutants adsorption is related to its type, composition, exchange adsorption capacity, characteristic and pH.

Molecular surface adsorption refers to small colloidal particles have huge surface area. The molecules on the surface often produce surplus force due to uneven stress. The energy produced by surplus force is called free energy or surface energy. Objects in nature all have the tendency to reduce their free energy and maintain the stability of dispersion system. And it is achieved by free energy consumption. This is one of the important mechanisms of soil colloid on chelate adsorption substances in molecular state of Cation exchange adsorption refers to adsorption and maintain effect on ionic substance in soil colloid. Actually it is the process that the cation on diffusion layer of colloidal dispersion and in soil solution exchanges each other to achieve balance. It has the following three characteristics.

① Cation adsorption process is a dynamic equilibrium in a reversible reaction. Higher the concentration of metal ions entering into the soil and higher the valence state is, more easily it is to be absorbed by colloid. When entering excess of heavy metal ions in the colloid surface, the adsorption capacity decreases and heavy metals may be desorbed.

② Cation exchange capacity is equal to the amount.

③ The adsorption affinity of colloidal substances in various cations is different. For the same price ion, the smaller of ion radius, more easily being absorbed. Colloidal metal ion adsorption capacity is often influenced by soil environmental conditions.

The different cationic colloidal substances can exchange different mount of cation. When the soil solution is neutral, the largest amount of adsorption is called cationic exchange capacity of this soil (CEC), which is also known as the maximum adsorption capacity of soil cation. The unit is mmol/100g soil. Soil cation exchange capacity (CEC) in our country from the south to the north, in order from low to high: Northeast black soil is 24. 44-34. 34; Cinnamon soil is about 16. 40 in North China; Yellow Cinnamon soil in Yangtze River valley is about 13. 23; the southern red soil was only 4. 77 and 4. 09. The tests of Minzi show that soil exchange capacity is the first factor of determining soil heavy metal adsorption. Followed is humus content, and the role of clay is not obvious. Sequential adsorption affinity of different clay minerals on the metal ion is not the same. For example:

Montmorillonite sequential adsorption of divalent metal ions:

$Ca^{2+} > Pb^{2+} > Cu^{2+} > Mg^{2+} > Cd^{2+} > Zn^{2+}$;

Kaolinite: $Pb^{2+} > Ca^{2+} > Cu^{2+} > Mg^{2+} > Zn^{2+} > Cd^{2+}$;

Illite: $Pb^{2+} > Cu^{2+} > Zn^{2+} > Ca^{2+} > Cd^{2+}$.

The adsorption and chelating effect of soil organic matter on heavy metal ions can occur at the same time. When the concentration of heavy metal ions is high, it is mainly exchange adsorption. When concentration is low, it is mainly chelation. Hasler pointed out that the sequential capability of organic matter in soil to combine metal ions is that Pb>Cu>Ni>Co>Zn>Mn>Mg>Ba>Ca>Hg>Cd. Jonasson believed that the order of humic acid and fulvic acid absorbing metal is: $Hg^{2+} > Cu^{2+} > Pb^{2+} > Zn^{2+} > Ni^{2+} > Co^{2+}$. Soil organic matter generally is only 5% of solid part. Its negative charge averagely is 21% (5%-42%) of total negative charge in soil. Therefore, the contribution that organic colloid absorb metal ions is smaller than that of inorganic colloid. But soil organic colloid has significant influence on pollutant, especially on the transformation organic pollutant and biological effect.

Soil colloid adsorption of heavy metals and pesticides played an important role in controlling their migration in the soil-plant system. For example, the activity of heavy metal in soil largely depends on absorption of soil. Soil clay minerals and humus have a strong capacity of adsorbing heavy metals. They can reduce the activity of heavy metal.

3. The composition and chelate of soil

Organic, inorganic ligands in soil can occur composition or chelation with metal ions, thus it can affect the migration and transformation of metal ion.

Soil organic ligands mainly are humus. Their surface have a variety of functional groups containing oxygen, nitrogen and other coordination atom. It can form composition or chelate with metal ions. The order of different ligands and metal ion affinity is: $—NH_2 > —OH > —COOH > C—O$. The carboxyl and phenolic hydroxyl functional groups are respectively accounted for 50% and 30% of the functional groups in humus. They become the main composition groups of humus-metal complexes. In the soil of our country, the content of hydroxyl groups of the humic acid (340-480mmol/100g), is lower than that of fulvic acid (700-800mmol/100g). The ratio of carboxyl and hydroxyl groups of the former is 0.7-1.9, the latter is 2.7-5.6. The composition and reducing capacity of fulvic acid is higher than that of humic acid.

Common soil inorganic ligands are Cl^-, SO_4^{2-}, HCO_3^-, OH^-, etc. They generate a variety of compositions with metal ions, such as $Cu(OH)^+$, $Cu(OH)O_2$, $CuCl^+$, $CuCl_3^-$, etc. The stability of metal compositions or chelates is related to ligands or chelating agents, metal ion species and environmental conditions. The order of composition or chelation ability of soil organic matter on metal ion is: $Pb^{2+} > Cu^{2+} > Ni^{2+} > Zn^{2+} > Hg^{2+} > Cd^{2+}$. The pH of soil medium has great influence on the stability of chelation. When the pH is low, H^+ competes with the metal chelator, so the stability of chelation is poor. When the pH is high, the metal ions can form insoluble inorganic compounds of hydroxides, phosphates or carbonates.

The influence of composition or chelation on metal ion migration depends on solubility of chelation. If the formed chelate is soluble in water, it is conducive to the migration of metal; On the other hand, it is conducive for metal to remain in the soil and its activity can reduce. Generally, the chelate of humic acid and metal ion is insoluble, while the chelate of fulvic acid and metal ion is soluble. Heavy metal and humus can produce stable soluble chelate and effectively prevent heavy metal precipitate as insoluble salt.

The formation of heavy metal and humus is soluble stable chelate, it can effectively prevent the heavy metal as the insoluble salt precipitation. For example, the chelate forming of humus and Fe, Al, Ti, V is soluble in neutral, weak acid or weak alkaline soil solution. It can migrate in the form of chelate. It has precipitated when lacking of humus.

The reaction and factor of inorganic, organic ligand and heavy metal ion have a detailed discussion in water pollution chemistry. The situation in soil is similar. For example, in the solution of soil surface, mercury mainly exists in the form of $Hg(OH)O_2$、$HgClO_2$. In saline soil of high chloride ion, it is based on $HgCl_4^{2-}$. Cd^{2+}, Zn^{2+}, Pb^{2+} can produce complex ion of $MClO_2$, MCl_3^-, MCl_4^-. Because the pH of saline soil is high, heavy metal can hydrolyze and form hydroxyl ions. As the same time, there is a competition between hydroxyl mating reaction and chlorine mating reaction. Besides, it can produce various complex ions, such as HgOHCl, CdOHCl, etc. Mating reaction between heavy metal ion, hydroxyl and chlorine can improve the solubility of insoluble metal compound. In addition, it can weak the adsorption of heavy metal in soil colloid, thus affect the migration and transformation of heavy metal in soil.

4. The redox properties of soil

A variety of organic, inorganic redox substances exist in soil undergoing redox reaction. Strength of redox reaction in soil can be expressed in redox potential (E). The main oxidants in soil are free oxygen, high valence metal ions. For reducing agent are low valence metal ions, soil organic matter and decomposition products formed under anaerobic condition. Besides pure chemical reaction, biological reaction is also soil redox reaction. If free oxygen is dominant in soil, oxidation is the main reaction. If organic is dominant, reduction reactions take place. The E is affected by ventilation and human measures. Indry farm, the E is high, so it is dominated by oxidation reactions. In deep ground, E is lowered to a negative value, and it is dominated by reduction reaction. For example, Hg^{2+} in soil can be reduced to Hg^+ by organic matter and microorganism, and then it can be reduced to HgO by disproportionation reaction. The E of upland soil is 8. 5-11. 84V. Hg is in forms of $HgCl_2$ and $Hg(OH)_2$; The E of paddy soil is −5. 1V and produces HgS precipitation by H_2S与Hg^{2+}. In addition, the E of soil is related to pH. If the pH decreases, the E increases.

The E of soil can affect the existing forms of organic and inorganic matter, thus it can affect their migration, transformation and the extent of poisoning crops in the soil. This is particularly important for price valence-variable elements. For example, under the reduction condition of low E, Cd^{2+} formed undissolved CdS precipitation. It is difficult to be absorbed by the plant. Its toxicity is decreasing. But after the paddy field is dry, CdS was oxidized to be soluble $CdSO_4$ which is easily absorbed by plants. If Cd^{2+} content is higher, it can affect the growth of plants. Therefore, toxicity of pollutants can be reduced by adjusting the E and the pH of the soil.

5. Acid and alkali of soil

The acidity and basicity of soil depends on the concentration of H^+ ions in soil solution, and H^+ mainly come from the hydrolysis of exchangeable aluminum in the soil:

$$Al^{3+} + H_2O \rightleftharpoons Al(OH)^{2+} + H^+$$

$$Al(OH)^{2+} + H_2O \rightleftharpoons Al(OH)_2^+ + H^+$$

$$Al(OH)_2^+ + H_2O \rightleftharpoons Al(OH)_3 + H^+$$

When CO_2 in soil dissolves in water to form carbonic, it can dissociate H^+. The content of inorganic acid and other saline matter in soil, as well as the type of positive ion on the colloid surface all affects pH value of soil. The latter shows when H^+ absorbed by colloid is replaced by cation in salt solution. Therefore, it is also known as exchangeable acid or potential acid. For example, in the interaction between soil and neutral salt solution such as KCl, H^+ is replaced into solution, equal amount of K^+ absorbed by colloid.

$$\text{colloid—H} + KCl \rightleftharpoons \text{colloid—K} + HCl$$

The more H^+ replaced from colloid, the higher value of soil exchangeable acid.

The normal pH of soil is 5 to 8. The pH of Acid soil may be less than 4 and pH of alkaline soil can be more than 11. In humid area with strong eluviation, soil is acidic or strongly acidic, $pH < 6.5$. In arid area with weak eluviation, most soil is alkaline or strongly alkaline, $pH > 7.5$. In the area where part is humid and part is dry, the pH of soil is between above two. Acid rain can cause soil acidification, which has been a worldwide environmental problem.

pH value is one of the important indicators of soil. Soil pH directly or indirectly affects the transition and transformation of pollutants in soil. ①pH value affects solubility of heavy metal plasma. ②pH value affects the oxidation of pollutant reduction system. ③pH value affects soil colloid adsorption of heavy metal ion. For example, the best range of pH value for silica gel to adsorb heavy metal ions is that Co^{2+}, 5~7; Cr^{3+}, 3.5~7. ④pH value affects obviously the activity of heavy metals in soil. For example, cadmium is more soluble in acidic soil and more toxic to plant, while in alkaline soil, it is less soluble and less toxic.

6. The biological properties of the soil

There are soil animal, protozoa and microbial community in soil. Soil microbe is the main part of soil organisms. It has various type and large amount. Especially on the surface of soil, there are a billion and billions of microorganisms, such as bacteria, Fungi, actinomycetes and yeast per gram of soil. They can produce various kinds of obligate enzyme, playing the main role in transformation and decomposition of soil organic matter. Soil microorganism and other organisms have strong self-purification capability to organic pollutants, called biodegradation. They can produce a variety of obligate enzyme, thus in the process of transformation of soil organic matter decomposition plays a main role. Soil microorganisms and other organisms have strong self-purification ability to organic pollutants, the biodegradation. Their detoxification and ability to update create decisive advantage for the material cycle of soil ecosystem. Besides, it also provides guarantee to keep soil fertility.

Pollutants can be absorbed by biology and accumulate in the body. Plant root absorb pollutants, which is the main way of plant pollution.

7. Self-purification of soil

The properties of the soil have a great influence to the migration of pollutants. For example, soil colloid can absorb all kinds of pollutants and reduce its activity. Microorganisms have special effect on the degradation of organic pollutants, which makes soil has good ability to self-renew without outside supports. The ability to renew itself is called self-purification of soil. The process of self-purification after pollutants entering soil is like this: Pollutants reduce its concentration by physical way like diffusion, dilution, volatilization, etc. Then it is degraded to be non-toxic or low toxic substances through biological and chemical way. Alternatively, it would be transformed into insoluble complex substances by chemical precipitation and mating reaction or chelation and oxidation-reduction. Besides, it can be absorbed by soil colloid, being difficult to be absorbed by plants. Then it would temporarily exit the micro-biological cycling, separated from food chain or into the air and water out of the soil.

The soil self-purification ability is related to the composition and other properties of soil, and the type and property of pollutants. The self-purification capacity of the soil is different from soil to soil. For one kind of soil, the self-purification capability is different to different pollutants. In general, the soil self-purification speed is slow. It is difficult to purify after pollutants entering into soil.

Section 4 Soil Pollution

The phenomenon that the pollutants produced by human activity enter into soil and accumulate to a certain degree leading to deterioration of soil is known as soil pollution. There are many differences between soil and water, air environment. Compared to water and air, the position of soil is more stable, which is good for gathering of pollutants. Therefore, soil is considered as the collective place of pollutants.

Pollutants enter into the soil through various ways. If the amount of pollutants in the soil is in the range of self-purification ability, normal ecological circulation can maintain. Soil pollution and purification are two processes which are contradictory but exist at the same time. If the amount of the pollutants produced by human activity and its speed entering soil is more than the speed of purification, it leads pollutants to continuingly accumulating in soil. Bad ecological effect and environmental effect will occur. Finally, the normal soil function loses, soil quality reduces, the growth of crops is affected and the productivity and quality of crops decrease. Soil pollution happens. Soil pollution can be identified from following two respects. ①One respect is whether underground water is polluted. ②Another respect is whether growth of crops is influenced.

When soil is polluted, the growth of plants is affected. At the same time, it affects the change of biota in soil and transformation of substances. Bad ecological effect happens. Soil pollutants will enter river and lake with surface runoff. When the concentration of pollutants in runoff is high, surface water is polluted. For example, overmuch N and P in soil, some organophosphorus and organochlorine pesticides and the move leaching of phenol and cyanogen always cause surface water pollution. Therefore, pollutants may cause secondary pollution on surface water and underground water after entering soil. Through soil-plant system and food chain, the pollutants in soil can finally affect human health. For example, itai-itai disease in Japan is a typical example for soil pollution indirectly damaging human health.

1. Soil pollution source

Soil pollution source includes natural pollution source and man-made pollution source.

(1) Man-made pollution source

Soil pollutants mainly are waste water and solid waste from industry and city, pesticide and fertilizer, dejection, residue of biology and fallout of air. Sewage irrigation and sludge are used as fertilizer, which makes soil polluted by heavy metal, inorganic salt, organic matter and pathogens. Solid waste of industry and city is randomly stacked, which causes leaching and releasing of harmful material, soil and underground water pollution. Modern agriculture uses large amount of pesticide and fertilizer, which may also cause soil pollution. For example, organochlorine insecticide such as three six and DDT remain in soil for a long time and concentrate in organism. Chemical fertilizer like nitrogen and phosphorus if it is not absorbed and used by plants or fixed by root layer soil, will accumulate under root layer soil, or enter into underground water, being potential pollutants to environment. Animal manure of livestock farm and waste of shamble the property of which is similar to human waste, can be used as fertilizer. If it is not properly dealt with, the parasite, pathogenic bacteria and virus can cause soil and water pollution. Sulfur dioxide and nitrogen oxides and particulate matter in the air can cause soil acidification by dry and wet deposition landing on the ground.

(2) Source of pollution of nature

In some ore deposits or enrichment center around the elements and compounds, due to the natural mineral decomposition and weathering, natural diffusion zone often forms, leading to the content of some elements in soil beyond normal content.

Soil pollution, according to its property, can be divided into three types. There are chemical pollution source, physical pollution source and biological pollution source, which is complex. Chemical pollution of soil is the most universal, serious, and complex one.

2. The main pollution of soil

There are many different kinds of soil pollutants, which generally can be divided into the following categories.

(1) Inorganic pollutant

It includes elements harmful for animals and plants and inorganic compounds, such as cadmium, mercury, copper, lead, zinc, nickel, arsenic and other heavy metals. Besides, nitrate, sulphate, fluoride, soluble carbonates and other compounds are common inorganic pollutants of soil. Overusing nitrogen or phosphorus fertilizer can cause soil pollution.

① Mercury.

The background level of mercury in soil is 0.01-0.15$\mu g/g$. Besides parent rock, pollution source is the main origin of mercury, such as application of mercury-containing pesticide and sewage irrigation. Therefore, the content of mercury in soil is quite different from place to place. Mercury from pollution source firstly enters surface soil. Soil colloid and organic matter has strong ability to absorb mercury. The mobility of mercury in soil is not good, which makes mercury often accumulate on the surface of soil and unevenly distribute in profile. Mercury in soil is not easy to lose with water but it is easy to vitalize into air. Many factors influence the vitalization of mercury. Mercury in soil can be divided into three types according to chemical speciation. There are metallic mercury, inorganic mercury and organic mercury. Mercury in soil exists as zero-valence mercury when pH is between normal value. Under the certain condition, mercury can transform among all kinds form. Some inorganic mercury entering soil can decompose to be metallic mercury. Under reduction condition, organic mercury can degrade to metallic mercury. Generally, the reaction that $Hg_2O \xlongequal{} Hg + HgO$ can occur in soil. Producing mercury may volatize. Under good ventilation in soil, mercury can stably exist in any forms. Some mercury can transform into soluble methyl mercury or gaseous dimethyl mercury.

Cationic state mercury are easily absorbed by soil. Many mercuric salts such as mercuric phosphate, mercuric carbonic acid and mercuric sulfide are difficult to dissolve. Under reduction condition, Hg^{2+} and H_2g produce insoluble HgS. Metallic mercury can be transformed into mercuric sulfide by sulfate-reducing bacteria. All of this prevent mercury move in soil. When oxygen is abundant, mercuric sulfide is slowly oxidized into sulfite and sulfate. Mercury in negative ion, such as $HgCl_3^-$ and $HgCl_4^{2-}$ can be absorbed by iron oxide with positive ion, ferric hydroxide and the edge of clay mineral. Mercury in molecular state like $HgCl_2$ can be absorbed by oxide of Fe and Mn. $Hg(OH)_2$ is difficult to dissolve so it can be largely held by soil. Because of the strong reaction between mercuric compound and soil composition, except metallic mercury which volatizes as vapour, mercury in other forms slowly transform in soil. Mercury diffuses as gas phase among soil sore. In general, mercury is more easily to transform than other toxic metal. When mercury is chelated with organic matter in soil, it will have horizontal and vertical movement.

Mercury is a harmful element to plant growth. If the content of mercury in soil is over high, it not only accumulates in plants but produce poison to plants. Usually, the compound of organic mercury and inorganic mercury as well as mercuric vapor can cause plants poisoned. For example, mercury does harm to growth and development of paddy. Hydroponic experiment of paddy in China Academic of science shows that when use inoculums the mercury content is 0.074$\mu g/mL$, the productivity of paddy decrease and the rate of blight grain increase. The root of the paddy has been damaged when dealt with 0.74 $\mu g/mL$. As concentration increases, root is more tortuous and become brown and rusty. When the content of mercury in medium is 7.4 $\mu g/mL$, rice leaves become yellow, tiller is restrained, the height of plant decreases and root is of dysplasia. In addition, with the concentration increasing, the content of mercury in every part of plant goes up. When the concentration of medium is 22.2μg/mL, paddy is seriously damaged. The lethal concentration for paddy is 36.5μg/mL. However, in the soil culture experiment of plants, although the content of mercury in soil is 18.5μg/mL, the productivity of paddy and wheat is not affected. Therefore, the effect of mercury on plant is closely related to environmental conditions. Different plants has different sensitivity to mercury. For example, soybean, sunflower and rose are very sensitive to mercury vapour. Paper birch, oak, ivy and reed strongly resist to mercury vapour. Peach and tomato are moderately sensitive to mercury vapour.

There two ways for mercury to enter into plants. One is through roots absorbing mercury ion. Under some conditions, it can also absorb methyl mercury and metallic mercury. Next are mercurial used on leaves, floating dust, mercury in rain and mercury vapour released by soil under because of contraction of temperature between day and night. It enters plant through leaves or absorbed by root. The mercury entering through leaves can be transported to other part of plant while the mercury absorbed by root often deposit on root after reacting with protein in root and hardly transform to ground.

The amount of mercury absorbed by plants depends on not only mercury content in soil but also its effectiveness. The effectiveness of mercury on plant are closely related to redox condition of soil, pH value and organic content. Different plants have different ability to absorb and accumulate mercury. Different organs of plant have different ability to absorb mercury. The absorption of plants on mercury is related to the forms of mercury in soil.

Different forms of mercury in soil have different effect on crop growth and development. The experiment about the effect of organic mercury and inorganic mercury in soil on crop growth and development suggests that when mercury concentration is same, the degree of harm that mercuric compound does to paddy's growth is like following. PMA$>HgCl_2>$ HgS. HgS is difficult to be absorbed by paddy. Although mercuric compound is same, it can exist in different forms when soil environmental conditions change. Therefore, its effectiveness on crops is different.

② Cadmium.

Cadmium abundance in crust is 5μg /g and the background level of cadmium in parts of our country is 0.15-0.20μg /g.

Cadmium pollution in soil is mainly from mine, smelt, sewage irrigation and the application of sludge. Cadmium can enter soil with using of phosphorus slag and calcium phosphate. With wind action, cadmium in waste gas spreads and settles in soil. In the soil beside road where traffic is busy, cadmium pollution often occurs.

Cadmium in soil includes available state, substation and insoluble state. Cadmium in available state exists as ion or complexation, which is easy to be absorbed by plant. What can be absorbed and transformed by clay and humus is cadmium in substation. Insoluble cadmium includes cadmium in sediment and insoluble chelation, which is difficult to be absorbed by plants.

Cadmium in soil can be absorbed by colloid. Absorbed cadmium usually accumulate at 0-

15 cm of surface soil. Under 15 cm of soil, the content of cadmium greatly decreases. The absorption rate of most soil to cadmium is between 80%-90%. Soil absorption of cadmium has positive correlation with pH value. Absorbed cadmium can dissolve and be transformed with water. The lower pH value is, the higher leaching rate is. If pH is 4, leaching rate of cadmium is more than 50%; If pH value is 7.5, cadmium is hardly dissoluble.

The transformation of cadmium in soil is related to the type, property and pH value and influenced by redox condition. Paddy land is a special soil in which potential redox is low. When paddy field is full of water, because of reducibility environment produced by shadowing effect of water, organic matter anaerobic decomposes into hydrogen sulfide. When using ammonia sulfate fertilizer, the function of sulphur reducing bacteria leads to reduction of sulfate radical, producing large amount of hydrogen sulfide. When it is drowned, cadmium exists as CdS and Cd^{2+} transformation is restrained. In oxide and leaching environment produced by draining off water, S^{2-} oxide or SO_4^{2-}, pH value decrease. Cadmium dissolves in soil, which is easy to be absorbed plants. PO_4^{3-} ion in soil can influence cadmium transformation. For example, Cd^{2+} and PO_4^{3-} produce dissoluble $Cd(PO_4)_2$, which is difficult to be absorbed by plants. Therefore, to manage cadmium pollution of soil, lime and phosphate fertilizer can be used to adjust soil pH value up to 5.0 and prevent cadmium damage.

In soil of dry farm, cadmium exist in the form of $CdCO_3$, $Cd_3(PO_4)_2$ and $Cd(OH)_2$. And $CdCO_3$ is the majority, especially in calcareous soil t the pH of which is more than 7. The reaction producing $CdCO_3$ is as follow.

$$Cd^{2+} + CO_2 + H_2O \Longrightarrow CdCO_3 + 2H^+ \qquad \lg K = -6.07$$

Cd^{2+} in soil can be led out.

$$-\lg[Cd^{2+}] = -6.07 + 2pH - \lg[CO_2]$$

If it is in the air of soil, CO_2 partial pressure is 0.0003 atm, so Cd^{2+} in soil can be led out.

$$-\lg[Cd^{2+}] = -6.07 + 2pH - \lg[CO_2]$$

Therefore, Cd^{2+} in soil of dry farm has negative correction with pH value.

Cadmium is a toxic element which endangers plant growth. Although the damage of cadmium to crops, when in low concentration, is not obvious symptom, it does harm to human health through food chain. When the concentration of cadmium in soil is high to some degree, it can remain in soil and affect the growth and development of plant. Pot experiment suggests that when cadmium content is 10 μg/g, it have bad effect on paddy growth. When cadmium content is 500 μg/g, it seriously damage paddy growth. The biological effect of cadmium on plant is related to its forms in soil.

Plant absorption and accumulation of cadmium depend on the amount and type of cadmium in soil, the activity of cadmium and type of plants. Many plants can absorb cadmium from soil and accumulate in body to certain amount. The absorbed cadmium amount in plant is not only related to cadmium content in soil but also its chemical forms. For example, the order of cadmium absorption and accumulation by paddy is: $CdCl_2 > CdSO_4 > CdS$. Different kinds of plants have difference on cadmium absorption. Different breeds of one plant also have big difference on cadmium absorption and accumulation. Crops like wheat, corn, paddy, oats and chestnut all can absorb cadmium by root. The absorption amount is corn > wheat > paddy > soybean. To the same crop, the distribution of cadmium in each part of plant is uneven. Its content is: root > stem > leaf > seed. Plant in different growth period absorb different amount of cadmium, among which growth period is the most. Therefore, factors that influence cadmium absorption is various.

Cadmium can does harm to human health through soil-plant system and food chain. Thus, environmental pollution is people's focus.

③ Lead.

The abundance of lead in crust is 12.5μg/g and the background level of lead soil is 15-20μg/g.

Soil in city, mine and surrounding of metal production plant contain more lead. Lead pollution caused by automobile exhaust concentrate in city and two sides of road. The shorter distance form road is and larger the traffic volume is, more serious the lead pollution is. For example, if lead content of soil near road is 809.6 μg/g, lead content of place 91m far away from road is 32.5μg/g.

Pb^{2+} entering soil is easy to be absorbed by organic matter and clay mineral. The lead absorption of different soil is as following: black soil (771.6μg/g) > cinnamon soil (770.9μg/g) > red soil (425.0μg/g). Humus' ability to absorb lead is stronger than clay mineral. Lead can produce stable metal complexes and chelat with ligand. Lead exists in forms of $Pb(OH)_2$, $PbCO_3$ and $PbSO_4$ solid in soil. However, the content of soluble lead is very low in soil solution. Therefore, the transformation of lead is not active and biological effectiveness is low. When soil pH decreases, part absorbed lead can be release, which improve the transformation of lead and biological effectiveness.

Lead absorption and accumulation of plant are depend on lead concentration in soil, soil conditions, type and position of plant as well as the size and shape of leaf. There are two ways for lead to enter into plant. One is being absorbed by root of plant and the other is by folia. Lead absorbed by plant and transformed to land surface depends on plant type and environmental condition but lead is mainly absorbed by root. Under different soil conditions, lead absorption of plant is not same. In acid soil, lead accumulation in plant is more than alkali soil. Other elements in soil can be competitively absorbed with lead by plant. For example, in calcareous soil, calcium competes with lead for absorption by plant root. When calcium exists, because of competition between calcium and lead, lead is absorbed to be on unimportant position of enzyme chemical structure. Although lead content in plant is high, toxicity is not heavy. Another example, when lead and cadmium exist in soil at same time, cadmium may decrease lead content in crops but lead will increase cadmium content. Therefore, factors to influence plant on lead absorption are complex.

Lead is not the necessary element for plant growth. The process of lead entering plant is metabolic. Lead is stable in environment and certain concentration of lead does no harm to crop growth. The damage degree of crop caused by lead depends on its sensitivity to lead. Usually, lead is believed harmful to plants. For example, soybean is very sensitive to the damage of lead. High concentration of lead in soil prevents paddy growth. There are many symptoms. Chlorophyll content in blade decreases which influence photosynthesis and postpone plant growth. Finally, productivity decreases. Generally, the increase of lead content in soil can cause the decrease of crop productivity. In serious polluted area, it even cause coverage area heavily decreases. Under other condition, plants growing in serious polluted area have strong ability to persist high concentration lead. Up to now, there is no certain conclusion on the relationship between lead content in soil and lead absorption of plants.

④ Chromium.

The abundance of chromium in crust is 200μg/g and the average background level is 100μg/g.

Chromium exists in four forms. There are tervalent ion Cr^{3+} and CrO_2^- as well as hexabasic negative ion CrO_4^{2-} and $Cr_2O_7^{2-}$, among which tervalent chromium is most stable. Soluble chromium in soil is only 0.01%-0.4% of total chromium. The transformation and migration of chromium is related to soil pH, redox potential and organic matter content.

When tervalent chromium enter into soil, 90% of it is absorbed and fixed by soil. When the compound of chromium and Fe hydroxide are seal in oxide, tervalent chromium can hardly migrate. In soil solution, the solubility of tervalent chromium depends on pH. When pH value is more than 4, the solubility of tervalent chromium will decrease. When pH is 5.5, it all precipitates, forming polyhydroxy compound of chromium in alkaline solution. In addition, chromium forms organic coordination in low pH solution and its migration ability is stronger.

Soil colloid's absorption to tervalent chromium has positive correlation to pH value. Cr^{3+} can even exchange Al^{3+} in lattice of clay mineral. Clay mineral can absorb 30～300 times tervalent chromium more than hexabasic chromium. Most part of hexabasic chromium in soil is free in soil solution, only 8.5%-36.2% of which is absorbed and fixed by soil colloid. Different types of soil and clay mineral has obvious difference on hexabasic chromium absorption. It is like this: red soil＞yellow brown soil＞black soil＞yellow soil, kaolinite＞illite＞vermiculite＞montmorillonite. The more organic matter soil has, stronger the electronegativity is and weaker the absorption force is.

Chromium migration and transformation in soil are greatly influenced by redox condition. Among common pH and E of soil, Cr(Ⅵ) can be rapidly reduced into Cr(Ⅲ). In different paddy field, reducing rate of Cr(Ⅵ) has positive correlation to organic carbon content. When organic carbon content is 1.56% or 1.33% in latosol, reducing rate are 89.6% and 77.2% respectively. Generally, reducing rate of Cr(Ⅵ) adds 30% with 1% increase of organic carbon. The reduction made by organic matter on Cr(Ⅵ) has negative correlation to soil pH. When content of organic matter in soil is very low, reducing rate of pH on Cr(Ⅵ) greatly increases. For example, when soil pH is 3.35 or 7.89, reducing rate of Cr(Ⅵ) is 54% and 20% respectively.

When waste water containing chromium enters into farm land, Cr(Ⅲ) is absorbed and fixed by soil colloid and Cr(Ⅵ) is rapidly reduced into Cr(Ⅲ) by organic matter, absorbed by soil colloid. However, under certain condition, Cr(Ⅲ) can be transformed into Cr(Ⅵ). For example, when pH is between 6.5-8.5, Cr(Ⅲ) can be oxidized to Cr(Ⅵ). The reaction is as follow.

$$4Cr(OH)^{2+} + 3O_2 + 2H_2O \longrightarrow 4CrO_4^{2-} + 12H^+$$

Besides, manganese oxide in soil can transform Cr(Ⅲ) into Cr(Ⅵ). Therefore, Cr(Ⅲ) is a potential danger.

In the process of growth, plants can absorb chromium from outside and it enters into plant body through root and leaves. Chromium content in plant is quite different among different types of soil and plants. The content of remaining chromium in plant has positive correlation to chromium content in soil. Most Chromium absorbed from soil accumulates in root, next stem leaf and seed least.

Microelement chromium is necessary to plants. The development of plants will be influenced lacking of chromium. Low concentration chromium stimulates plant growth but excess chromium accumulated in plants is harmful which directly and indirectly endangers human health. For example, Cr(Ⅲ) between 20 and 40μg/g in soil boosts growth of maize seeding. Cr(Ⅲ) prevents its growth when it is 320μg/g. Another example, when Cr(Ⅵ) is 20μg/g, it boosts growth of maize seeding. When Cr(Ⅵ) is 80μg/g, prevention to growth is notable.

High concentration of chromium not only does harm to plants but also influence plant absorbing other nutrient. For example, when soil contains more than 5μg/g chromium, it prevents upper part of plant to absorb calcium, potassium, phosphorus, boron and copper. As a result, the top of damaged soybean will wither.

Toxicity of chromium in soil to plant is related to following elements.

(a) Chemical speciation. For example, Cr(Ⅵ) is more toxic than Cr(Ⅲ).

(b) Soil property. Soil colloid has strong ability to absorb and fix Cr(Ⅲ). In acid environment, Cr(Ⅵ) also can be easily absorbed by soil colloid. Organic matter in soil has functions of absorption and chelation. Besides, it can reduce soluble Cr(Ⅵ) to insoluble Cr(Ⅲ). Therefore, the content of soil clay and organic matter can influence the chromium toxicity to plants.

(c) Redox potential of soil. In same concentration of Cr(Ⅲ), there are more chromium in available form in dry land than paddy field.

(d) Soil pH. Cr(Ⅵ) is more toxic in neutral and alkaline soil while Cr(Ⅲ) is more toxic in acid soil.

In general, chromium does little to prevent plant growth because the ability of chromium to migrate in plant is weak. Rice cultivation experiment shows that the order of heavy metal migrating in plants is Cd>Zn>Ni>Cu>Cr. Therefore, chromium is one of the most difficult elements to be absorbed. The reason may be: (a) The reduction of tervalent chromium to bivalent chromium can hardly occur in plant. (b) Hexabasic chromium is of effectiveness but plant absorption to hexabasic chromium is influenced by negative ion like sulfate radical.

⑤ Copper.

The average value of copper in crust is 70$\mu g/g$. Copper content in soil is 2-200 $\mu g/g$. Copper content in soil is 3-300$\mu g/g$ in our country. The copper content in soil of most area is 15-60 $\mu g/g$ and average value is 20 $\mu g/g$.

Soil copper pollution mainly from copper mine and waste water from metal production plant. In addition, industrial dust, city sewage and copper containing pesticide all can cause soil copper pollution. For example, nearby copper mine in southern China, copper content in polluted soil is 1730-2630$\mu g/g$, which is 91-138 times more than comparison soil. The land area polluted by copper is about 30430hm^2, which is about 80% of total heavy metal pollution area. The average copper content in soil is 1000 $\mu g/g$ in Watarase-gawa area. The top value is 2020 $\mu g/g$, soluble copper of which is 250 $\mu g/g$.

The forms of copper in soil are as following: (a) Soluble copper. It is about 1% of total copper in soil. (b) Replaceable copper, It can be absorbed by soil organic matter and inorganic colloid and replaced by other positive ions. (c) Non replaceable copper. It refers to copper closely absorbed by organic matter as well as that in original mineral and secondary mineral, which can be replaced by neutral salt. (d) Insoluble copper. Most of them are salt that can dissolve in acid but cannot in water. For example, CuO, Cu_2O, $Cu(OH)_2$, $Cu(OH)^+$, $CuCO_3$, Cu_2S, $Cu_3(PO_4)_2 \cdot 3H_2O$, etc.

Humus in soil can produce chelation with copper. Soil organic matter and clay mineral has strong ability to absorb copper ion. The ability of absorption is related to its content and composition. The ability of clay mineral and humus to absorb copper ion is: humus>montmorillonite>illite>kaolinite. The ability to absorb copper of main soil in our country is: black soil>cinnamon soil>red soil.

Soil pH has big influence on copper migration and biological effect. Free copper has negative correlation with soil pH. In acid soil, copper easily migrate and its biological effect is stronger.

Copper which widely distributed in all plants is a necessary element to biology. When using copper fertilizer in soil of copper deficiency, crop productivity greatly improved. For example, copper sulfate is a common copper fertilizer, which can be used as base fertilizer, seed manure and top dressing. Besides, it can be used to deal with seed. But too much cooper endangers plant growth. For example, copper content in soil is 200 $\mu g/g$, wheat die . When copper content is 250$\mu g/g$, paddy die either. Another example is that when using solution whose copper content is 0.06$\mu g/mL$ to irrigate farm land, paddy productivity decreases. When concentration increases to 0.6$\mu g/mL$, productivity decreases 45.1%. If copper concentration is 3.2$\mu g/mL$, paddy has no harvest. Research shows that copper toxicity to plant is influenced by other elements. If there are 1 $\mu g/mL$ copper sulfate in culture medium, barley growth can be prevented. However, when adding other kinds of salt, although copper content is 4$\mu g/mL$, barley won't stop growing.

In body of plants growing in copper polluted soil, copper can accumulate. Copper accumulation in plant does not have obvious relation with total copper in soil but it is closely related to content of copper in effectiveness. Effective copper includes soluble copper and replaceable copper absorbed by soil colloid. The content of effective copper in soil is directly related to soil pH, organic matter content, etc. Different plant has different ability to absorb and accumulate copper. Copper distribution in different part of one plant is also different.

⑥ Zinc.

Zinc content in soil is 10-300$\mu g/g$. Average is 50$\mu g/g$. Zinc content in soil is 3-70$\mu g/g$ in our country and average is 100$\mu g/g$.

When using waste water containing zinc to irrigate, zinc enter into soil as forms of Zn^{2+} as well as complex ion $Zn(OH)^+$、$ZnCl^+$、$Zn(NO_3)^+$ and it is absorbed and accumulated by soil colloid. Sometimes it forms hydroxide, carbonate, phosphate and sulfide precipitation. Or it combines with organic matter in soil. Zinc mostly enrich in soil surface.

According to the studies of L. M. Shuman, the content of zinc in various parts in soil are that clay>iron oxide>organic matter>silt sand>sand>exchange state. Most of the zinc in soil exists in binding state, or as organic compounds and all kinds of minerals which are generally not easily be absorbed by plants. Plants can only absorb soluble or replaceable Zn. Migration ability and effectiveness of zinc depends mainly on soil pH, then the adsorption and fixed capacity of soil followed. Overall, the concentration of Zn in soil is higher than the effective concentrations of other heavy metals, and effective Zn is of 5% to 20% of the total zinc amount at the average.

Migration of zinc in the soil depends on the pH. When the soil is acidic, zinc adsorbed by clay minerals is easy to desorbs, and insoluble hydroxide zinc reacts with acid then converse into Zn^{2+}. Therefore, zinc is easy to migrate in acidic soils. When zinc in the soil is present with Zn^{2+} based, it is easy to migrate or be absorbed by plant. Therefore, zinc deficiency often occurs in acidic soil.

Due to rice fields is flooding, in a reduced state, sulfate-reducing bacteria will transform SO_4^{2-} into H_2S, Zn^{2+} and S^{2-} will product small solubility ZnS. Besides, soil zinc accumulation occurs. Zinc interacts with organic matter, forming soluble or insoluble complexes. Therefore, organic matter in soil will have a greater impact on the migration of zinc.

Zinc is an essential element for growth and development of plant. Zinc sulfate is often used as the trace element fertilizer, but an excess of zinc will harm plant roots. Therefore, productivity and quality of crop are impacted. Increasing soil acidity will increase the hazards of zinc to plant. For example, by adding 100 $\mu g/mL$ zinc solution in neutral soil, the growth of onion is normal; when adding 500 $\mu g/mL$ zinc, onion leaf yellowing; but in acidic soil, when adding zinc solution, 100 $\mu g/mL$, the growth of onion is retarded, while adding 500 $\mu g/mL$ zinc, onions almost stop growing.

Plants tolerate zinc concentrations greater than the other elements. Various plants have different sensitivity to high concentrations of zinc poisoning.

In general, the zinc concentration in the soil, will inevitably lead to the accumulation in the plant, and zinc accumulation in plants is closely related to soil zinc. For example, the amount of zinc in brown rice and zinc content in soil is of linear correlation. Other elements in the soil can affect the plant's absorption of zinc. For a instance, the overuse of phosphate fertilizer can form insoluble zinc phosphate and fixed, with plant's absorption of zinc reduced, or even cause zinc deficiency. Temperature and sunlight also have an impact on plant's absorption of zinc. Different plant absorption and accumulation of zinc is quite different. Usually, plants zinc content is 10-160 $\mu g/g$, but some plants zinc absorption are strong. Zinc accumulation in plant is up to 0.2-10 mg/g. Zn's distribution in various parts of the plant is not uniform. Such as in rice and wheat, zinc content distribution is: root>stem>fruit.

⑦ Arsenic.

The average arsenic content in the earth's crust is 2 $\mu g/g$, soil arsenic is generally about 6 $\mu g/g$. In our country, the average amount of arsenic in soil is about 10 $\mu g/g$.

Chemical valence of Arsenic is variable. Arsenic in soil exists in trivalent or pentavalent state. Its presence forms can be divided into soluble arsenic, adsorptive, replaceable arsenic and insoluble arsenic. Soluble arsenic is mainly AsO_4^{3-}, AsO_3^{3-} and other anions, which is generally only 5% to 10% of the total amount of arsenic. In our country, soluble arsenic in soil is less than 1%, and the total amount is less than 1$\mu g/g$. Therefore, even if the soluble arsenic into the soil, it is easy to translate insoluble arsenic and accumulate in the surface soil.

Migration of arsenic in soil is related to the content of iron, aluminum, calcium, magnesium and phosphorus, as well as soil pH, redox potential and action of microorganisms.

Soil colloid has adsorption effect on AsO_3^{3-} and AsO_4^{3-}. Ferric hydroxide and aluminum hydroxide, positively charged, and aluminum ions on the surface of aluminum silicate clay mineral can absorb arsenic anions, but organic colloid has no obviously adsorption to arsenic. Different clay minerals or anions have different adsorption on arsenic. Studies have shown that arsenic adsorption capacity of clay minerals, saturated with Fe^{3+}, is 620-1172 $\mu g/g$; Adsorption strength is that montmorillonite>kaolinite>dolomite; Adsorption amount of clay minerals saturated with Ca^{2+} is: 75-415μg / g; adsorption strength as following: kaolinite>montmorillonite>dolomite.

Arsenic can form insoluble arsenic compounds with iron, aluminum, calcium, magnesium and other ions. It can also produce co-precipitation with amorphous iron and aluminum hydroxide, so the arsenic in the soil may be fixed by iron, aluminum, calcium and magnesium, etc., making it difficult to migrate. Solubility of arsenic (V) compounds is that $Ca_3(AsO_4)_2>Mg_3(AsO_4)_2>AlAsO_4>FeAsO_4$. Therefore, Fe^{3+} has the strongest ability to fix AsO_4^{3-}. The order of arsenic adsorption capacity of several soils is as follows. Red soil>Latosol>Yellow brown>Black Earth>Earth>Loess.

Adsorbed arsenic in soil can be converted to soluble arsenic. The process is related to soil pH and redox conditions. If *E* of soil lowered and pH value increased, the solubility of arsenic will increase significantly. In alkaline conditions, the positive charge reduced in soil colloids, and arsenic adsorption capacity will reduce. The content of soluble arsenic will increase. Because AsO_4^{3-} is easily fixed and adsorbed by soil than AsO_3^{3-}, if arsenic in the soil exists in AsO_3^{3-} state, the relative solubility of arsenic will increase. Transformation between AsO_4^{3-} and AsO_3^{3-} in soil depends on redox conditions. Because dry land soil is in oxidized state, AsO_3^{3-} could be oxidized into AsO_4^{3-}; And if paddy soil is in reduced state, the majority of arsenic is in AsO_3^{3-} form. Solubility and the effectiveness of arsenic increase relatively, so arsenic victims will increase. In addition, AsO_3^{3-} damage crops more than AsO_4^{3-}.

Soil microbes can also promote morphological changes of arsenic. Fifteen lines heterotrophic bacteria was isolated, which can be oxidized AsO_3^{3-} to AsO_4^{3-}. Soil microorganism can also play a role of gasification of removing arsenic. Pot experiment found that there is a big difference between the arsenic amount applied and the sum of rice absorb arsenic and arsenic residues in the soil. That is because arsenic mold has gasification effect on arsenic, making this part of the arsenic reduced to AsH_3 and other forms, gasifying from the soil out of the soil. In addition, soil microbes can also convert inorganic arsenic to organic arsenic.

Phosphorus compound and arsenic compound have similar properties. Therefore, the presence of phosphorus compound will affect migration and biological effects of arsenic. Generally, phosphorus adsorption capacity of the soil is stronger than arsenic, leading to phosphorus seizures fixed position of arsenic in soil. Then arsenic solubility and bioavailability relatively increase. Gile has mentioned the problem of arsenic absorption in soil that phosphorus can be absorbed by iron and aluminum in soil. Iron plays a main role in arsenic absorption. In addition, the affinity of aluminum for phosphate is far more than the affinity for arsenic, so aluminum adsorbed arsenic can easily be replaced by phosphorus.

Therefore, the property of arsenic is contrary with cadmium, chromium, etc. When the soil is in an oxidized state, it is not that harmful; when waterlogged soil in a reduced state, AsO_4^{3-} was reduced to AsO_3^{3-} and arsenic harm to plants increased. Therefore, in practice, to arsenic polluted paddy soil, improving the soil redox potential and adding some substance can weaken the damage that arsenic does to crops.

Generally, Arsenic is not an essential element for plant. Low concentration of arsenic stimulates the growth of many plants while high concentration of arsenic is harmful. Arsenic poisoning will hinder growth and development of crop. Studies have shown that: when soil arsenic is 25μg/g or 50μg/g, wheat can yield increasing 8.7% and 20%, respectively; when arsenic is up to 100μg/g, the growth of wheat was seriously impacted; When containing 200-1000μg/g arsenic, all wheat died. Different arsenic compounds have different effects on the growth and development of crop. Such as organic arsenic is easily absorbed by the rice, which is more toxic than inorganic arsenic. As for inorganic arsenic, AsO_3^{3-} has huge harm to crops than AsO_4^{3-}.

Crop's absorption and accumulation of arsenic is related the amount of arsenic to in soil. The ability of different plants to absorb and accumulate arsenic is very different, and arsenic accumulation in different parts of the plant is different. The main pathway of arsenic to flow into the plant is the absorption of root and leaf. Plant roots can absorb arsenic from the soil, and then it migrates to various parts of the plant. After the organic arsenic being absorbed by plants, organic arsenic can be gradually degraded to inorganic arsenic in the body. Like heavy metals element, arsenic could eventually enter the human body through the soil-plant system and the food chain.

In summary, the soil pollution of heavy metal is mainly from sewage irrigation, sludge application and atmospheric dust; waste and municipal solid waste piled up anywhere can also cause soil pollution of heavy metal. High concentration of heavy metals in the soil endangers the growth of plant and affects the productivity and quality of agricultural products. Harmful levels of heavy metals on plant growth depend on the content of heavy metals in soil, especially the effective form of content. The main factors affecting transformation and biological effects of heavy metals in soil are: colloid adsorption of heavy metals, mating reaction and chelation of various organic and inorganic ligands, the redox state of the soil, effects of the soil pH and coexisting ions, as well as the soil microorganisms. Therefore, the factors affecting heavy metal migration and biological effects in soil of are multifaceted.

Heavy metals eventually enter the human through the soil-plant system and food chain, and affects human health. Heavy metals cannot be degraded by microbes. At same time, because colloid has a strong ability to absorb heavy metal ions etc, making it hard to migrate. Therefore, once soil suffered heavy metal pollution, it is difficult to be completely eliminated. It is believed that the soil is the of "sink" heavy metals pollution, so a series of pollution-prevention measures should be adopted to effectively combat heavy metal pollution.

(2) Organic pollutants

It includes chemical pesticides, herbicides, petroleum organic compounds, detergents and phenols. Pesticide is the major organic matter in soil. There are 50 kinds of commonly used pesticides.

(3) Radioactive substances

The weighed mean by area of radionuclides ^{238}U, ^{226}Ra, ^{232}Th and ^{40}K in the soil of 28 provinces in China are 39.5Bq/kg, 36.5Bq/kg, 49.1Bq/kg and 580.0 Bq /kg, respectively. The content of ^{238}U, ^{226}Ra, ^{232}Th is accordance with spatial differentiation, whereas ^{40}K presents no significant changes. In addition, the radionuclides show closely relationship with soil forming rock and distribution of uranium resources. The radioactive pollutants in addition to can harm human body directly, can also enters the body through biological chain and food chain, produce internal radiation in the body, damage the body tissue cells, causing diseases such as cancer, leukemia and genetic disorders, etc.

(4) Pathogenic microorganisms

Pathogenic microorganisms in the soil, such as intestinal bacteria, bacillus anthracis, intestinal parasites, mycobacterium tuberculosis, etc., mainly come from human and animal manure, sewage, waste, industrial waste water, hospital sewage and waste residue. Once the pathogenic microorganisms enter into the soil from the outside, they will multiply and cause soil quality degradation. The ecological balance will be destroyed. Plant makes disease or death directly. If human directly contacts with soil containing pathogenic microorganism, human health will be affected. If a man eats the vegetables, fruits and so on, which polluted by soil, people will get sick.

Section 5 Soil Pollution Control

Pollutants can enter the soil through a variety of ways, which cause changes in the normal function of the soil, then affect the normal growth and development of plants. However, the soil also plays the role of cleaning up pollutants. Especially organic pollutants in the soil can be purified through diffusion, dilution, volatilization and photochemical degradation, biochemical degradation, chemical degradation. If contaminants flow into the soil over purification capacity of the soil in the number and speed, which means exceed the environmental capacity of the soil, it will eventually lead to disorders of normal function of the soil, and hindering normal growth of crops.

Environmental capacity of soil refers to the maximum amount of pollution load of soil can hold without pollution, at certain environmental unit, within a certain time limit for compliance with environmental quality standards, and ensure the quantity and quality of the product.

To some extent, the soil is the "sink" of environmental pollutants. Soils are closely linked with plant life activities; contaminants ultimately affect human health through the soil-plant system and food chain. Therefore, soil pollution control is very important. First, we must control and eliminate sources of pollution. All effective measures should be taken to eliminate contaminants in soil, or controlling migration and transformation of pollutants in soil, making it can hardly enter into food chain.

1. Control and eliminate soil pollution source

Control and eliminate the soil pollution source is fundamental measure to prevent pollution. Control of soil pollution source means control the volume and speed of contaminants entering soil, so that it could naturally and slowly degrade in the soil to avoid soil contamination.

(1) Control and elimination of industrial "three wastes" emissions

Clean technology should be vigorously promoted to reduce or eliminate pollution source. Industrial "three wastes" and municipal waste must be dealt with and recovered, namely waste resource. On the emissions of "three wastes", purification treatment should be taken, and the number and concentration of pollutants emissions should be controlled.

Water source is lacking and distributes unequal in our country. Water pollution has been more and more serious recent years, causing increasing shortage of water resources.

Agricultural water has been severely strained. Therefore, China has developed sewage irrigation in many areas. On the one hand, it solved the farmland water problem. On the other hand, sewage contains a considerable number of nutrients, which also leads to soil contamination. Therefore, before using sewage for irrigation and the applying sewage sludge, regional irrigation water quality standards and agricultural sludge application standards should be developed at first according to the environmental capacity of the soil. And we should always observe the composition, content, and dynamic of pollutants in sewage. The amount of irrigation and sewage sludge fertilizer must be controlled in order to avoid blindly abuse irrigation water which causes soil contamination. In addition, industrial waste cannot be stacked arbitrarily.

(2) Control the use of chemical pesticides

We can prohibit or restrict the use of highly toxic pesticide residues, such as organic chlorinated pesticides; or develop high-effective, low poison and low residual pesticides, such as pyrethrin, nicotine and other natural plant ingredients pesticides; vigorously carry out studies of microbial pesticides and hormones pesticides. Microbes can make insects to die of infection, such as nuclear polyhedrosis virus has better effect in prevention and control mulberry caterpillar. Pesticides have insect hormone (insects gland secretions), molting hormone (molting hormone testosterone is used to prevent moth larvae), juvenile hormone (Antheraea Yamamai oil juvenile hormone makes insects can't survival) and pheromones, like drosophila, gypsy moth, pink bollworm of cotton and other sexual attractant. In addition, pesticide containing amino acids, fatty acids, nucleic acids and other ingredients which constitute biology can be used. They could be degraded easily. To explore and promote the biological way to control pests, we carry out biological predators on prevention act, such as the applying of insects, bacteria, mold, viruses and other microbes as natural enemies of pests and diseases. Pest control act of infertility should also be carried out.

With heavy use of pesticides for long time, various problems gradually exposed, mainly in the following aspects.

① Refractory property and wide range of pollution.

Chlorine pesticides especially DDT, has seeped into almost all organisms including humans on Earth, which endangers global ecological and human health.

② Low utilization rate.

It is estimated that after the trial, about 10% to 20% of the pesticides attached to the crop stems, leaves and fruit. More than 80% scattered in the soil and air during spraying. After wide spreading through wind and rain, it serious pollutes surrounding water, and even cause fish death.

③ Poor Selectivity.

Chemical pesticides kill pests, but also do destruction to the natural enemies and other beneficial organisms at the same time. It also destroys ecological balance, causing pests soared rampant.

④ Endangering birds.

In order to protect seeds and shoots of crops from pests, farmers often use pesticides to dress seed, resulting in a lot of birds died of pecking seeds. Meanwhile, it can also cause death and disease after the birds peck the food sprayed pesticides.

In nature, birds and insects are in a contradiction relationship. Birds are the natural enemies of insects, playing a regulatory role in the number of insects and altering their distribution area. Production and reproduction of insects is quite fast. If there is no control of birds, earth probably would have already been flooded by insects. Therefore, birds are our assistant in pest control. Most birds has benefits of forestry and animal husbandry industry. When humans use chemical pesticides to control pests, some birds also are damaged with decreasing number. Therefore, the number of pests and use of pesticide increase, forming a vicious cycle in which the ecological balance of nature has been seriously destructed.

⑤ Pest resistance.

Pest resistance continues to deepen and expand. There are least 504 kinds of insects becoming resistant to pesticides all over the world, and some pest resistance up to ten thousand times, which may lead to resurgence of object pest and secondary outbreaks of pest. Therefore, the amount and frequency of application of pesticides increase, resulting in a vicious cycle of chemical pesticide application. Finally, it leads to environmental pollution caused by the use of chemical pesticides becoming more serious.

⑥ Endangering human health through the food chain.

Organism enrichment of pesticides is quite remarkable. Pesticides can accumulate in crops, grain and livestock, then do harm to human health through the food.

In addition, poisoning and suicides caused by pesticide occurred frequently with social consequences. According to incomplete statistics, about 100 million people get disease each year due to pesticide poisoning worldwide. At least 20000 people died of pesticide contamination. According to information from the Ministry of health food hygiene supervision and inspection, in 2000, there are 41 cases caused by pesticide poisoning nationwide. 1300 people poisoned and 35 people died which is one fifth of the total annual number of deaths. What a shocking figures which is enough to alarm us. From the 1960s, government and public began to focus on chemical pesticides and their hazards posed above, and started looking for safer and more effective methods to control pests. In the 21st century, the hazards of chemical pesticides have aroused great attention, and people began to find solutions.

(3) Developing biological pesticide

Although China joined the WTO, but developed countries have adjusted and improved the product imported technological barriers for China's WTO accession, forming a "green barrier." The US Food and Drug Administration often detain food imports from China for the reason of not meeting FDA standards for pesticide residues. EU, Japan, South Korea following US, set up obstacles to China's exports of agricultural products. Bio-pesticides is an important guarantee to achieve safety of agricultural products and smooth export to production. The reality has brought opportunities and challenges for the development of biological pesticides.

Bacillus thuringiensis (called Bt), as a biological pesticide, its green and environmental protection has become widely accepted. It is mainly used to kill more than 200 kinds of pests on grain, cotton, vegetables, fruits, green and trees. It is high selective, not kill natural enemies of pests, environmental friendly, and harmless to humans and animals. It can meet people's demand for green food, and be in accord with the direction of sustainable development of agriculture. Therefore, it has been rapidly developed.

In 1970s, People began to produce Bt in large scale. However, because of backward level of fermentation resulting in higher costs, coupled with people's weak environmental awareness, Bt biological pesticide was unwelcomed under the strong competitive pressures among chemical pesticides. Many Bt biological pesticide manufacturers closed down. However, the pace of science did not stop. China had begun to invest the funds from the "seven five" plan for Bt biological pesticide research. After the 30 years development, Bt biological pesticide has made great progress in terms of production costs, technological content, production technology and theoretical research. Today, Bt bio-pesticides has a new look. With the enhancement of people's environmental awareness and demand for green food, Bt biological pesticide meet a rapid development period.

Bt biological pesticide has the following advantages.

① It has high effect for pest control and can maintain quality of agricultural products.

② It has strong pest specificity without killing natural enemies of pests and beneficial animals. It is non-toxic to humans and animals.

③ Variety of factors and composition work together. Pests and pathogens are difficult to produce resistance.

④ It maintains ecological balance.

⑤ No residue. It is conducive to sustainable development.

Bt biological pesticides still have some problems currently, mainly in the following aspects.

(a) The prices of Bt biological pesticides were flat with chemical pesticides, but need to reduce costs further to improve market competitiveness.

(b) Poor light stability. It is easy to lose activity under sunlight. It is effective in field for only about 5 to 7 days. New anti-sunlight Bt biological pesticide has come out and is badly in need to spread.

(c) Only prevent leaf-eating pest generally.

In order to overcome the shortcomings of Bt biological pesticide, scientists are innovating constantly. In the next five years, it is expected that Bt price will be lower than chemical pesticides, showing significant market competitiveness.

2. Soil improvement and repair

Increasing organic matter content in soil, mixing clay and sand and improve sandy soil, can increase or improve the properties of soil colloids. These increase the adsorption capacity and amount of toxic substances in the soil. Therefore, the soil environment capacity is increased and the purification capacity of soil is improved.

Analyzing, separating or developing new breed microbial can increase the microbial degradation of organic contamination. It is an important step to improve soil purification. Progress has been made in this aspect.

(1) Applying chemical amendments

Chemical amendments include inhibitors and strong adsorbent. Inhibitors commonly used are lime, phosphates and carbonates, etc. They can form insoluble compounds with heavy metals to impede heavy metals transfer into the crop. Application lime in acidic contaminated soils can increase soil pH, which enables cadmium, copper, zinc, mercury and other heavy metals to form hydroxide precipitate. According to tests, after using lime, cadmium content of rice can be reduced by 30%. Application of calcium and iron phosphate can be effectively suppressed activity of heavy metal like Cd, Hg, Pb, Cu, Zn. For example, Cd^{2+} and Hg^{2+} could form insoluble $Cd_3(PO_4)_2\downarrow$, $Hg_3(PO_4)_2\downarrow$ with phosphate respectively, which is significant for the elimination of Cd, Hg pollution in soil.

Applying strong adsorbent can make pesticide molecular lose activity, and also could reduce the damage of pesticides on crops. If 0.4% of activated carbon was added, Ehrlich dose pea absorbed from soils can be reduced by 96%. Organic, green manure, montmorillonite and so has a similar effect.

(2) Biological Improvement Measures

Enrichment of plant, including planting crops with strong absorption capacity to heavy metal, can exclude part of pollutants. For example, yellow-throated snake grass is 10 times higher than in rice on heavy metal absorption. Planting non-edible crops can remove heavy metals in soil to some extent. In addition, planting resistant plants, which means plant with poor absorption capacity on heavy metal, is also one biological way to improve.

In order to reduce the damages pollutants do to crops, plowing the soil is a choice. It also called deep ploughing, which digs and mixes the soil up and down and reduces the levels of pollutants in the surface soil. This method needs less amount of ground-breaking, but it is inadvisable in seriously polluted area.

In recent years, chemical and bio-remediation of soil and groundwater contamination has become a hot topic, and has been developed. Chemical remediation of soil organic contamination is cleaning organic matter in soil with a surfactant or organic solvent; Bio-remediation is the process using microorganisms in soil to degrade toxic and harmful organic pollutants into harmless organic matter (CO_2 and H_2O) . Degradation can change the physical and chemical conditions of the soil (including pH, humidity, temperature, aeration and added nutrients), and can also be inoculated with special built the engineered microorganisms, in order to increase the degradation rate.

Characteristics of bio-remediation of contaminated soil are as follows: lower cost than heat and physical and chemical methods; without destroying the soil that environment plants need to grow; completely oxidation of pollutants, no secondary pollution; good effect, 99% removal rates to low molecular weight contaminants; situ treatment, simple operation.

3. Development of green food, Promotion for human health

It is the goal of the people to develop green food and promote the human health. In order to achieve this goal, the commonly used measures are as follows.

① Use efficiency and low toxicity of chemical pesticides.

② Use Green Pesticide.

③ Biological deinsectization Technology.

④ Trapping deinsectization Technology.

⑤ Water-cover deinsectization Technology.

⑥ Steam deinsectization Technology.

⑦ Weed deinsectization Technology.

⑧ Nitrogen deinsectization Technology.

⑨ UV lamp deinsectization technology.

Questions

1. Describe soil formation and composition.
2. What are the basic characteristics of the soil?
3. What is the capacity environment of the soil?
4. What is soil pollution and how to distinguish?
5. What are the main pollutants in the soil? What are ways for them to enter the soil?
6. What are the methods to degrade chemical pesticides in the soil?
7. What is the common soil bio-remediation technology?

Chapter 5
Disasters And Environmental Protection

Key Point: Earthquake cause analysis and warning; The formation mechanism of volcano, eruption conditions and environmental impact; Mudslides cause analysis and disaster prevention; The cause and hazard of EI Nino phenomenon; The formation and classification of hurricane; The main components of biological diversity, protective measures and importance; The characteristics, habits, toxicity and protective measures of five kinds of deadly pest. The five kinds of deadly pest are bark scorpion, fire ant, tick, black widow spider and African killer bee; Food safety, food quality, food safety incidents, cause analysis and solutions; Many environmental problems of urbanization, the underlying causes behind environmental pollution, exploring new path and measures of environmental protection.

Section 1 Geological Hazard

1. Earthquake

(1) Definition and hazards of earthquake

Earthquake is seismic waves resulting from mediums violently eruption in earth. It is rapidly shack of earth surface, which leads to land shake in certain area. It is a manifestation of tectonic movements. It is also a manifestation of energy release of the earth crust and rock. It is estimated that around 5000000 earthquakes occur each year. Ninety-five percent of them occur in deep underground and their energy is very small, so people feel nothing, and it can only detect with special instruments. When the earthquakes are larger than magnitude 3, people begin to have feelings. But the lower earthquakes have no hazards of human safety and health. On average, the great damage earthquakes occur a dozen times every year all over the world.

Earthquake is the largest natural disaster that threatens people's safety and ecological environment. Powerful earthquake is greatly destructive. They can instantaneously damage a wide range of cities and villages all in ruins, causing serious disaster. China-Taiwan and the Japanese archipelago lie in the circum-Pacific seismic belt. The west and southwest of China lie in the Mediterranean-Himalayan seismic belt.

A large part of territory in China lies in active seismic zone, so earthquakes always occur in China. Such as the Xingtai earthquake in Hebei province in March 1966, the Haicheng earthquake in Liaoning province in February 1975, the Tangshan earthquake in Hebei province in July 1976, the Nantou earthquake in Taiwan in September 1999, the Wenchuan earthquake in Sichuan province May 2008, the Yushu earthquake April 2010, the Yaan earthquake in April 2013. These serious earthquakes bring great damage to people at earthquake region.

(2) Causes of earthquake

So far, there are three theories of the cause of earthquake. The first theory is "the elastic rebound theory", proposed by Reid in 1911. It said that when continuing accumulation of strain energy in earth exceeds rock strength, fault is produced. After the formation of faults, rock has elastic rebound and tendency to recover former state. Therefore, accumulated energy suddenly releases, causing earthquake.

The second theory is "magma impact hypothesis", proposed by Matsuzawa Takeo of Japan in 1955. This theory holds the view that earthquakes are correlated with magma impact. Rocks inside the earth have uneven thermal conductivity, so the volume of parts of melt expand and squeeze the wall rock. Finally, the wall rock ruptures causing earthquakes. The third theory is "phase change hypothesis", proposed by Percy Bridgmam of America. This theory holds the view that under the condition of hugh temperature and pressure, underground substances change from one crystal to another. Earthquake happens because of sudden change of substances volume. Though imaginative, the three theories can not reasonably explain the about 5000000 earthquakes happened each year. There are still other theories, none of them have been acknowledged by authority.

(3) Earthquake warning

According to the different mode of transmission, earthquake wave can be divided into three types: longitudinal wave, transverse wave and surface wave. The propagation velocity of longitudinal wave in the earth's crust is 5.5-7.0km/s. Firstly the longitudinal wave spreads from the epicenter, then makes the ground shake up and down. It has the lighter destructiveness. The propagation velocity of transverse wave in the earth's crust is 3.2-4.0km/s.

Secondly, the transverse wave spreads and makes ground shakes back and forth, left and right. It has the stronger destructiveness. The superposition of longitudinal wave and transverse wave form surface wave, which arrives finally. It is the mixed wave that arouses by longitudinal wave and transverse wave encountering in the surface. It has long wavelength and large amplitude. It only can propagate along the surface. So it is the main factor which seriously destroys the builds.

Earthquake warning uses the principle that electric wave is faster than earthquake wave, and issue warning before the earthquake wave arrives under the premise of the earthquakes had occurred. Earthquake warning can make people leave the disaster scene early and creates chance to handle with any possible secondary disasters, so that casualties and secondary disasters can be reduced. The key of earthquake warning is to estimate accurately magnitude and epicentral location using the data of a few seconds before the earthquake wave. It also can quickly estimate the impact of the earthquake on warning target and so on.

United States, Japan, Mexico, Romania and other countries are actively developing earthquake warning system. And they also have completed a certain scale and number of earthquake warning system or automatic emergency disposal system. China is an earthquake-prone country. Earthquakes cause many casualties and huge economic loss. Daya Bay Nuclear Power Station (GNPS) has established the earthquake warning system that combines monitor network with human decision-making. But there are no other major construction projects of earthquake warning system except GNPS. The relevant studies have just begun.

2. Volcano

(1) Formation mechanism of volcano

The volcanic formation involves a series of physiochemical process. Rocks in outer mantle partially melt under certain conditions of temperature and pressure. The melt separates from the parent rock and travels up through pore and fracture. Then magma chambers are gradually formed in certain parts. With the supply of magma continuously, the excess pressure of magma in chamber gradually increases.

When earth's crust of overburden cannot stop magma going upward, magma go up to the earth' s surface through the weak band. Because of the decreasing pressure in the rising process, the volatile content in magma gradually melts and forms bubble. When the bubble volume is more than 75%, the bubble locked in the liquid comes out, and it leads to explosive eruption. The magma viscosity suddenly drops after gas coming out, and it forms turbulence. If the magma viscosity is higher or the volatile content is less, it forms the quiet type overflow. The physiochemical process from the earth' s mantle and rock melt to explosively eruption is varied. So it forms different volcanic activities.

(2) Conditions of volcanic eruption

The formation of volcano in a place should meet the following five conditions. ①There should be the formation of partial melting. ②There should be the concentration of magma in the earth's crust. ③There should be physiochemical process of rock melting and gas evolution in magma chambers. ④There should be the tension field of tectonic activity. ⑤There should be the double driving force of pressure gradient and buoyancy.

(3) Environmental impact of volcanic eruption

There are amount of volcanic ashes and gases belched by volcanic eruption, It causes great impact on the climate. Volcanic eruptions inject gases which contain sulfur and poisonous sulfur dioxide. The volcanic ashes and gases are sprayed into the air, and then spread with winds far away. The volcanic ashes keep the sun out. As a result, temperature drops and harvests reduce. What is worse, it also causes some species to extinction. Sulfur dioxide is oxidized in the atmosphere. It forms acid rain and destroys vegetation. The forces of volcanic ash and rain combine cause mudslides. Mudslides can destroy roads and bridges as well as flood the countryside and city, making millions of people homeless. The mud formed with soil and rock debris floods the whole city. The volcanic ash is different from ordinary ash. The volcanic ashes are hard granules and insoluble in water. Breathing in volcanic ashes injuries respiratory tract and lung, causing respiratory disease. The volcanic ashes also influence equipment. They can damage transport like cars and airplanes.

(4) Disasters of volcanic eruption

Disasters of volcanic eruption are mainly mud stream similar to mudflow. Hot pyroclastic flows, lava flow and volcanic surge flow in the process of flow, met a lot of surface water, forming the hot volcanic mudflow accumulation. These deposits caused by heavy rain or earthquake collapse, flow into the rivers or lakes, forming cold volcanic mud flow accumulation. Lahars can submerge vast region and destroy buildings. Take eruption of Nevado del Ruiz and Pinatubo for example, volcanic mud stream produced by the former eruption submerged whole valley and town, causing 2500 death. The accumulation of unconsolidated volcanic ash caused by the latter eruption, washed and corroded by rainstorm. As a result, it caused accumulation of mud stream, which submerged large area and cost heavy economic loss.

(5) Benefits of volcanic eruption

The benefits of a volcanic eruption mainly include volcanic geothermal energy, hot springs, and fertile soil. Magma that can stay under ground is a good source of underground water is heated by energy produced by magma. We call this kind of energy geothermal energy, large scale of which forms geothermal field. Volcanoes can create land. Seawater accounts for 71% of the earth surface. Magma coming from underwater volcanic eruption cools and solidifies repeatedly until it reaches to the surface of water, and an island is formed. Hawaiian Islands and Iceland are formed in this way. Up to now, there is active volcano to spurt magma on these two islands. In addition, clouds of volcanic ashes will cover sunlight after volcano eruption, leading to average temperature going down. Therefore, volcano eruption can spontaneously adjust earth climate to balance.

3. Debris Flow

(1) Definition and property of debris flow

In mountainous and rugged area, because of stimulation of water from rainstorm, snow and ice or broken frozen lake, loose soil accumulated in valley pours to expansive downstream religion with large amount of mud and rock. Debris flow forms. Debris flow often occurs in some areas and has less predictability. It has good concealment, unexpectedness and destructiveness. The whole process of debris flow is only a few hours usually. Rarely, it is a few minutes. Because of emergency, short duration, great impact and terrifying force, debris flow has destructive force on land, environment and society. Debris flow is the result of mountain degrading, surface erosion increasing, water and soil losing. Its formation and development are mainly impacted by large-scale comprehensive environmental factors such as geology, geomorphology, hydrology, climate and so on. Unreasonable and unwise destructive activities of human increase the frequency and intensity of debris flow.

(2) Causes of debris flow

① Natural factors.

Landform is the basic conditions to debris flow. Debris flow usually occur in area where there are high mountain and deep valley, precipitous landform, high drop of gully bed, "two peaks flanking a valley", and stream easily forms. The landforms of debris flow can be divided into three main parts: forming region, circulating region and accumulating region. The landforms of upper forming region mostly are "three sides by mountain, one side by exit" scoop-shaped or funnel-shaped. This structure can accumulate energy. The landforms of middle circulating region mostly are narrow and deep valley. The valley bed has deep longitudinal gradient. This structure can increase current velocity and make a sheer drop of debris flow. The landforms of lower accumulating region mostly are flat piedmont plain or valley terrace. This structure can have accumulation. These open terrains mostly are farmlands or farmhouses. If debris flow occurs, they will suffer great losses.

Incompact substance contributes to the formation of debris flow. Debris flow mostly occurs in area characterized by complex geologic structure, developmental faults and folds, intensive new structure movement and frequent seismic activity. Surface rocks breaking, collapse, edge failure and sliding provide nourishing solid material for debris flow. In addition, rocks subject with loose structure, weathering, joint development and soft hard alternate layers structure provide fragmental material for debris flow.

Water t is a trigger for debris flow. Water is an important component of debris flow, and dynamic source of transporting medium. Water source of debris flow includes heavy rainfall, snowmelt, continuing raining for a long time, collapse of reservoirs, etc.

② Human Factors.

Irrational human activities are intrinsic factors that induced debris flows. To some extent, these activities intensify the frequency and degree of harm when debris flow occurred. Driven by the interests, the people in some areas would take such unwise activities as deforestation, overgrazing, destruction of vegetation on steep slopes, land reclamation, constructing transportation and water conservancy facilities casually, discarding slag waste randomly, etc. These activities have undermined the underlying conditions of surface runoff, and provided detritus for debris flows, exacerbated soil erosion, increasing the risk of debris flow.

(3) Hazards of debris flows

① Debris flows are more disruptive than the flood.

Debris flow, containing a lot of mud, sand and rock, pours from the height with other things accumulated in. Thus its energy is far greater than the flood. Because the process is short, no time was left for people to evacuate. It often leads to badly losses of lives and property.

② Debris flow is of large impact and strong destruction.

Debris flow is ferocious, wherever it goes through, houses collapsed, roads cut off, as well as roads, railways, stations buried, bridge embankment, farmland destroyed, trees smashed down. Zhouqu County Debris Flow (Gansu Province) engulfed half the town instantly. 2. 5 million square meters has been razed to the ground.

(4) Prevention of Debris Flows

Debris flows happened in different regions have different characteristics, because of which appropriate measures are required to be taken. Commonly used control measures include engineering treatment measures and non-engineering treatment measures. Engineering treatment measures include water storage, water diversion project, blocking, exhaust engineering and dredging projects. Non-engineering treatment measures include reforestation, planting vegetation, water conservation, reduced surface runoff, prevent soil erosion, restore ecosystem functions, etc. In debris flows area where collapses and landslips are of intense activity, engineering measures should be mainly used along with non-engineering measures. In a slope erosion and gully erosion debris flow area, non-structural measures should be dominate, supplemented by engineering measures. While in debris flow areas with both slope erosion and gravity erosion, comprehensive treatment works best.

Section 2 Weather Disasters

1. El Nino

(1) What is El Nino

El Nino refers to seawater of coastal areas in Peru and Ecuador in South America suddenly become warm before or after Christmas in some years. Specifically, Peru Current, flowing through the South American coast, is blown by the southeast trade winds almost parallel to the coast of Peru. While the surface waters flow off shore, the deeper water upwells with nutrient salts to surface, thus plankton becomes flourish. It attracts a large number of Peruvian sardine and other cold-water fish to breed and habitat in this area, making the region well known as the Southeast Pacific Fishery. But in some years, the southeast trade winds temporarily weakened. South tributary of the Pacific equatorial countercurrent crossed the equator south along the coast of Ecuador, so that the temperature of water along the coast of Ecuador and Peru increased rapidly. Because they did not adapt to the new environment, cold-water plankton and fish suffered a lot of death. Due to the temperature of coastal water rise much more intensively before and after Christmas, the birth date of Jesus, Peru residents called this seasonal phenomena of sea temperature rising El Nino (El Nino is Spanish transliteration, meaning El Nino or Son).

(2) The causes of El Nino

The earth is a spheroid composed by core, mantle, and crust. The core accumulates heat by constantly thermonuclear reaction, while the crust and mantle also release heat in two major ways. One is volcanic eruptions, earthquakes and mountain building. The other is earth's mantle thermal convection. In the ocean, hot magma flows from the central ridge and brings out the heat. After passing surface heat to the waters, it condensed into oceanic crust and went to both sides. And then it moving to the edge trench of the continental shelf and being completely cold, it sinks to beneath of the continental crust and melts in the mantle, and then starts a new round of circulation. This theory is called sea-floor spreading hypothesis, which is very similar to the liquid flow in boiling pot. Right now, the theory has been proved correct by paleomagnetism and it is successfully applied to explain some phenomena of volcanoes, earthquakes along the ocean edge. Distributed? in north-south direction, the Central Pacific Ridge divided the Pacific Ocean into two parts.

The central Pacific Ridge is not in the middle of the Pacific Ocean exactly, but in the east of the Pacific Ocean. From the central ridge to the east of the American continent, there is only a narrow piece of the ocean with simple ocean floor structure and thin oceanic crust. Finally, it dives beneath the American continent at a larger speed of about 10cm one year. The Pacific region of western equatorial is broad and of complex structure. Its crust here finally drives into the mantle at the speed of about 4-6cm one year, which is slower than the east. Due to the different subduction structures and speeds of the oceanic crust to mantle, the temperature difference of the ocean surface exits.

When the earth' s rotation speeds up, the crust summon up. The asthenosphere material in the interior of the earth will flow in the direction of the equator. Meanwhile, the material is flowing from the earth to the direction of the rotational axis. According to conservation of angular momentum, the earth' s rotation should turn slowly. This event happened later than seawater warming. Because there is a lag process for warming of seawater to pass to the surface, the El Nino occurs when the earth slows down or simultaneously. Westward type appears easily in the western ocean of Peru and Ecuador, which has obvious relationship with the Luni-solar tidal force in this region. Therefore, when the earth' s interior is unstable (the western ocean of Ecuador is three fracture point where geothermal materials have a huge upwelling potential and the upper materials has tensile fracture structure), this small solid tide can trigger deep undersea thermal mass pours to seabed, and warm the seawater, causing El Nino the phenomenon.

(3) Hazards of El Nino

The effects of El Nino are significant and widespread. Firstly, El Nino is an abnormal ocean event, which has a direct impact on local fisheries. Secondly, El Nino affects the local climate through the ocean-sea interaction, and the affection will extend to a wider range by the flow of air. Meanwhile, the climate anomalies will lead to a series of related problems, such as rainstorm, flood disaster, drought, forest fire, dust storm, warm winter, and various diseases. Therefore, the El Nino has extensive impact on two aspects, including the broad range of space (it is not only affects the low-latitude regions, but also the middle-high latitude regions) and various fields (atmosphere, oceans, ecology, agriculture, human health, etc.).

El Nino is a strong signal of climate anomalies. However, El Nino has different influence in different region. According to the degree of influence, the earth will be divided into three regions worldwide. The first area is hardly affected or less affected by El Nino, such as Europe, especially northern Europe. The second area is occasionally affected by El Nino. The influence is serious but the impact has not yet been uncertain, such as China, South Africa. The third area is often affected by El Nino, such as some countries in South America, Western Pacific and Indian Ocean region, Australia, etc.

2. Hurricane

(1) Definition and Classification of Hurricane

Tropical cyclone in Atlantic Ocean, Gulf of Mexico, the Caribbean and the eastern North Pacific is called hurricane. Hurricane usually occurs in the summer and early autumn, along with thunder and lighting. In just one day, the hurricane will be able to release large amount of energy, and this energy is sufficient to meet the demand of electricity of the entire United States for about six months. According to the maximum average wind near the center of the tropical cyclone, the World Meteorological Organization make the definition and classification standards of tropical cyclones. Tropical cyclone is divided into four levels. The wind less than level 8 is tropical depression. The wind between level 8 to 9 is tropical storm. The wind between level 10 to 11 is strong tropical storm. The wind more than level 12 is typhoon or hurricane.

Hurricanes refers to tropical cyclones, the wind speed of which is more than 33 m/s. Due to geographical differences, their names are different. Intense tropical cyclone, which appears in the western North Pacific and the South China Sea, are called "typhoons" . Intense tropical cyclone which occurs in the eastern part of the Atlantic Ocean, Gulf of Mexico, Caribbean Sea and the North Pacific, are called "Hurricane." Hurricanes and tornadoes can't be confused. The latter is of short period, and it is an instant outbreak no more than a few hours. In addition, a tornado has generally emerged together with hurricane. The most important feature of tornado is that there often has been one or more like "elephant nose" kind of funnel cloud column when a tornado occurs, accompanied by storm, lightning or hail. When tornado passes through the water, the water rise to form a water column and connect with the cloud, which known as "dragon water" . Passing through the land, tornado often rolls down the houses, even sucks people into the air.

Hurricanes are divided into five category levels. One level hurricane center has sustained wind speed of 64 to 82 sea miles per hour, or 119-153 kilometer per hour. Two level hurricane center has sustained wind speed of 83-95 sea miles per hour, or 154-177 kilometer per hour. Three level hurricane center has sustained wind speed of 96-113 sea miles per hour, or 178-209 kilometer per hour. Four level hurricane center has sustained wind speed of 114 to 135 sea miles per hour, or 210-249 kilometer per hour. Five level hurricane center has sustained wind speed over 135 sea miles per hour, or 249 kilometer per hour. According to the definition of the United Nations World Meteorological Organization, the sustained wind speed is the average wind speed in 10minutes.

(2) The Formation of Hurricane

Warm waters, moist atmosphere and ocean winds are three conditions for formation of hurricane. A series of thunderstorms have been formed at the beginning of the storm, and then a warm and humid air is wandering in the tropical ocean. When hot air rises, the water vapor condenses into a droplet, releasing latent heat that heat the surrounding air and sent hot air to the upper atmosphere (latent heat is one of the factors in the formation of tropical cyclones) . If there is a sufficient amount of storm clouds nearby, warm and humid air will form an "air column" from the ocean leading to the border of upper stratosphere vertically. Meanwhile, in the rising process, low pressure in the bottom absorbs a lot of air around. The release of latent heat, in the form of condensation of water vapor, makes the air column get more energy to travel over the sea. With the constantly rising of air column, the bottom pressure is further down. If this phenomenon occurs in the equatorial latitudes of more than 4°, a composite of acceleration acts on the air flows within the column, and turn faster. Finally, whirlwind forms.

With advantages, tropical cyclones could obtain energy rapidly in the formation process, and it turns from low pressure into the storm, and finally forms a hurricane with an amazing energy. National Oceanic and Atmospheric Administration have calculated that in one day, sustained typical hurricane could release 2×10^{14} joules of energy, which is equivalent of 1 million atomic bombs invested in the Hiroshima. Fortunately, most of the energy within the hurricane distributed to the air, rather than to the side. Faced with the destruction caused by hurricanes, climate warming can only make the situation worse. The wet air column is generally formed only when the sea surface temperature is above 26℃. When the oceans are warm, most of the sea surface temperature exceeds this threshold. Each 1℃ exceeds the threshold temperature can enhance the strength of hurricanes. When "Katrina" hurricane occurred, the surface temperature of Mexico Gulf is about 30℃. This is a powerful evidence to prove that temperature rise will leads to more and stronger hurricanes.

Section 3 Biodiversity and Lethal Pests

1. Biodiversity

(1) Definition

Biodiversity refers to the ecological complexes which are formed by biology and biological environments, as well as a variety of ecological processes related to it, including animals, plants, microorganisms and their genes, as well as the complicated ecosystem formed by them and their living environments. Biodiversity is substantially consisted of genetic diversity, species diversity and ecosystem diversity.

As time goes on, people are coming to realize that a very close relationship exists between either all kinds of species in nature or biology and surroundings, when carrying out the practices of nature conservation. Thus, it is difficult to obtain satisfying effect if nature conservation is restricted to the protection of species. To save the rare and endangered species, not only their wild populations but also their habitats should be carefully protected. In other words, effective protections are necessary to the whole ecosystem where the species live in. The definition of biodiversity emerges under such backgrounds.

(2) The conservation of biodiversity is imperative

In the later period of the 20th century, all fields of life science, especially the molecular biology, had achieved huge developments which brought about a revolution for the status of life science in natural science. Many scientists believed that bioscience would become the leading subject in natural science. They even predicted that the 21th century would be a biological century. More and more professionals engaged themselves into the research of life science. Among many branches of bioscience, the conservation of biodiversity is not only one of the most urgent missions in bioscience, but also one of the focus issues concerned by global field of biology. According to reliable data, over 100 species on earth die out every day. Many of them are extinct before known by humans, which is undoubtedly a kind of sadness and disaster to humans. Thus, actions for the conservation of biodiversity are imperative and extremely urgent.

(3) Main compositions of biodiversity

① Genetic diversity.

Genetic diversity is an important composition of biodiversity. The generalized genetic diversity refers to the totality of genetic information carried by the biology on earth. Genetic information is stored in the gene of individual organism. Thus, genetic diversity is also the diversity of biological gene. Any specie or individual organism is considered as a gene bank, because it conserves abundant genes. More abundant genes one species includes, stronger ability of acclimatization as well as larger contribution to life evolution and species differentiation it has.

The narrow genetic diversity mainly refers to the intraspecific change of gene, including the genetic variation between significantly different populations in one species, and the genetic variation in one population. Moreover, genetic diversity can perform at multiple levels, such as molecule, cell or individual. For most species of sexual reproduction in nature, there is usually no completely same genotype exists among individuals in one population. And these multiple individuals with different genetic structures form the population.

During the long-term evolution of biology, change (or mutation) of genetic material is the fundamental cause of genetic diversity. Mutation of genetic material mainly includes two types, that is, change of amount and structure of chromosomes, and change of nucleotides inside genetic locus. The former is called as distortion of chromosomes, and the latter is called as gene mutation (or point mutation). In addition, genetic recombination can also lead to genetic variation of biology.

② Species diversity.

Species diversity refers to the abundance of biological species, including animals plants, microorganisms and so on. Species diversity contains two aspects. The first aspect refers to the abundance of species in a certain area, which can be called as regional species diversity. And the second aspect refers to the uniformity of ecological distribution of species, which can be called as ecological diversity or community species diversity. Species diversity is an objective indicator utilized to evaluate the abundance of biological resources in a certain region. Regional species diversity is an indicator utilized to evaluate the abundance of biodiversity in a country or region. Regional species diversity contains: (a) sum of species, meaning the species amount of specific groups in a designated area; (b) density of species, meaning the species amount of specific groups in unit area; (c) proportion of endemic species, meaning the proportion of one endemic species of specific groups to the total amount of species in a certain area.

③ Ecosystem diversity.

Ecosystem refers to the natural complex formed by variety of organisms and their surroundings. All species are the composition of ecosystem. Diversity of ecosystem mainly refers to the diversity of composition and function of ecosystem on earth, and the diversity of various ecological processes, including many aspects such as diversity of habitat, diversification of biomes and ecological process. Among them, diversity of habitat is the formation foundation of ecosystem diversity, and diversification of biomes can reflect the type diversity of ecosystem.

(4) Significance

Biodiversity is the material basis for the survival and development of human society. Biodiversity provides us with food, fiber, wood, medicines and lots of industrial raw materials. Biodiversity also play an important role in preserving the soil fertility, ensuring the water quality, regulating climate and so on. Furthermore, biodiversity is important to the regulation of many aspects, such as the ingredient of atmosphere, the temperature on surface of earth, the oxidation-reduction potential and pH of the surface sediment. Maintaining the biodiversity is beneficial to the preservation of some rare and endangered species.

Once any species extinct, it can never regenerate, and humans will lose these valuable biological resources forever. The protection of biodiversity has great strategic significance to scientific research. Thus, it is extremely urgent.

(5) Protective measures

The first is in situ conservation. In order to protect biodiversity, a certain area of divisor land or water which contains protective objects shall be conserved and managed. For example, natural conservation areas are established to conduct in situ conservation. Natural conservation area is a typical natural system and a natural distribution area for rare and endangered wildlife, including natural relics, land, terrestrial water, sea and other different types of ecosystem. Natural conservation area is also equipped with important function for scientific research, popular science propaganda and ecological tourism.

The second is ex-situ conservation. Ex-situ conservation means establishing zoos, botanical gardens, arboretums, wildlife parks, seed banks, gene pools, aquariums and other different forms of conservation facilities at places out of biodiversity distribution, to protect the relatively precious species and the species or their genes with appreciative value under the assistance of artificiality. The purpose of ex-situ conservation is to find a temporary living space for the species on the verge of extinction. When recovering from extinction and possessing the ability to survive in nature, they will be sent back to ecosystem.

The third is to establish gene bank. People have been starting to establish gene banks in order to realize the desire of conserving species. For example, establish the net of global gene banks to protect the cultispecies of crops and their wild relatives that may be extinct. Most gene pools have stored the seeds of main crops such as cereal, potato and legume.

The fourth is to establish legal system. To conserve biodiversity, people also need to utilize legal means and improve the relative legal systems. For example, enhance the assessment and approval for the import of exotic species, as well as implement the uniformity of supervision and management. In addition, to provide financial support for practical work, fund institution should be established, which guarantees national earmarks as well as trying to obtain the donation and assistance from individuals, society and international organizations.

2. Lethal pests

Insects are one of the oldest organisms on earth, which have evolved for millions of years. The quantity of insects exceeds that of all other organisms. Insects can fly, crawl, slither, aggregate, sting and bite. They can raid without any warning and cause great threat to mankind. Millions of people have died of latent diseases transmitted by insects up to now. Therefore, insects are lethal pests.

(1) Lychas mucronatus

① Morphological characteristics.

Lychas mucronatus from Arizona are about 7-8 centimeters long, of which the outline is similar to pipa. Highly corneous hard skins are on the surface of their whole body which is segmented apparently. The trapezoidal cephalothorax is consisted of six sections. The backside is covered by semitransparent and hard carapaces. A pair of black median eyes is located on the centre of back and three lateral eyes are located on both sides of front. Insects have six pairs of appendages, of which the first pair is intact limbs utilized to assist eating, the second pair is long, thick and crab-claw-shaped tentacles utilized to prey, touch and defend, while the last four pairs are walking limbs. The mouth is located on the bottom of abdominal antechamber. Insects are dioecious and have different appearance with sex. The praeabdomen is wide and consisted of seven sections. The postabdomen is flexible, narrow-long and consisted of five sections as well as a tail barb. The first section of postabdomen has a genital which is covered by a gonopore.

② Life habit.

These scorpions prefer moist air and slant warm temperature, so they frequently appear in the living places of people. The Arizona lychas mucronatus are easily found be trapped in indoor water channel and bathtub, closet or sunless area of storeroom. They also scramble wall or adhere to celling. They usually live in seam of wood and bricks outdoors. They even aggregate to form a large group and live under barks during their hibernation. The Arizona lychas mucronatus prefer living at poplar riparian areas and phoenix tree gardens, because these places contain enough moisture and humidity to support the survival of insects and other prey species. The wide distribution of irrigated lawns increases the environmental humidity of living quarters, which has already led to explosive reproduction of Arizona lychas mucronatus in some regions. Lychas mucronatus are nocturnal and fond of ambushing prey. They like eating insects with medium or small size, such as small crickets, cockroaches and spiders. Lychas mucronatus are distributed in southwestern Arizona, California coast of New Mexico and Colorado River, northern Sonora and southeastern Utah.

③ Toxicity.

The stinger of Arizona lychas mucronatus usually cause partial pain and slight swell, lymphangitis and regional lymph node swelling, temperature rise of skins around wound, and haphalgesia. The venom of centruroides exilicauda can immediately cause pain as well as numbness and tingling of the affected part, but usually no swelling. Children may appear to be jittery and waggle randomly due to the abnormity of head, neck and eyes, while adults are mainly exposed to tachycardia, high blood-pressure, accelerated breathing, weakness and dyskinesia. Adults and children may all be exposed to dyspnea and salivation in excess as well. The stinger of brown scorpion may cause the death of children under six years old and hypertensive.

(2) Ticks

① Morphological characteristics.

Tick, also called Bishi and Biechi, is known as Caopazi, Goubie, Caobiezi, Niushi, Yinchichong, Caopichong, Goudouzi, Niubiezi, Caopazi (Scutiger is also called Caopazi in northeast areas) . Stick hibernates in the grass and plants of shallow hills or boards in the fur of domestic animals. When stick doesn't suck blood, it is very small like a mung or rice grain. When it sucks blood, its size will become bigger like a soybean, even like a fingernail. Anaplasmosis caused by stick bite is kind of infectious disease. People are generally injected with this illness. Medical workers and companions are especially in dangerous when intimately contact with severe patients or their blood without any protection.

This parasite extremely likes the animals with dense fur, especially cattle. They are always found under the neck and in the legs of cattle. They usually gather in group, and are difficult to get rid of. It is easy to find them in the countryside of Sichuan, Yunnan, Guizhou province.

Adults have strong chitin scutum on the back. They are generally called hard ticks which belongs to ixodidae. The ticks without scutum are called soft ticks which belongs to argasidae. There are about 800 kinds of ticks all over the world including 700 kinds of ixodidae, 150 kinds of argasidae and one kind of nuttalliellidae (only found in Europe) . There are about 100 kinds of ixodidae and 10 kinds of argasidae recorded in China. Ticks are temporal parasites in the surface of many kinds of spine animals. They are the media and reservoir host of the disease spreading from animals to human.

② Hemophagia habit.

Both larva and adult, male and female of ticks suck blood. The hosts of ticks include mammalia, birds, reptiles and amphibians. Some of them bite human. Ticks' host have a wide range. For example, 200 kinds of mammalia, 120 kinds of birds and few kinds of reptiles are Ixodes persulcatus' host, and they attack human. It is significant in epidemiology. Ixodid ticks often bite host by daylight. It is a long time, generally few days, for them to bite hosts. Soft ticks often bite host at night. The time of sucking blood is short, generally 1 to 60 minutes. Tick can suck a lot of blood and get bigger several times or more. Female ixodid ticks can even get 100 times bigger.

③ Relationship of host.

The parasitic site of ticks is selectivity. Ticks often choose thin fur or the undisturbed place. Such as neck, ear back, armpit, inner thigh, private parts, groin and so on. Canestrinis often parasitize on the neck caruncle and breast of cow, and next is scapular. Argas persicus often parasitize on the wings, legs and axillary of fowls.

Ticks' sense of smell is sensitive. They can distinguish animals' bromidrosis and CO_2, when they are 15m away from the host. They can jump on the host's body when hosts are coming. Such as Ixodes persulcatus inhabiting in forest often gather in grass across the road and the top of bush when they find food. They can climb onto the hosts when they pass by. Hyalomma asiaticum kozlovis inhabiting in desert zone are often active on the ground and initiative to find food. Ticks inhabiting in the livestock housing often find host on the ground, walls, and stakes.

④ Deadly ticks.

Ticks live in forest or rotten flag. Therefore, ticks are rarely seen in the big city where forest coverage is low. There are a dominant difference between hard ticks and soft ticks. The mouth of hard ticks for sucking blood is bigger and their shells are hard after sucking blood. Hard ticks are smaller than soft ticks. But hard ticks are more frightened to human. They bury their whole head in flesh and won't get out even they finish sucking blood. It is not easy to fetch them out because their body will break into two parts. Even if its body is be divided into two parts, the part in body cannot be fetched out, so it is more frighten.

⑤ Diseases.

(a) Forest encephalitis: Forest encephalitis caused by tick-borne encephalitis virus is a kind of acute infectious disease of nervous system and natural source disease in forest. In China, the major vector is ixodes persulcatus. Virus can be reserved in ticks for a long time and passage. This disease spreads widely in the forests of Heilongjiang and Jilin May to August. Timber jacks are easily got this disease. What's more, forest encephalitis can also be found in Sichuan, Hebei, Xinjiang, Yunnan provinces and municipalities.

(b) Xinjiang hemorrhagic fevers: Xinjiang hemorrhagic fevers caused by a kind of RNA virus through ticks is a kind of acute infectious disease. It is a natural disease in pastures of desert. The major vector is Hyalomma asiaticum kozlovi. Pathogene can live in ticks' body for few month and multiply by eggs. Besides, sheep blood and fresh blood of acute stage patients can cause this illness.

(c) Tick-born relapsing fever: Tick-born relapsing fever also called epidemic relapsing fever is a kind of nature spirochete disease spreaded by turicata. Its major clinical features are irregularity intermittent fever. This disease is prevalent in Xinjiang. The pathogene is Iran borrelia of southern xinjiang rural type. The vectors are Ornithodoros papillipes. The northern xinjiang wild type's vectors are Laplace borrelia and blunt edge tick. Infection source can be murid and patients.

(d) Lyme disease: In the summer of 1985, lyme disease is firstly found in Meilin county Heilongjiang province. The pathogene is Burt's Borrelia. It is a kind of natural focal disease spreaded by hard ticks and prevalent in spring and summer. The major vectors are ixodes persulcatus in China. Their major hosts are wild small rodents. This disease spread wildly in about 20 countries of five continents. Lyme disease is prevalent in 20 areas in China.

(e) Q hot: The pathogen is Bay's rickettsia. Clinical characteristic is rapid attack. This disease is prevalent between wild animals (rodent) and livestock. The major vectors are cattle and sheep. The pattern of infection is through respiratory tract, digestive tract, bite of ticks, and fecal pollution. The pathogen can live in ticks' body for a long time and multiply through spawn, for example, the pathogen can be kept in ornithodoros papillipes for 2 to 10 years. This disease spreads all over the world. Q hot has been found in dozens of provinces, cities, and municipalities in China.

⑥ Prevention methods.

(a) Environmental prevention

People use rotation, isolation to destroy ticks in prairie, combining with reclaim wasteland, cleaning weed, cleaning colony house of livestock, sealing the hole and killing rodent to kill ticks.

(b) Chemical prevention

We can spray DDVP, malathion and sumithion in the areas where ticks live in or overwinter. Livestock can have medicated bath regularly to kill tick.

(c) Medical prevention

Smearing and spraying grass bouquet oil of Luofu mountain on skin and room can effectively prevent ticks' bite.

(d) Biological prevention

Ticks inhabit in grass and forest. We can spray biopesticide in the ground where ticks live in. This method can works for a long time and is harmless to human and animals. Algae cover (0.12% propylene glycol alginate) can also be used. Physical control has no color, smell, toxic and pollutions to human.

(e) Personal prevention

We must wear five tight clothing, long stockings and boots, and helmets as well when we are in the areas of ticks. Spray repellent on the exposed skin. Check with each other and not let ticks out of epidemic areas.

(3) Solenopsis invicta

① Solenopsis invicta's morphological characteristics.

Solenopsis invicta is similar to ant and its size is 3 to 6 millimeters. To protect themselves, although they don't have inner skeleton, they have strong outer skeleton. They use weasand to breathe. Their bodies contain three parts including head, breast and abdomen. They have three pairs of feet and a pair of tentacle. Like ants, their worker-priests and soldier ants are gynes who have no ability to have children. Queen ant produces and hatches eggs. Among them, there are a few males because they just need one male in the process of copulation. Therefore, male ant is much less than female ant. Solenopsis invicta can't live in the environment of the temperature below 17℃.

② Life habit of Solenopsis invicta.

Adults' diets are wide. They kill insects, earthworms, frogs, lizards, birds and small mammals. They also collect seeds. They attack big size birds' eyes and other vital organs. Solenopsis invicta always bring eco-catastrophe. They are the archenemy of biodiversity protection and agricultural production. Solenopsis invicta is in single or multiple groups of post-production. Queen ant can lay up to 800 eggs a day. A nest having few queen ants can product 2000 to 3000 eggs a day. The amount of eggs could be the largest when foods are adequate. There are 240 thousands of ergates in a mature nest.

Solenopsis invicta has "five fierce": They like attacking, biting human, eating seeds, sting animals and drilling electric box. They can invade house, schools, lawns and other places. They have many chances to contact with people and often bite people. There are two diffusion paths of Solenopsis invicta: active diffusion and passive diffusion. Active diffusions (natural diffusion): They can migrate for 3 to 5 thousands of meters and diffuse with rain and drainage ditches. Passive diffusion (man-made diffusion): Horticulture pollution, turf pollution, the moving of waste soil pollution, compost pollution, farming machines pollution, empty containers pollution, vehicles pollution and so on.

③ Solenopsis invicta's hazards.

Hazards to animals and plants: Solenopsis invictas eat seeds, shoots, roots and fruits causing crops losses. They not only destroy the environment, but also prey many small animals such as newly hatched birdies, arthropods, earthworms and so on.

Hazards to infrastructures: Solenopsis invictas destroy infrastructures, power units, telecommunication equipments, dams and so on. Biting electric wires often causes short circuit even a small fire.

Hazards to human: The venom discharged in their tails can cause anaphylaxis. Once people were bitten by Solenopsis invicta, they would feel persistent heartburn-like pain. Partial skin forms erythema, blisters, hard swelling and itching. The broken blisters may cause the second bacterial infection. A few physical sensitive people may have severe anaphylactic shock. The main symptoms are systemic anaphylaxis, bronchoedema, dyspnea, pale complexion, cold limbs, fall of blood pressure and other circulatory failure. What's more, patients may swirl, lack in strength, fidgets, conk, and twitch, of gatism and so on because of cerebral anoxia and encephaledema.

④ Prevention of Solenopsis invicta.

The prevention of Solenopsis invicta is based on season and technology including physical prevention, chemical prevention and biological prevention.

(a) Seasonal prevention: Solenopsis invicta is active in the wet season of spring and summer. To kill them, baits work well. In winter when weather is cold, they are weak and almost inactive. Irrigating the nest by agentia is effective.

(b) Technical prevention: Physical prevention is disposing nest with boiling water and water logging. Boiling water disposing is generally pouring boiling water directly into the visible nests. Almost 60% Solenopsis invictas can be killed with this method. Boiling water must be poured into whole areas of nests. Single disposing can hardly success. We must do it again every 5 to 10 days and continue 3 to 4 times. However, disposing areas are often recurrent. We should note security, avoid being scalded or hurt other plants. Adding some soap or detergent is much better. Water logging is a method that flood nests to kill them. This method need to dig the whole nest and soak it for at least 24 hours.

Solenopsis invicta chemical control technology is the type of control methods which is currently the most studied, thorough and widely used. At present, pesticide formulations in the prevention and control of Solenopsis invicta at home and abroad mainly used are emulsifiable concentrate, suspension concentrate, micro-emulsions, wettable powders, dusts and granule bait. Pesticides to control Solenopsis invicta are used for spraying, hole irrigation, bestrewing, and so on; pesticides which direct contact may cause death are bifenthrin, permethrin, fenvalerate, cypermethrin, deltamethrin, permethrin chlorofluorocarbons, heptafluorobutyric permethrin, pyrethrin and so on; Pesticides used to put booby-traps of bait are sulfluramid, fipronil, fluorine hydramethylnon, spinosad, acephate, methoprene, pyrethrins and so on. Those methods in the prevention and control of Solenopsis invicta are generally divided into two categories. One is spraying, hole irrigation, bestrewing for anthill. These methods that use more tag pesticides has quick effect, but because of Solenopsis invicta are social insects that can transmit information generating shift nest phenomenon. Another important method is baiting to trap and kill Solenopsis invicta. Although effects are relatively slow, it is the favored products in other countries.

There is a variety of biological factors to inhibit Solenopsis invicta. In its origin, South America, the natural enemy of Solenopsis invicta includes at least 18 species of parasitic flies, more than 10 kinds of pathogenic microorganisms, at least three kinds of parasitic nematodes, one kind of parasitic wasps, one kind of parasitic ants and dozens of symbiotic creatures. In this region, the main reason for no outbreak of Solenoids is that the existence of strong competition and a variety of species of predatory and parasitic predators.

(4) African killer bees

① The life habits.

Years ago, 26 Queen Bee was unexpectedly taken into Brazil from Tanzania. And it hybridized with local bees, forming today's brutal African killer bees. It is a bred hybrid bee species produced by man-made accident. The species of insects with severe toxicity can kill people in a short time. Tens of thousands of people suffer every year. These bees blooms up in the wild to attack humans and animals, and scientists believe that killer bees living in Africa, where many natural enemies, if it do not take the initiative to attack, will be destroyed other animals. In a difficult career, those aggressive groups of bees survive through natural selection and breed. They work as groups and ferociously drive. Many animals will fled when met them. Even the lion cannot deal with them. The fatal nature of African killer bees lies in its extremely sensitive to outside defense system. What more frightening is that even if the victim has already been driven back and away from the hive, killer bees are still pursued and kill them ruthlessly.

African killer bees have large populations and strong mutation capacity. If people use chemical pesticides for long term, its increasing resistance may weaken its toxicity. These killer bees also are beneficial insect to environment and ecological construction.

They often eats pest in forest, which makes big contribution to controlling plant disease and pest. Killer bee is the murder for people's life, but at same time, it is the defender for forest and farmland. We should take these all to consideration when judging killer bee.

② Mechanism of toxicity.

The deadly substance of "Killer bees" is heart toxin, and what substance can lead to die for people in the body of wasps? Bee venom is pale yellow liquid secreted by worker bees. Chemical composition of venom is extremely complex. Besides a lot of water, it contains a variety of peptides, enzymes, biogenic amines, choline, glycerol and other substances and 19 kinds of free amino acids and so on. In the composition of bee venom peptide substance, the highest content of melittin accounts for about 50% of dry venom. It is this component that becomes a lethal disease in patients with the culprit. Since melittin is strong heart toxins that not only has the function of the contraction of blood vessels, but also highly solubility in blood, it does great damage to hearts.

③ The prevention countermeasure.

Queen can secrete a substance called fromen. Once bees sense the smelling, they suddenly became docile, and they will stop fighting. This artificial synthetic substance used in human self-protection.

(5) Black widow spiders

① The characteristics.

Black widow spiders that also known as latrodectus mactans and often referred to as Black Widow which is a spider with a strong neurotoxin. It may be fatal, if a man is bitten. The abdomen of adult female black widow spider is shiny black, and has an hourglass-shaped red spot. This record is usually red, and some may occasionally between white and yellow or red to orange. For some species, spot may be separated two points. Female black widow spider including legs' length is approximately 38 mm long. The body is about 13 mm long. Male black widow spider is only about half-size of the female spider, or even smaller. They have a longer legs and a smaller abdomen relative to their body size. They are usually dark brown, and have a yellow stripe and a yellow hourglass spot in mind. Adult male spiders can be distinguished to underage female spiders with more slender body, longer legs and larger limbs. Adult males have no venom.

Black Widow's venom is 15 times more severe than the prairie rattlesnake. Its worth is thousands time more valuable than gold in the American market and one. One ounce is at least $2.3 million. In addition, its silk strength exceeds all known natural fibers, even the famously tough steel wire and Kevlar fiber cannot compete with it. Its toughness can extend 27% before the snap, and strength is 2 times more than that of ordinary spider.

② The life habits.

Black widow spiders usually live in temperate or tropical regions and distribute in the southern Soviet Union, southeastern Europe and Central Asia, the Americas, and the Chinese island of Hainan, Xinjiang and other places. They tend to feed on insects, but occasionally they also prey lice, millipedes, small centipedes, scorpions and other spiders. When the prey entangled in the web, black widow spider may attack quickly from habitat and wrap prey with strong net, and then pierce prey and inject toxin. Toxin will work in 10 minutes or so, and the prey always dominated by spiders closely in that times. When the prey activity stops, the spider will inject digestive enzymes into the wound. Then the black widow spider will take the prey back to habitat.

③ The toxicological characteristics.

Although the toxicity of Black Widow toxin is very strong, but the dose is small, and there are few death cases resulting from Black Widow biting. Black widow spider' s venom will promote release of the neurotransmitter acetylcholine, which acts on muscles and cause muscle contraction. In general, the black widow spider toxin is a greater threat to children and the infirm.

④ Prevention.

Black widow spider generally will not take the initiative to attack humans without frightened. They often hide in the dense bush, so people should look around if there are traces of the black spider. It is best not to be barefoot when working or resting in the grass. In addition, Russia has six types of wasps which are natural enemies of black widow spiders. Wasps can be introduced to control the number of black widow spider.

Section 4 Food Safety

With the development of economy and society, food safety has been a global concern. Food security has become the basis to protect human life and health, to improve the quality of human life, to promote food trade and maintain the world peace and development. And governments and academia are constantly exploring effective ways to solve the increasingly serious problem of food safety.

1. Food safety and food quality

Currently, the concept, carried out by FAO/WHO in the To Protect Food Quality and Safety- the Guide to Strengthen National Food Control System in 2003, is widely accepted. Food safety refers to acute or chronic endanger to consumer' s health, which must be eliminated. Food safety is nonnegotiable.

In reality, food quality and food safety are often confused. Food quality refers to the sum of all the features affect the value of a food, including: ①the sensory quality, such as the appearance, texture, flavor, etc. ; ②nutritional quality, for example , protein, fat, vitamins, etc. ; ③security-quality, such as chemical residues, heavy metals, microbial toxins; ④ethical qualities such as animal welfare, environmental friendly.

As can be seen, food safety refers only to the property of food, which may does damage to human health. It is only a part of the food quality. In addition, the Food Safety is also different from "food security" or "food security in quantity" which is the quantity of food protection for a country or society.

Animal welfare generally is that animals, especially controlled by human, should not be subject to unnecessary suffering, even if it is for people's need to use for food, working tools, company or research. This position requires moral sentiment in human behavior. It is not as animal advocator said that the status of animal should be equal to human and it can pursue more rights in politics and philosophy.

Animal welfare concept consists of five basic components: the physiological benefits, without the worry of hunger and thirst; environmental benefits, is to let the animals have suitable housing; health benefits, mainly to reduce animal injuries; conduct welfare, should ensure that the animal's nature of free expression; psychological benefits that reduce animal fear and anxiety.

2. Food safety incidents

(1) The melamine-contaminated milk powder incident

On September 11, 2008, media reported that there were 14 babies suffering from kidney stones in Gansu, and then the provincial health department started to investigate. That night, Sanlu Group acknowledged that infant milk powder contaminated with melamine. Up at 8: 00 am on September 21, 2008, the number of babies in hospital who drank infant milk powder contaminated by melamine was more than ten thousand. Four death cases were confirmed by official. On September 16, 2008, Sanlu Group party secretary was dismissed from office. Meanwhile, four persons were deposed, there are the deputy mayor of Shijiazhuang, the chief of municipal animal husbandry and fisheries bureau administration, chief of city food and drug administration and the chief of quality and technical supervision bureau. On September 22, 2008, the state administration of quality supervision, inspection and quarantine announced the limitation of melamine which allowed 1 mg of melamine in 1000 grams of infant formula milk powder.

(2) The sudan red events

A food company imported five tons of red chili powder from Indian in 2002, which was used in the production of various brands of Worcester sauce for many downstream grocers or hotel use. UK Food Standards Agency finds out this chili containing carcinogens, and therefore subject to arrest. In February 2005, the British Food Standards Agency published a notice on the official website: products belonging to 30 enterprises, such as Heinz, Unilever, may contain carcinogenic industrial dye Sudan Red I. Subsequently, a massive action banned Sudan I swept the globe.

Just after the days when UK Food Standards Agency issued this notice, Beijing Municipal Food Safety Office notified to public about the situation in Beijing such as Red Duck Eggs etc. Sudan Red I is a kind of red dyes, used to grace for the solvent, oil, wax, gasoline, and shoes, floor, etc. The study showed that Sudan Red I is carcinogenic, both our country and the European Union have banned for use it in food production.

(3) Other food safety incidents

In the "special action" in March 15, 2011, CCTV exposed the incident that ShuangHui added "clenbuterol" into fodder to feed pig. Clenbuterol is an ingredient which can increase the amount of lean meat of animals. The animal fed with clenbuterol can be on the market in advance and decrease cost. However, long term use of clenbuterol which has strong toxicity may cause chromosome mutation and lead to therioma.

In November 9, 2011, General administration of quality supervision inspected part of the products of Unilever and showed that the rare earth content of "Lipton Tieguangyin" exceeded three times than criterion. The impact of rare earths on human health depends on its concentration and excessive intake does harm to human body. In October 19, 2011, "SiNian" confirmed the contamination of golden staphylococcus in its seafood and pork-stuffed dumplings; besides, in November 7, the golden staphylococcus was detected in the cabbage and pork dumplings of "SanYuan"; moreover , in November 17, the golden staphylococcus was found in wonton noodle soup of "MongTsai Wharf" . According to the database above, the three frozen food brands are all caught in the problems of bacteria. In April of 2011, some deli and noodle restaurants turned pork into beef through adding beef extract in the port to make profit in Hefei, Nanjing and other places. Experts point out that using food additive in a certain safe dose will not cause threaten, but if excessive and long-term intake will cause a harm even may cause cancer. In the end of June 2011, "Xinhua View" preliminarily revealed the shady industry chain of "waste oil" in Beijing, Tianjin and Hebei where there were "waste oil" processing factories. It had high-tech processing technology and a huge industrial chain. Besides these products have the access to supermarket in the form of small packages.

In August 2011, industrial and commercial bureau of Zhejiang province found that the content of nitrite in red cubilose was 350 times than criterion. These red cubilose mostly imported from Malaysia and other countries though Guangdong, Xiamen. In the early April 2011, some medias exposed that HuaLian supermarket and its food counters in Shanghai sold three kinds of steam bread such as corn steam bread, sorghum steam bread and black steam bread which are produced by the same company. The company added some pigment and essence into the out of date steam bread to produce these new ones. In April 2011, the Municipal Public Security in Huanggu, Shenyang province, attacked six processing sites of soybean sprout and they seized 25 tons of soybean sprouts which was illegally added some additive. It is verified that sodium nitrite, urea, enrofloxacin, 6-benzyladenine hormones and other harmful substances are detected in these soybean sprouts. In December 2011, it was reported by some media that there was a black workshop producing duck blood in Nanjing. They used 1. 5 to 2 kilograms of raw material to produce more than 10 kilograms duck blood by adding additive. The yield could magnify 5 times through this process.

With the development of China's food industry, food safety issues are more and more serious. These issues appear in all aspects such as food source, processing, production, circulation and consumption, which is from farm to table. Therefore, we will ask a question why food safety issues occur so frequently in our country.

3. Analysis about the causes of food safety incidents

(1) The apparent factors

Judging from the apparent, the reasons why food safety incidents occur so frequently are following. ①the contamination of water, soil, air and other environmental resources of agriculture. ②harmful chemical residues in agricultural products such as fertilizers, pesticides, growth hormones in the process of farming and aquaculture production. ③the illegally and excessively use of food additives (preservatives) in agricultural products processing and storage process. ④foodborne illness caused by microbial contamination. The food safety risks of new materials and technology, such as genetically modified foods. ⑤new challenges in food safety from scientific and technological progress. ⑥production operators counterfeit in food.

(2) The reasons of regulations and management

There are many flaws in the food safety management system in China. First, the legal system is not comprehensive. Although there have been lots of laws and regulations related to food safety, but they are decentralized and the adjustment range is narrow. So the laws and regulations cannot meet the needs of consumers' demands for food safety in new time and the punishment of food safety issues is not serious. Besides, there are many reasons as following: "Segmented supervision" lead to the fragmentation of the regulatory system (many departments unable to manage a pig) . The duties and responsibilities of local government is inaccurate; the lack of professional management department; the information between producers and consumers is serious asymmetry; food safety system lacks incentives for innovation; the consumers have a limited rationality; the producer has characteristic of opportunism and the implementation system is ineffective etc.

(3) Deep Analysis

Chemical pollutants play an important role in food production, processing, circulation, marketing, consumption and other aspects. Physical properties, chemical properties, biological properties, toxicological properties, toxicity, clinical manifestations, potential risks, the rule of migration and transformation, analytical techniques, damage repairing technology, removing technology and other aspects of chemical products should be strengthened.

4. Solutions

(1) Building a food safety system

The theory and practice research on food safety system indicates that it is of great significance to build a systematic and powerful food safety control system. National food safety system should include regulation system, management system, monitoring system, scientific support system and other aspects; the establishment of food safety system should involve all stakeholders, so that they work together for a common goal and build a common approach to ensure food safety. To achieve maximum effect on existing resources, it is allowable to make comparison and balance.

(2) Improving government supervision and regulation

Research on food safety supervision and regulation of the government shows that because of the externality, limited rationality and information asymmetry of comprehensive research on awareness, causes and solution for food safety, the market itself cannot solve all kinds of food safety problems at present, so these problems must be solved by government regulation. Government regulations provide useful ideas for solving the problem of food safety management from the supply. Chins has realized that government regulation is very important to solve the problems of the food safety. It has been improved in eliminating information asymmetries, strengthening the role of external regulation of resources, achieving cost minimization of government. Therefore, government regulation should draws more attention.

(3) Management model to control food safety

The current study focused on the model of HACCP which is focusing on the controlling to the risk factors, internationally. In our country, while researching and applying the HACCP, they also began to research on food safety control model from the perspective of the supply chain. Besides, they also discussed the model of quantitative classification and information tracking. But in author' s opinion, the existing documents on food safety control mode are fragmented and the theme of them is not clear. The HACCP model valued by the majority workers of theory, emphasis too much on food safety control to the production and processing aspects. It cannot solve the food safety issues " from farm to table" . Therefore, the research on the management system of food safety is not perfect, and mode or method for controlling the food supply chain still needs further study.

Section 5 Environmental Protection

1. Urbanization has brought many environmental problems, so environmental situation of our country is very serious

Many environmental problems which happened in last hundreds of years in developed countries, emerged in our country in just only 30 years with the rapid development. The phenomenon forms environmental problems of the structure type, compression type, complex type. New environmental problems turn up while old ones have not been solved yet. There is still a gap between improvement of environmental quality and the people' s expectation.

(1) Air pollution and atmospheric environment deterioration

Since January 1, 2013, 74 cities in our country began to monitor and publish the data of $PM_{2.5}$ according to the new air quality standards. In the 74 cities which are the first ones to achieve air quality testing according to the new air quality standards, more than half of the data, which are conducted by 80 national network monitoring sites in Beijing, Tianjin and Hebei seriously, continuingly exceed the standard. One third of the data, which is conducted by 129 national network monitoring sites in the regional cities in Yangtze River Delta, seriously exceed the standard. The data, which is conducted by monitoring sites in other municipalities and the capital city, also exceed the standard in different degree. At the beginning of 2014, affected by extremely adverse weather conditions, haze with long time, large scope, heavy pollution always appears in China's central and eastern regions. According to the new air quality standards, 70 percent of our country is substandard. The problem of $PM_{2.5}$ caused public concern. In fact, long-time haze and poor air quality are not new phenomenon. In recent years, after winter coming, this phenomenon always appears in the mid-east region. It is attribute to meteorological and pollution emission. Fog is natural, but haze is man-made. Sustained haze is the result of the damage to nature by industrial civilization and is also the evidence that humans must take the responsibility for pollution. Large amount of pollution sources discharge like coal, motor vehicles, industrial and dust are the root of serious pollution.

(2) Water pollution and deterioration of water environment

Although total water resources of China ranks the sixth in the world, the per capita availability ranks No. 109, so China is listed as one of the world's 13 water-poor countries. Water shortage is especially serious in the north and northwest of China. Water problems have severely restricted the sustainable development of the regional economy. Water pollution are still serious. Nowadays, half of the wastewater discharged by industry still cannot meet the drainage standard even though these water has been disposed. Among 1200 rivers under monitoring, 850 rivers are polluted in different degree. Those rivers flowing through the city are more seriously polluted. Almost 85 percent of the rivers seriously exceeded standard in their city reaches and nearly 50 percent of water supply in key towns do not meet drinking water standards. In 2012, the Grade Ⅴ water is 10.2% among 10 Great Basin and 24 lakes are in Grade Ⅲ among the 61 Key Lakes. 10% of underground drinking water source is substandard and there are still 298 million people cannot drink water under the security guarantee in countryside. The capacity of sewage treatment is lower.

(3) Ground Collapse

The frequent ground collapses have become potentially dangerous in city, which is related to the urbanization process. There are electricity, coal gas, oil, communication and other else in the underground of the city in each city. The water companies, the electricity sector, the gas companies, oil companies and communication companies laid the pipes and cable, which made the underground network very complicated. However, with the development of technology, the time of each pipeline being laid is different so that this road is in a circulation of excavation, landfill and excavation. Ground collapse happens with heavy rain, which is threat to city safety.

(4) Ecological protection and environmental protection in rural areas

Nowadays ecosystem degradation and soil erosion are serious. There are nearly 600000 administrative villages without environmental protection infrastructure. Every year 90 million tons of sewage and 280 million tons of waste are produced. Pollution census results show that emissions of chemical oxygen demand (COD), total nitrogen, and total phosphorus from agricultural sources account 44%, 57% and 67% of total emission amount respectively. The environmental risks are increasing.

2. The underlying causes of environmental pollution

(1) Insufficient knowledge of sustainable development strategy

Nowadays with the guide of GDP, some local governments are keen to bring in projects with high and fast returns which can increase the GDP through stimulating the economy and increasing employment. However, the development model which is eager for success makes the local government value more development but less environment. The development direction, means and thinking are accumulated behind the environmental issues. The environmental protection is a particular test of the "strategic vision" for the local government. There may be short-term economic growth at the expense of environment, but the consequence and harm of long term pollution will cost several generation's health even life.

(2) Lax enforcement of environmental protection laws, low investment

Policy from the multiple divisions, a policy does not perform and a policy is difficult to perform. Law is generally considered the most effective means of regulation for Government. As early as in 1979, the standing committee of the National People's Congress passed the Environmental Protection Law (Trial) . This is one of the first law enacted after the reform and opening up. Management of country should rely on law and the punishment should be specific to individual. Hardships of environmental protection in China for over a decade shows that if the environment law enforcement remains on the fine and limited on toward "public", it is not effective and efficient. Only if people who deliberately polluted environment are punished by law, the pollution will not continuingly happen. The reason for failure to implement the legislation ahead lies in that the investment on environmental protection accounts a small proportion of total economy. To maintain the status, the environmental protection investment should be accounted for 2% of gross domestic product (GDP), and to improve the pollution situation, investment should reach 3% of gross domestic product (GDP) .

(3) Officials emphasis on economic growth but neglect environmental protection

Local government and officials claim that they "pay high attention to" environmental protection laws and regulations. However, when the environmental protection is inconsistent with local economic development, environmental protection is obviously at a disadvantage. When questioned by superior departments, media reporters and local people about the investigation and treat of local factories' over emission, local government and official are uncover that they stress more on the importance of economic growth instead of environmental protection. Actually, local official cannot be blamed for peruse higher GDP as social progress and people's livelihood improvement need the financial support provided by the economic development. What should be blamed is the evaluation system which treasures GDP only even at the expense of land, environment and the interests of the people. Under nearly irrationally pursuit of GDP index and the corresponding evaluation system, it is difficult to realize the overall coordination, scientific development. City government do not have the advanced ideas to manage the city as a city manager. They are only in pursuit of economic interests, and ignore or d not attaches great importance to the urban ecological environment protection and pollution prevention. The result is that the urban environment problem becomes more serious. The image of city is greatly affected, hindering economic development.

(4) The lack of effective supervision system of environmental accountability

Environmental monitoring system of accountability is to ensure the smooth conduct of the environmental protection accountability. The foundation of sustainable development in environmental accountability is to establish long lasting environmental accountability system and effective supervision. China's long-term environmental accountability system is imperfect. It does not take the responsibility to censure not environmental friendly conduct or ignore serious problems existing in government. Therefore, the environmental accountability loses its authority, which contributes to the spread of violation of environmental laws.

3. Actively explore new environmental method

China's environmental policy has changed over time in the aspects of the concept and goals of environmental policy. It has changed from local environmental improvement to treatment after pollution to polluting while treating and then into environment and protection first and sustainable development. From the aspect of environmental policy tool, it simply adopted executive management first, and then market and public has been added.

China has step into the new period where economic development is improved by environmental protection. We cannot afford to take western countries' developing method that treatment after pollution and sacrificing environment for economic development. Some area in China set economic development apart from environmental protection, even treating after pollution at the cost of ecological environment. Lessons teach us that we must not take the old way but actively explore a new environmental way which meets the basic rudiments of low cost and emission, efficient, and sustain. It supports social and economic development with minimum damage of ecological environment, bringing three-win situation of economic benefits, social benefits and ecological benefits. The goal of new environmental way is to accelerate transformation of environmental management. It aims at improving environmental quality, constructing suitable systems of ecological environment protection and efficient preventive treatment, perfecting systems of environmental regulation system and science and technology standard, improving systems of environment management and law enforcing supervision as well as citizen involvement.

4. Strictly control $PM_{2.5}$ pollution

$PM_{2.5}$ pollution is hot in environmental protection. The state council executive meeting carried 35 measures in the Plan to Control Air Pollution, including following six aspects.

(1) Accelerating improvement and upgrade of industrial structure and cutting down overcapacity

It is required to strictly control the added capacity of high-energy consumption and high pollution industrial. For the steel, electrolytic aluminium and coke industrial, new project of them should be equivalent capacity or reduce capacity by exchange. The assignment of eliminating backward capacity should be speed up, finishing elimination in 21 key industrials in advance for a year. Construction of illegal projects in overcapacity industrials should be cleaned up, cutting down overcapacity.

(2) Adjusting the energy structure and controlling total consumption of coal

We should increase natural gas, coal seam methane supply and develop nuclear power, hydropower, wind power and biomass energy as well as promote the proportion of clean energy. Besides, the total amount of energy consumption should be controlled. The coal consumption in Beijing-Tianjin-Hebei, Yangtze River delta and Pearl River delta and Shandong province achieve negative growth. The projects consuming coal adopt other substituted energy.

(3) Strictly controlling pollution caused by vehicle and improving quality of fuel

We should practice the National Five Standard and accelerate to eliminate high exhaust emission vehicle. Besides, the step to decrease sulphur in fuel need speed up by implement better quality higher price. The development of public transportation is in the first place and walking and riding system also need be constructed.

(4) Controlling muti-pollutants at same time and strengthen comprehensive control

Take comprehensive measures such as making the coal-fired power plant desulfurization, denitration and dust removal, controlling industrial boiler furnace pollution, comprehensive treatment of volatile organic compounds, dust environmental management, dining lampblack pollution treatment, pollution control in warm season the north and so on.

(5) Establishing a monitoring and early warning system and properly dealing with heavy polluted weather

Establish pollution weather monitoring and early warning system as soon as possible, release monitoring and warning information timely, make and improve the emergency plan. According to warning levels, adopt measures to limit production, emissions and production for heavily polluting enterprises, stop constructing earthmoving operations, motor vehicle travel restrictions and other measures.

(6) Strengthen organizational leadership, implementation measures of safeguard

Establish a regional coordination system. Make full use of tax advantage, increase financial investment, and guide social capital to control air pollution. Accelerate modify the Environmental Protection Act and Air Pollution Control Act to crack down on environmental violations. Clarify the government social corporate responsibility to construct the great pattern of universal action.

5. Environmental Protection "Contribution in the contemporary" and "Benefit future generations"

(1) The adjustment of industrial structure, energy structure adjustment is the inevitable course

Adjust industrial distribution and structure. Rely on scientific and technological progress to promote clean production, and compression, eliminating the industry and product of low technical content, high energy consumption and heavy pollution, to improve the level and optimize the industrial products. Combining with the urban transformation, gradually adjust industrial layout. Move polluting enterprises out of the city, resolutely implement off closed, shutdown, merger and transfer industry, vigorously promote clean production to improve ecological agriculture, the proportion of green plants in the agricultural and industrial industry. Strengthen and promote the advanced concept of urban management, scientific planning, strict management, include the whole process of urbanization in national and regional sustainable development strategy and implement the strategy of sustainable development of urbanization.

(2) Adjust the important function of China's society in different parts of the configuration

To improve the urban environment, we must start from the adjustment of Chinese society in all kinds of important functions in different parts of the configuration, removing parts of function from function-centralized area.

It is a drastic way to avoid the superposition of humanistic effect and geographical climate conditions, alleviate pollution and its negative impact on human life. At this point, a new round of urbanization bears a special assignment. If you can achieve the optimal distribution of population and social activities, it can lay good foundation for the control environmental pollution.

(3) Fiscal policy boost environmental protection

First of all, it is to adjust the enterprise behavior of prevention and control of environmental pollution through credit lever. The fund for enterprise to prevent and control environmental pollution as well as of technical renovation project, can be applied from bank with low interest rates, interest loans. For the polluting enterprises, we can carry out the credit control by not to accepting the loan application or improve loan interest rates. Furthermore, polluting industries need to pay a certain amount of margin to the insurance company. Once environmental pollution accident occur, the insurance company should provide polluters with pollution damages and pollution control fund.

(4) Further improvement of the environmental legislation

To obtain the ideal effect of environmental law enforcement, perfecting environmental legislation can not be ignored. At present, when increasing environmental protection law enforcement, we should accelerate further improvement of China's environmental protection legislation. Through legislation, modification, cancellation and others, those regulation which do not meet the need of the current environmental protection situation need to be adjusted, making our country's environmental legislation to keep pace with the times. Eliminating incoordination in general principles of the Civil Law and the Environmental Protection Law of the related content, is one of the legislation problems need to solve at present.

(5) Accountability must be systematic

According to the law, accountability starts from top to bottom. Accountability is not only to quell the disturbances or cover up the problem, but to solve the problem completely. Therefore, accountability mechanisms should be a system. Reasonable accountability system should make that all parties have anticipation of the consequences of their actions. Let them know what things, what behavior cannot be done because of responsibility.

6. The development trend of China's Environmental Management Policy in the next 10 years

By 2025, the level of economic development of our country will enter the ranks of upper-middle income countries, and gradually move into late industrialization. The urbanization rate will reach 60% and the peak consumption of resources and energy are expected to continuingly arrive. These factors not only provide a strong driving force for China's environmental protection work, but also become the basic motivation of environmental protection work stress. With the continuous improvement of living standards, the people's requirements on environmental quality have accordingly increased. It will be more difficult to achieve environmental quality levels and in 2020 China well-off society. The urgent requirement of environmental protection work in the new situation is to pay more attention to and speed up practice of environmental policy. To set out environmental policy with global vision and strategic thinking and to optimize the regional distribution of environmental carrying capacity, China will put more efforts to promote in the following 10 areas in the next 10 years. ①Strengthen the policy of cutting down total emissions of pollutants; gradually incorporate carbon reduction into the schedule. ②Environmental policy will be gradually shifted to the improvement of environmental quality and environmental risk prevention and control of environmental policy for the target type of pollution control policies. ③Increase investment in environmental protection and promote the enforcement of environmental economic policy. ④Implement regional differences in environmental policies. ⑤Promote cooperative environmental pollution control policies. ⑥ Know the national environmental protection responsibility and improve the environmental policy execution. ⑦ Support public participation, construct "government - enterprise - the public" corporate governance mode. ⑧Improved performance appraisal index system, improve the capacity of local government regulation. ⑨Comprehensively deepen economic and environmental source control system to improve the quality of China's economic development. ⑩Comprehensively promote the construction of ecological civilization.

Questions

1. Please describe formation reason of earthquake, volcano and debris flow.
2. How earthquake and hurricane form?
3. What are common fatal creatures?
4. How to avoid food safety incidents?
5. What are latest environmental protection measures?

References

[1] Ronald A. Bailey, Herbert M. Clark, et. al. Chemistry of the Environment, Academic Press, 2005.

[2] Liu Zhaorong, Chen Zhongmin, Zhao Guangrong, et al. Course of Environmental Chemistry, Chemical Industrial Press, 2004.

[3] Thomas G Spiro, William M Stigliani. Chemistry of the Environment, Second Edition, Tsinghua University Press, 2003.

[4] Eldon D Enger, Bradley F Smith. Environmental Science - A Study of Interrelationships (Seventh Edition), Mc Graw-Hill, Tsinghua University Press, 2000.

[5] Wang Xiaorong. Environmental Chemistry, Najin University Press, 1993.

[6] Dai Guishu. Environmental Chemistry (21 century course), High Education Press, 2002.

[7] Yu Yufu, Ye minlv, Zheng Zhijian. Introduction of Environmental Chemistry, Fudan University Press, 1997.

[8] Deng Nansheng, Wu Feng. Course of Environmental Chemistry, Wuhan University Press, 2000.

[9] Ye Changmin, Wang Chunxia, Jin Longzhu. Environmental Chemistry of 21 century, Science Press, 2004.

第一章 绪论

本章要点：环境化学的定义；环境化学的分类；环境化学的研究内容；环境化学的特点；环境化学的发展方向；环境污染；世界上曾发生过的八大公害事件；近年来世界瞩目的公害事件；面临的主要环境问题；人们对环境问题的认识；我国环境污染防治面临的主要问题及解决方案。

第一节 什么是环境化学

一、环境化学的定义

环境化学是一门研究潜在有害化学物质在环境介质中的存在、行为、效应以及减少或消除其产生的科学。环境化学的任务有三项：第一，研究有害化学物质在环境介质中的存在、化学特性、行为和效应；第二，研究控制污染物的化学原理和方法；第三，用化学的基本理论和方法发现和解决环境问题。它是环境科学的核心组成部分，也是化学学科的重要分支。

二、环境化学的分类

环境化学主要包括环境分析化学、环境污染化学及污染控制化学。

① 环境分析化学研究如何运用现代科学理论和先进实验技术来鉴别和测定环境中化学物质的种类、成分形态（包括状态、结构）及含量。基于环境污染物分析的种类多、浓度低、干扰严重等特点，通常需要采用灵敏度、选择性、准确度与精密度俱佳的方法。自动化连续监测与卫星遥感技术不断融入环境分析化学之中，充实了该项技术对污染物的实时监控能力。环境分析化学包括环境有机分析化学和环境无机分析化学。

② 环境污染化学主要研究化学污染物在环境介质中迁移、积累和转化过程中的化学行为和变化规律。环境污染化学包括大气环境化学、水环境化学、土壤环境化学和污染生态化学。

③ 污染控制化学是论述化学污染物分离和控制的科学，它侧重研究污染控制和治理过程中的化学问题，为化学污染物的治理提供理论和方法。污染控制化学包括大气污染控制化学、水污染控制化学和固体废弃物控制化学。

三、环境化学的研究内容

环境化学的研究内容包括以下 5 点内容。①检测有害物质在环境中存在的浓度水平和形态。②研究潜在有害物质的来源、分布、迁移转化与归宿，它们在个别环境介质中和不同介质间的环境化学行为。③探究有害物质对环境和生态系统以及人体健康产生效应的机制和风险性。④寻找有害物质已造成影响的缓解和消除方法。⑤寻求防止产生危害的方法和途径。

四、环境化学的特点

环境化学的特点是从微观的原子、分子水平上，研究宏观的环境现象与变化的化学机制及其防治途径，其核心是研究化学污染物在环境中的化学转化和效应。

1. 所研究的物质在环境中数量庞大，并且体系复杂

环境中的化学污染物质品种、数量繁多，组成复杂；同时，各种污染物质在环境体系中可以发生多种机制的化学反应和物理变化。这就决定了环境化学研究对象是一个组成多样、形态多变、机制复杂的体系。

2. 污染物浓度极低，一般处于μg/mL或μg/L数量级

化学污染物质在环境中的浓度水平很低，一般处于μg/mL或μg/L数量级，而与这些污染物共存的其他化合物的浓度则处于常量水平。为了对这些污染物进行准确的定性和定量分析，常常需要使用现代分析技术手段，如：红外光谱技术、核磁共振技术、色质联用技术和同位素示踪技术。

3. 环境化学研究时间跨度和空间跨度大、综合性强

环境中的化学反应不容易控制，从事环境化学研究需要很大的时间跨度和空间跨度。环境化学研究化学污染物质在环境生态系中的分布、迁移、转化和归趋，尤其是它们在环境介质中的积累、相互作用和生物效应等问题。

五、环境化学的发展方向

随着环境问题的日益严峻和人们认识的提高，对环境化学的研究出现了新的发展趋势。

1. 环境分析化学

环境分析化学的研究方向主要集中在环境样品的前处理技术完善和开发方面。分析对象从以无机化合物为主转变为以有机化合物和有机金属化合物为主。仪器分析方法的改进集中于提高仪器分析的灵敏度和重现性方面。

2. 环境污染化学

环境污染化学研究化学污染物在大气、水体和土壤环境中的形成、迁移、转化和归趋过程中的化学行为和生态效应。大气环境化学的研究对象涉及大气颗粒物、酸沉降、大气有机物、痕量气体、臭氧耗损及全球气候变暖等，其空间尺度从室内空气、城市、区域环境远至全球。

水环境化学主要研究河流湖泊和水库，其次是河口、海湾和近海海域，地下水污染；集中研究水体界面化学过程、金属形态转化动力学过程、有机物的化学降解过程、有机物的化学降解过程、金属和准金属甲基化等方面。

土壤环境化学主要研究农用化学品在土壤环境中的迁移转化和归趋及其对土壤和人体健康的影响，包括：有机污染物在土壤中的降解、土壤中温室气体的释放、污染物在固-液界面上的化学过程和土壤污染的化学与生物修复等内容。

环境生态化学主要研究农用化学品在生态系统中产生效应的化学过程，如稀土农用的环境化学行为及生态安全等。

3. 污染控制化学

污染控制目前有两种模式：一种是传统的末端控制；另一种是污染预防与清洁生产。末端控制对各国污染控制技术的发展和环境污染治理起着积极的作用，但末端控制只能减少或阻止污染物排放，不能阻止污染物产生。末端控制中的化学研究内容包括：大气污染控制；水体污染控制；固体废物污染控制及资源化研究；污染控制材料和技术研究，包括絮凝剂、吸附离子交换剂、膜材料、催化剂、分离技术等的研究。

鉴于末端治理的局限性，20世纪80年代中期，欧美国家提出了控制模式从终端污染控制模式（End-of-pipe Control）向全过程控制模式的转变，提出了污染预防（Pollution-Prevention）、清洁生产（Cleaner-Production）、绿色化学（Green Chemistry）和工业生态学（Industrial Ecology）等一系列污染控制的战略思路，希望生产过程中不产生有害中间产物和副产品，实现废物和排放物的内部循环，达到污染最小化和资源、能源利用最大化的目的。这是环境管理战略的一次重大转变。

第二节　环境污染事件

环境问题是指各种环境退化及对生物圈的不利影响的现象，主要包括酸雨、温室效应、臭氧空洞、淡水资源缺乏、资源短缺及生态退化。

一、环境污染

近年来，由于人为因素使环境的构成或状态发生了变化，环境素质下降，从而扰乱和破坏了生态系统和人们正常生活和生产条件。

环境污染具体表现在以下四个方面：有害物质对大气、水、土壤和动植物造成污染并达到致害的程度；生物界的生态系统遭到不适当的干扰和破坏；不可再生资源被滥采滥用；固体废弃物、噪声、振动、恶臭、放射线等造成对环境的损害。造成环境污染的因素有物理的、化学的和生物的三个方面，其中由化学物质引起的污染占80%～90%。

二、世界上曾发生过的八大公害事件

1. 比利时马斯（Meuse）河谷烟雾事件

1930年12月1～5日，比利时马斯河谷工业区内13个工厂排放的大量烟雾弥漫在河谷上空无法扩散，使河谷工业区有上千人发生胸疼、咳嗽、流泪、咽痛、呼吸困难等，一周内有60多人死亡，许多家畜也纷纷死去，这是20世纪最早记录下的大气污染事件。

2. 美国洛杉矶光化学烟雾事件

从20世纪40年代起，已拥有大量汽车的美国洛杉矶城上空开始出现由光化学烟雾造成的黄色烟幕。它刺激人的眼睛、灼伤喉咙和肺部、引起胸闷等，还使植物大面积受害，松林枯死，柑橘减产。1955年，洛杉矶因光化学烟雾引起的呼吸系统衰竭死亡的人数达到400多人，这是最早出现的由汽车尾气造成的大气污染事件。

3. 多诺拉烟雾事件

1948年10月26～31日，美国宾夕法尼亚州多诺拉镇持续雾天，而这里却是硫酸厂、钢铁厂、炼锌厂的集中地，工厂排放的烟雾被封锁在山谷中，使6000人突然发生眼痛、咽喉痛、流鼻涕、头痛、胸闷等不适，其中20人很快死亡。这次烟雾事件主要由于二氧化硫等有毒有害物质和金属微粒附着在悬浮颗粒物上，人们在短时间内大量吸入了这些有害气体，以致酿成大灾。

4. 伦敦烟雾事件

1952年12月5～8日，伦敦城市上空高压，大雾笼罩，连日无风。而当时正值冬季大量燃煤取暖期，煤烟粉尘和湿气积聚在大气中，使许多城市居民都感到呼吸困难、眼睛刺痛，仅四天时间内就死亡了4000多人，在之后的两个月时间内，又有8000人陆续死亡。这是20世纪世界上最大的由燃煤引发的城市烟雾事件。

5. 四日市哮喘事件

1955年日本第一座石油化工联合企业在四日市上马，1958年在四日市海湾打的鱼开始出现有难闻的石油气味，使当地海产品的捕捞量开始下降。1959年由昭石石油公司投资186亿日元的四日市炼油厂开始投产，四日市很快发展成为“石油联合企业城”。然而，石油冶炼产生的废气使当地天空终年烟雾弥漫，烟雾厚达500m，其中漂浮着多种有毒有害气体和金属粉尘，很多人出现头疼、咽喉疼、眼睛疼、呕吐等不适。从1960年起，当地患哮喘病的人数激增，一些哮喘病患者甚至因不堪忍受疾病的折磨而自杀。到1979年10月底，当地确认患有大气污染性疾病的患者人数达775491人，典型的呼吸系统疾病有：支气管炎、哮喘、肺气肿、肺癌。

6. 痛痛病事件

19世纪80年代，日本富山县平原神通川上游的神冈矿山实现现代化经营，成为从事铅、锌矿的开采、精炼及硫酸生产的大型矿山企业。然而在采矿过程及堆积的矿渣中产生的含有镉等重金属的废水却直接长期流入周围的环境中，在当地的水田土壤、河流底泥中产生了镉等重金属的沉淀堆积。镉通过稻米进入人体，首先引起肾脏障碍，逐渐导致软骨症，在妇女妊娠、哺乳、内分泌不协调、营养性钙不足等诱发原因存在的情况下，使妇女得上一种浑身剧烈疼痛的病，叫痛痛病，也叫骨痛病，重者全身多处骨折，在痛苦中死亡。从1931年到1968年，神通川平原地区被确诊患此病的人数为258人，其中死亡128人，至1977年12月又死亡79人。

7. 水俣事件

从1949年起，位于日本熊本县水俣镇的日本氮肥公司开始制造氯乙烯和醋酸乙烯。由于制造过程要使用含汞（Hg）的催化剂，大量的汞便随着工厂未经处理的废水被排放到了水俣湾。1954年，水俣湾开始出现一种原因不明的怪病，叫“水俣病”，患病的是猫和人，症状是步态不稳、抽搐、手足变形、精神失常、身体弯弓高叫，直至死亡。经过近十年的分析，科学家才确认：工厂排放的废水中的汞是“水俣病”的起因。汞被水生生物食用后在体内被转化成甲基汞（CH_3HCl），这种物质通过鱼虾进入人体和动物体内后，会侵害脑部和身体的其他部位，造成脑萎缩、小脑平衡系统被破坏等多种危害，毒性极大。在日本，食用了水俣湾中被甲基汞污染的鱼虾的人数达数十万。当时中毒者有283人，其中60多人死亡，到1991年，有2248人被认定为公害病患者，其中1004人已死亡。ChiSS公司为此赔偿了上百亿日元，40多年来问题仍未彻底解决。这是水污染的一个典型事件。

8. 米糠油事件

1968年日本九州爱知县一个食用油厂在生产米糠油时，因管理不善，操作失误，致使米糠油中混入了在脱臭工艺中使用的热载体多氯联苯，造成食物油污染。由于当时把被污染了的米糠油中的黑油用去作鸡饲料，造成了九州、四国等地区的几十万只鸡中毒死亡的事件。随后九州大学附属医院陆续发现了因食用被多氯联苯污染的食物而得病的人。病人初期症状是皮疹、指甲发黑、皮肤色素沉着、眼结膜充血，后期症状转为肝功能下降、全身肌肉疼痛等，重者会发生急性肝坏死、肝昏迷，以致死亡。1978年，确诊患者人数累计达

1684人。

三、近年来世界瞩目的公害事件

全世界平均每年约发生200多起比较严重的公害事件。比较令人瞩目的环境污染事件有以下10起。

1. 意大利塞维索化学污染事故

1976年7月10日意大利北部塞维索地区的一家农药厂爆炸，导致剧毒化学品二噁英的污染，使许多人中毒，附近居民被迫迁走，几年内当地畸形儿的出生率大为增加。

2. 美国三里岛核电站泄漏事故

1976年3月28日，美国宾州三里岛压水堆核电站发生了堆心熔毁的严重事故，一座反应堆大部分元件烧毁，一部分放射性物质外泄。事故持续了36h，给人们留下了终生不灭的影响，直接经济损失达10多亿美元。

3. 墨西哥的液化气爆炸事故

1984年11月19日，墨西哥国家石油公司所属的液化气供应中心发生爆炸，死亡1000多人，伤400多人，3万多人无家可归。

4. 印度博帕尔农药泄漏事故

1984年12月3日，美国联合碳化物公司设在博帕尔市的农药厂的剧毒化学品异氰酸甲酯罐爆裂外泄，受害人数达20万，死亡2000人以上。

5. 前苏联切尔诺贝利核电站泄漏事故

1986年4月26日，位于基辅地区的切尔诺贝利核电站四号反应堆爆炸，造成重大放射性污染，周围十多万居民被疏散，伤数百人，死亡31人。

6. 莱茵河污染事故

1986年11月1日，瑞士巴塞赞德兹化学公司的仓库起火，使大量有毒化学品随灭火用水流进莱茵河，造成了西欧10年来最大的污染事故。

7. 海湾战争造成的环境污染

1990年年底爆发的海湾战争历时42天，期间油井大火昼夜燃烧，是迄今为止历史上最大的石油火灾及海洋石油污染事故，也是人类历史上最严重的一次环境污染，其污染程度超过切尔诺贝利核电站发生的核泄漏事故。这次战争所造成的环境污染是灾难性的，已给世界带来了影响。

8. 苏丹红污染事件

苏丹红（一号）是一种红色染料，用于为溶剂、油、蜡、汽油增色以及鞋、地板等的增光，是一种禁止作为食品添加剂使用的工业染料。2005年2月2日，英国第一食品公司（Premier Foods）向英国环境卫生部门报告，该公司2002年从印度进口的5t辣椒粉中含有苏丹红一号染料，并且已经生产为辣椒酱等调料销往众多下游食品商。2月18日，英国食品标准署确认了这次污染，并追查了使用Premier Foods公司供应的原料的食品商，分四批列举了575种含有苏丹红一号的食品，并警告消费者不要冒险食用以减少可能导致癌症的概率，同时要求这些食品必须在2月24日12时之前在监督下从货架上全部撤除，对已销售的清单上的产品需要提供无条件退货服务。英国食品标准署称这是英国历史上最大规模的食品召回事件。在英国发出食品警告之后，2005年2月23日，中国国家质量监督检验检疫总局发布了《关于加强对含有苏丹红（一号）食品检验监管的紧急通知》，要求清查在国内销售的食品（特别是进口食品），防止含有苏丹红一号的食品被销售及食用。3月29日，中国紧急制定了食品中苏丹红染料检测方法的国家标准，开始正式实施。我国在许多食品中发现了苏丹红成分，包括肯德基的“新奥尔良烤鸡翅”、“香辣鸡腿堡”、“劲爆鸡米花”等五种食品，亨氏食品的“桂林辣椒酱”、“坛坛香”辣椒制品等。

9. 三聚氰胺污染乳制品事件

2008年9月8日，甘肃岷县14名婴儿同时患有肾结石病症，引起外界关注。至2008年9月11日甘肃全省共发现59例肾结石患儿，部分患儿已发展为肾功能不全，同时已死亡1人，这些婴儿均食用了三鹿18元左右价位的奶粉。而且人们发现两个月来，中国多省已相继有类似事件发生。中国卫生部高度怀疑三鹿牌婴幼儿配方奶粉受到三聚氰胺污染，三聚氰胺是一种化工原料，可以提高蛋白质检测值，人如果长期摄入会导致人体泌尿系统膀胱、肾产生结石，并可诱发膀胱癌。

2008年9月19日，伊利、蒙牛及光明三品牌液态奶被中国质检总局公布含有三聚氰胺，当局抽检蒙牛121批次产品中，有11批次被检出三聚氰胺，检出值在每千克0.8～7mg。伊利81批次产品中，有7批次检出三聚氰胺，检出值在每千克0.7～8.4mg。光明93批次产品中，6批次检出三聚氰胺，检出值在每千克0.6～8.6mg。不过在三元和雀巢产品中并未发现三聚氰胺。2008年9月17日，中国国家质检总局宣布取消食品业的国家免检制度，所有已生产的产品和印制在包装上已使用的国家免检标志不再有效。

10. 福岛核电站核泄漏事件

北京时间2011年3月11日13时46分，日本发生9.0级地震并引发高达10m的强烈海啸，导致东京电力公司下属的福岛核电站一、二、三号运行机组紧急停运，反应堆控制棒插入，机组进入次临界的停堆状态。在后续的事故过程当中，因地震的原因，导致其失去场外交流电源，紧接着因海啸的原因导致其内部应急交流电源（柴油发电机组）失效，从而导致反应堆冷却系统的功能全部丧失并引发事故。

2013年7月22日，在事故发生两年多之后，东京电力公司表示，核电站内的放射性污水正泄漏流入太平洋，当地渔民与核子监督机构的专家早先就怀疑会发生这个问题。而东京电力公司先前坚决否认这个问题的存在。因此，日本首相安倍晋三命令政府介入处理这个紧急问题。8月20日，核电站又发生一起事件，多达300t的高辐射浓度污水从污水储存槽外泄。这污水足以对附近的工作员工造成健康危害。这次污水外泄事故被评为国际核事件分级表中的第三级。

第三节　人们对环境问题的认识过程

一、面临的主要环境问题

环境问题是指各种环境退化及对生物圈的不利影响的现象，主要包括酸雨、温室效应、臭氧空洞、淡水资源缺乏、资源短缺及生态退化。

二、人们对环境问题的认识

20世纪60年代，人们把环境问题只当成一个污染问题，没有把环境问题与自然生态、社会因素联系起来，低估了环境污染的危害性和复杂性，未能追根寻源。

1972 年，联合国在瑞典斯德哥尔摩召开了人类环境会议，第一次把环境问题与社会因素联系起来。这次会议是人们对环境问题认识的一个里程碑。但是，它没有从战略高度上指出防治环境问题的根本途径，没有明确环境问题的责任，没有强调需要全球的共同行动。

20 世纪 80 年代，人们对环境的认识有了新的突破性进展，提出了持续发展的战略，指明了解决环境问题的根本途径。

1992 年，联合国在巴西里约热内卢召开了联合国环境与发展大会，这是 20 世纪人类社会的又一重大转折点，树立了人类环境与发展关系史上新的里程碑。

2002 年，联合国在南非约翰内斯堡召开了联合国可持续发展大会。

我国分别在 1973 年、1983 年、1989 年和 1996 年召开了四次全国环境会议，推动了我国环境保护事业的发展和人们对环境认识的发展。

随着环境问题的日益严重，人类对环境问题的认识也在不断提高。在当今世界上，人们越来越清楚地看到，环境问题已不再是某个国家和局部地区的事情。"酸雨"、"温室效应"、"臭氧层破坏"等已远远超出国家和地区的局限而成为全球性的问题。这些问题严重威胁着人类的生存和发展，成为世界各国关心的大问题，也成为国际社会关注的热点。

三、我国环境污染防治面临的主要问题及解决方案

1. 资源利用与经济发展的问题

从人均的角度来看，我国人口众多、资源相对不足、环境承载能力较弱，是世界上人均资源占有水平较低的资源小国之一。从总量的角度来看，我国地域辽阔，资源丰富，是世界上少数几个资源大国之一。按自然资源总量的综合排序，我国是世界第四位。其中矿产资源探明储量潜在价值居世界第三位；水能、太阳能和煤炭资源分别居第一、第二、第三位，石油资源居世界第八位；耕地、草原面积居世界第四位，森林面积居世界第六位。

我国自然地理条件复杂，资源种类齐全，但资源质量不高，自然生产力低，我国资源分布不平衡，资源之间不匹配，资源分布与生产力布局不协调。众多人口对有限资源的压力将愈来愈大。后备资源供给不足，潜伏着更大的资源危机。如果不能在地质勘探方面强化投入并取得突破性进展，我国石油、天然气、关键矿产资源将严重衰竭。

人口持续增长，耕地不断减少，供水能力紧张，能源紧缺愈加深重，矿产资源不足，后备资源基础薄弱，资源总需求迅速扩大，各类资源供应长期紧缺，是我国人口与资源、经济增长与资源供给矛盾的基本格局，资源短缺将长期成为遏制我国经济发展的瓶颈。

2. 环境破坏与资源开发的问题

随着我国国民经济的高速发展、人口的急剧增加和人民生活水平的提高，对以矿产资源为能源和原材料的消耗量将成倍增加，我国煤炭资源作为所占比例最大的能源，所占比例达 75%左右，并且在 21 世纪将是我国的主要能源，但矿产开发，特别是煤炭资源的开发过程中，在给国家提供大量能源、原材料、创造巨大的物质财富、为国民经济的快速发展提供巨大能源保障的同时，却给矿区环境带来了严重的破坏和污染。

在以矿产资源开发为主的地区，矿藏的开采、洗选、冶炼等过程都需要大量用水，因此，会排放出大量的污水、废水，造成严重的水污染。采矿活动对水资源的破坏极其严重，随着航道掘进，需要排出大量的矿井水，由于大量地排水，疏干了采矿区上方或矿坑周围地层含水层，形成地下水降漏斗区，造成地下水位的下降，其影响范围可达 40～50km 的半径，造成这一区域水文环境的破坏。

3. 经济发展与环境污染的问题

随着经济的进一步发展，我国的环境污染日益加剧，并且严重威胁着我国城镇居民的身体健康，甚至在将来会严重制约我国经济的合理发展。

我国环境污染严重，其主要原因是来自工业的排放，随着我国经济的快速增长及居民生活水平的提高，能源消耗巨大，废气、废水与固体废弃物日益增加。在治理污染的道路上，任重道远，需要依法办事，制止环境污染事件发生，保护人类赖以生存的环境。随着社会经济的快速发展、城市化进程的加速和人民生活水平的迅速提高，今后10～20年，我国环境保护和污染防治工作将面临更大的挑战：①7%的经济增长——污染防治的压力越来越大；②小康社会生活水平——环境质量的要求越来越高；③发达国家技术进步——国际环境标准越来越严格。

俗话说“解铃还需系铃人”，化学物质造成的环境问题最终绝大部分还得靠化学的方法和手段来解决，环境化学就是这样一门学科，它一方面研究造成污染的原因，另一方面又研究治理污染的对策，在解决各种环境问题的过程中起着举足轻重的作用。同时，由于环境因素的复杂性，环境化学又与物理学、气象学、生物学、地学、医学、工程学和统计学等都有着不可分割的关系，是一门综合性和交叉性都很强的学科。污染治理之路，发达国家走的是先污染后治理之路，我们不想走，能否实现，路很艰难，路在我们脚下。

思考题

1. 什么是环境化学？如何分类？它有哪些特点？
2. 什么是环境科学？它的研究领域和任务是什么？
3. 环境化学有哪几个组成部分？
4. 什么是环境污染物？
5. 历史上有哪些重大的公害事件？
6. 近年来世界瞩目的公害事件有哪些？
7. 人们认识环境的过程是怎样的？
8. 我国环境污染防治面临的主要问题及解决方案是什么？

第二章 大气环境化学

本章要点：大气层的垂直结构及分布特征，对流层与平流层的特点；地球的辐射平衡、温室效应；自由基反应和光化学烟雾形成机制；酸沉降化学机制；城市空气质量评估方法。

第一节 地球大气的演化

我们居住的地球周围环绕着厚厚的大气层。距地球表面高度 1000km 以外仍然存在着稀薄大气，不过大气圈质量的 99%集中在近地表 30km 的范围内。地球大气的主要成分是氮气和氧气，这是我们这颗星球上能够出现和存在高等生物的最重要的条件之一。空间探测发现，地球大气的化学组成在太阳系的九大行星中非常特殊，离地球最近的两颗行星——金星和火星的大气化学组成与地球完全不同。金星的大气密度是地球的 90 倍，其中 90%以上是二氧化碳，几乎不存在氧气，由于高浓度 CO_2 的强温室效应，使得金星表面的温度极高，达 750K；而火星周围的大气则极其稀薄，其密度仅为地球大气的 5‰，且主要成分也是 CO_2（占 80%以上）。由于火星的大气非常稀薄，它的表面温度比地球表面低，且有较大的日变化。

如果只考虑地球在太阳系中的位置、轨道这些空间参数及地球自身的大小、质量等因素，由金星和火星的大气成采用内插法可以推算出一种“理论上的”地球大气组成，它与实际地球大气的组成差别非常大。实际地球大气的 CO_2 浓度不足“理论”地球大气的1/1000，而 O_2浓度高出 700 多倍，N_2的浓度也高出 26 倍。这样一种特殊的地球大气是怎样形成又是怎样维持的？这曾是长期令科学家们困惑的一个问题，目前较为普遍的一个观点是认为地球上生物圈的出现和发展改变了其原始大气的组成，并且生物圈的活动仍然控制着当今地球大气的基本组成。

地球是一个特殊的星球，能够孕育生命完全依赖于她所拥有的特殊条件。第一，她的质量适中，质量过小则留不住大气，质量过大则会引来许多彗星的撞击；第二，地球有自传，会产生磁场，磁场的存在能够驱散来自太阳的高能射线（太阳风）的轰击，有利于保护生命；第三，地球有一个月球作卫星，月球是地球的一个自旋制动系统，会减轻地球的自转，从而减少地球上火山和地震发生的频度；第四，地球上存在水，水是生命之源，为生命的诞生创造了条件；第五，地球上存在水蒸气和二氧化碳，因此，有温室效应，温室效应减小了昼夜的温差，有利于生命的存在；第六，地球大气的演化使大气中氧气成分增加，在紫外线的照射下，逐渐形成臭氧层，吸收了对生命有伤害的紫外线，使动植物得到保护，进而推进了大气的演化过程。

今日大气圈的形成，经历了一个漫长的演化过程。在地球最初形成的时候，还处于熔融体的状态，原始大气的主要成分是 H_2、He、$H_2O(g)$、N_2和 CO_2等，随着地球逐渐冷却凝固，地表火山活动十分频繁，大气也发生着剧烈的变化。这一阶段的大气组成主要是 H_2、CH_4、NH_3、$H_2O(g)$、N_2、Ar 和 H_2S 等，仍是一种还原性的气氛。由于适当的日地距离，地球形成固体核心以后保持了适中的地表温度，加上适当的地表尺度，在一定阶段出现了液态水。

在水出现之前，由于太阳光强烈的紫外线辐射，生命无法存在。直到水圈形成之后，在水面下 10m 深处，短波紫外线被吸收了，而生命所需要的 $\lambda>290\text{nm}$ 的太阳辐射仍能保持足够的强度，使得水体吸收的一些还原态的气体成分在光能的作用下产生了简单的有机质，生命就从这里开始了。

这段时期只有水的光解作用可以产生少量的氧。

$$H_2O(g)+h\nu \longrightarrow H_2+1/2O_2 \tag{2-1}$$

大气中氧气的浓度还非常低，只有现在的 1‰，所以最初出现的这些原始的生命只能从化学和光化学过程所形成的有机质的发酵过程中获得能量。到了后来，它们有足够的能力通过光合作用制造碳水化合物并释放出 O_2：

$$CO_2+H_2O+h\nu \longrightarrow \{CH_2O\}+O_2(g) \tag{2-2}$$

由此，大气中 O_2 的浓度开始明显增加，到距今大约 6 亿年前，达到现在浓度的 1%，人们把这一浓度看做是生物发展史上的第一个关键浓度。因为在这一浓度条件下，由下列反应产生的大气臭氧浓度明显增加。

$$O_2+h\nu(\lambda<240\text{nm}) \longrightarrow O+O \tag{2-3}$$

$$O_2+O+M \longrightarrow O_3+M \tag{2-4}$$

反应（2-4）称为三体反应。式中的 M 表示第三体分子，其作用是吸收反应产生的多余能量而使产物稳定下来。M 可以是体系中大量存在的任何物质，在大气体系中一般是 N_2 或 O_2 分子。与此同时，也发生着 O_3 的清除反应。

$$O_3+h\nu(\lambda<290\text{nm}) \longrightarrow O_2+O \tag{2-5}$$

$$O_3+O \longrightarrow 2O_2 \tag{2-6}$$

尽管这时所生成的臭氧量还不多，但已经使到达水面的阳光紫外线大大减少，生物可以在水面生长，并通过光合作用释放出更大量的 O_2。到距今大约 4 亿年前，大气中氧气的浓度达到了现在的 1/10，这是生物发展史上的第二个关键点，O_3 的浓度也因此显著升高，并且逐渐形成了较为稳定的臭氧层，位于距地面约 15～30km 的高度。

臭氧层对太阳紫外辐射的屏蔽作用使地球的面貌发生了巨大的变化。生物从海洋发展到了陆地，绿色植物开始大量生长，光合作用产生的 O_2 大大增加，成为现代大气的主要成分，大气也由此转变成为一种氧化性的气氛。

总结大气演化的全过程，可以概括出以下几个特点。

① 整个演化过程的趋势是：较轻的气体如 H_2 和 He 逐渐从大气层中逃逸，CO_2 的含量大大减少，O_2 的含量则显著增加。N_2 由于化学性质稳定，且为生命必需元素和主要营养元素，在大气中逐渐累积并趋于稳定。现代大气是一个相对稳定的体系。

② 演化规律是从原始大气的还原性气氛转变成了现代大气的氧化性气氛。

③ 地球的特殊条件为生命的诞生奠定了良好的基础。生命的出现和生物圈的形成对地球大气的演变起到了重要作用，并且至今仍在调控和维持着这种“特殊”地球大气的稳定。

第二节　大气层的垂直结构分布特征

地球大气的总质量为 5.2×10^{15}t，相当于地球本身质量的百万分之一。由于万有引力的作用，大气的密度随着高度的增加而降低。

研究发现，在距离地面的不同高度范围内，大气的温度、化学组成及其他性质呈现出不同的变化规律，据此，人们将大气圈划分为若干层次，其中最常见的是按照气温的垂直分布特点来划分，称为热分层法。热分层法一般将大气分为对流层（troposphere）、平流层（stratosphere）、中间层（meosphere）、热层（therosphere）和逃逸层（stratopause）共五个层，如图 2-1 所示。

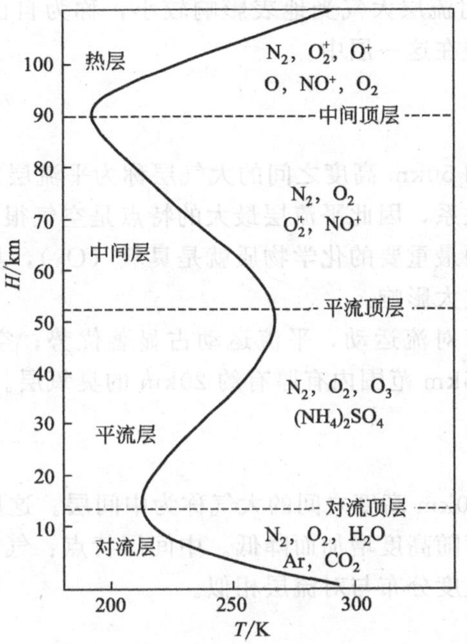

图 2-1　大气的热分层结构

一、对流层

对流层是最贴近地球表面的一层大气，它的上界是随纬度和季节而变化的。在赤道附近，对流层的厚度约为 16～18km，在中纬度地区约为 10～12km，而在两极附近只有约 8～9km。对流层一般夏季较厚，冬季较薄，对流层最显著的特点如下。

1. 气温随高度的上升而下降，大约每上升 100m，气温降低 0.6℃

贴近地表的空气因吸收地面辐射的热量温度较高，越向上吸收越少，温度越低。但底层空气受热会膨胀上升，而高层的空气冷缩后会下降，这样在垂直方向上就形成了强烈的对流运动，“对流层”的名字也就由此而来。

大气的这种上下对流，极有利于地面排放出的污染物的稀释和扩散。但在夜间，地面温度下降，容易出现低层大气下冷上热的稳定状态，这种现象称为“逆温”。这种情况一般在每天凌晨都会不同程度地存在，太阳出来后，阳光照射到地面，底层大气温度升高，逆温现象会很快消失。然而若遇寒冷的季节，阴天或静风的天气条件，逆温现象会持续较长时间。

在逆温条件下，排放到大气中的污染物扩散不出去，容易发生污染事件。

2. 密度大，水汽含量丰富，天气现象多

对流层从厚度上讲只是大气圈的一小部分，但它的质量占整个大气圈质量的 75%以上，水汽则占到 90%以上。因对流层顶部的温度已降到 −50℃以下，来自地面的水汽在到达这里之前就早已凝结成冰，因此大气中的水汽几乎都聚集在对流层中。

对流层大气因受地表影响的状况不同，在实际研究中常划分为两个部分。

① 在 1～2km 以下，大气与地表物体间的机械和热力作用强烈，通常称为摩擦层或边界层。排入大气的污染物绝大部分在这一层中活动，地表的建筑物、森林、山地等对大气中污染物的扩散和传输有很大影响。

② 在 1～2km 以上，对流层大气受地表影响较小，称为自由大气层。主要的天气过程如雨、雪、雹等的形成都是在这一层中。

二、平流层

从对流层顶向上至大约 50km 高度之间的大气层称为平流层。与对流层不同，平流层大气的温度与高度呈正相关关系，因此平流层最大的特点是空气很少是上下方向的对流，而是以平流运动为主。平流层中最重要的化学物质就是臭氧（O_3），臭氧层的变化会对对流层大气质量及地表生物圈产生重大影响。

平流层特点：空气没有对流运动，平流运动占显著优势；空气比对流层稀薄得多，水汽、尘埃含量甚微；15～35km 范围内有厚有约 20km 的臭氧层。

三、中间层

自平流层顶向上到约 80km 高度之间的大气称为中间层。这里空气更为稀薄，因为没有吸收辐射的物质，因而温度随高度增加而降低。中间层特点：气温随高度的增加而降低，顶部可达－92℃左右。垂直温度分布与对流层相似。

四、热层

指距地面高度 80～500km 之间的大气层。这一层大气中的氧强烈吸收波长小于 180nm 的阳光远紫外辐射的能量，使大气随高度增加而迅速升温，层内温度很高，且昼夜变化大。

五、逃逸层

热层以上的大气层，空气极为稀薄，密度已接近太空密度，因此也常称为外层大气。

在以上各层中，与人类关系最为密切的是对流层和平流层，即距地面高度 50km 以下的部分，又称为低层大气。

第三节　地球的辐射平衡

一、物体的最大辐射波长与温度的关系

入射到地球表面的阳光可看成是平行光束，入射到地球大气层外界的阳光总强度用太阳常数 S_0 表示。太阳常数定义为：在与光传播方向垂直的平面上单位面积接受到光的总量，其平均值为 $1368W/m^2$。

太阳的表面温度大约为 6000K，高温炽热气体以电磁辐射的形式放射出能量。太阳光谱几乎包括了整个电磁波谱，其中红外部分占总能量的 50％，可见光部分约占 41％，X 射线、γ-射线和紫外线大约占 9％。

太阳每天给地球提供巨大的能量输入，同时，地球也以同样的速度摆脱这个能量，因

此，维持了地球上温度的恒定。地球以辐射光的形式失去能量。当然地球不会像太阳一样发光。热的物体以一定的波长范围释放辐射。随着温度的增加，辐射波长会向短波方向漂移。这就是一个铁块在炉子里是红色的，当温度增加时铁块发白。地球释放的波长太长了，我们的肉眼无法看到。

太阳和地球的辐射光谱分布见图 2-2。这两条曲线很规范，是期望的随着太阳和地球温度而变化的“黑体”辐射曲线。黑体的光谱是平滑的，而来自太阳和地球的实际光谱是崎岖不平的，因为特殊的原子和分子跃迁对发射有贡献。

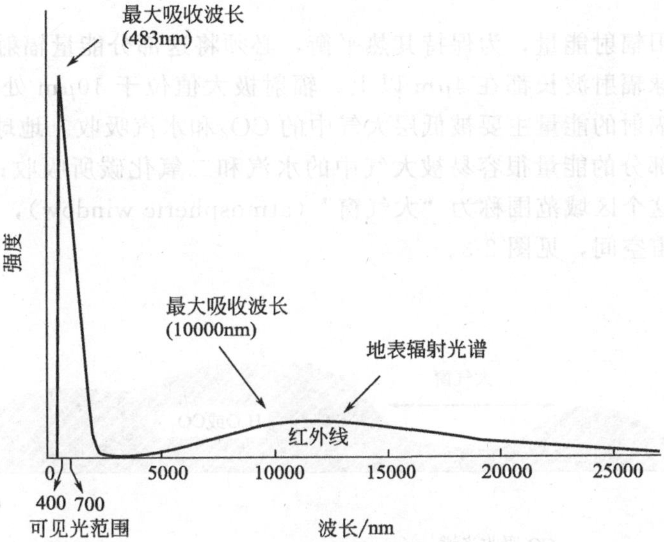

图 2-2 太阳和地球的辐射光谱分布图

根据 Wein 位移定律，黑体最大辐射能力所对应的波长 λ 与热力学温度 T 成反比，其数学表达式为：

$$\lambda_{peak}(\mathrm{nm}) = 2.9 \times 10^6 (\mathrm{nm} \times \mathrm{K}) / T(\mathrm{K}) \tag{2-7}$$

太阳是一个非常热的物体，它的峰值波长为 483nm，对应的温度是 6000K。它辐射的大部分波长落在 400～700nm 之间，这是可见光的区域。这个波长之所以是可见的，是因为我们的眼睛已经演化能适应太阳光。地球向外辐射，峰值波长约为 10000nm，对应的温度是 288K。因此，当地球主要吸收了可见光区的辐射（这是太阳表面高温的特征）后，地球以红外线区（这是地球冷表面的特征）的辐射把能量发射出去。

二、大气成分对太阳辐射的吸收

太阳光在穿过大气时，由于大气对其的吸收和散射作用而减弱，使投射到大气上界的辐射不能全部到达地面。太阳辐射通过大气层到达地面的过程中，大气组分如 N_2、O_2、O_3、H_2O 和 CO_2 等能吸收一定波长的太阳辐射。波长小于 290nm 的太阳辐射被 N_2、O_2、O_3 分子吸收，并使其解离：

$$N_2 + h\nu(\lambda < 120\mathrm{nm}) \longrightarrow N + N \tag{2-8}$$

$$O_2 + h\nu(\lambda < 240\mathrm{nm}) \longrightarrow O + O \tag{2-9}$$

$$O_3 + h\nu(220\mathrm{nm} < \lambda < 290\mathrm{nm}) \longrightarrow O + O_2 \tag{2-10}$$

故波长小于 290nm 的太阳辐射不能到达地面，而 800～2000nm 的长波辐射则几乎都被水分子和二氧化碳所吸收。因此，只有波长为 300～800nm 的可见光能透过大气到达地面，这部分约占太阳光总能量的41%。

云对太阳辐射产生反射作用，云越厚则反射率越大；云的全球平均反射率约为20%。由于大气的吸收、反射、散射作用，太阳辐射只有部分到达地面。经过大气减弱后的太阳辐射到达地面后，并不全部被地球吸收，而要被地面反射一部分。反射率取决于地表的性质和状态。例如，森林的反射率为15%，耕地为20%，沙漠为28%，雪地为80%，海洋为6%；地球的平均反射率为29%～34%。

三、地球的热平衡

地球吸收了太阳辐射能量，为保持其热平衡，必须将这部分能量辐射回太空，这一过程称为地球辐射。地球辐射波长都在4μm以上，辐射极大值位于10μm处，即主要是红外长波辐射。地球表面辐射的能量主要被低层大气中的CO_2和水汽吸收。地球辐射的波长在4～8μm和13～20μm部分的能量很容易被大气中的水汽和二氧化碳所吸收；而8～13μm的辐射被吸收得很少，这个区域范围称为“大气窗”(atmospheric window)，这部分长波辐射可以穿过大气到达宇宙空间，见图2-3。

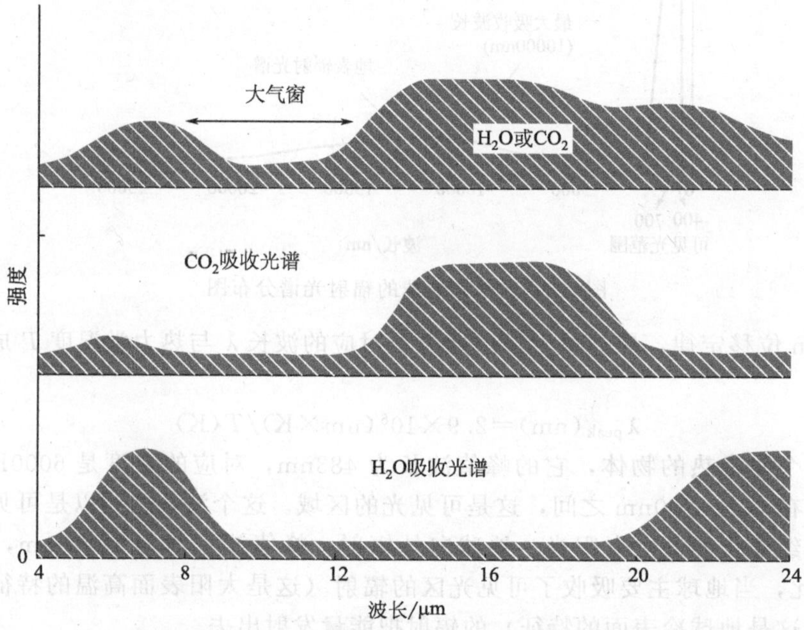

图2-3　水和二氧化碳对红外辐射的吸收

CO_2和H_2O吸收地面辐射的能量后，又以长波辐射的形式将能量放出。这种辐射是向四面八方的，而在垂直方向上则有向上和向下两部分，向下的部分因与地面辐射方向相反，称为“大气逆辐射”。由于大气逆辐射的作用，一部分地球辐射又被返回地面，使实际损失的热量比它们长波辐射放出的热量少。因此，大气对地表保持在适宜的温度范围起了重要作用。

由于在很长时期内地面的平均温度基本上维持不变，因此可以认为入射的太阳辐射和地球的长波辐射收支是基本平衡的。

由此可见，发生于地球和大气间的能量得失过程与化学物质的光化学、光吸收作用密切相关，尤其是O_3、水汽和CO_2等。所以，大气中这些成分的变化会对地球的能量平衡产生很大的影响。如近地面大气中水汽和CO_2量增加，它们会吸收地面长波辐射，在近地面与大气层间形成绝热层，使近地面热量得以保持，并导致全球气温升高，直接影响人类的生活和安全。这就是所谓的“温室效应”（greenhouse effect）。

通过大气的实际能量流是相当复杂的。图 2-4 显示了地球以每年 10^{20}kJ 为单位的能量输入和输出。大约 54.5 单位的太阳能量撞击地球及其大气，大约 16.3 单位（30%）被反射到太空，对地球的热平衡没有施加影响。这种光的大部分被云和大气所反射，较少量（2.2 单位）被地球表面反射。剩余的 38.1 单位（70%）被吸收，其中 13.0 单位（24%）被大气和云所吸收，25.1 单位（46%）被地球表面所吸收。

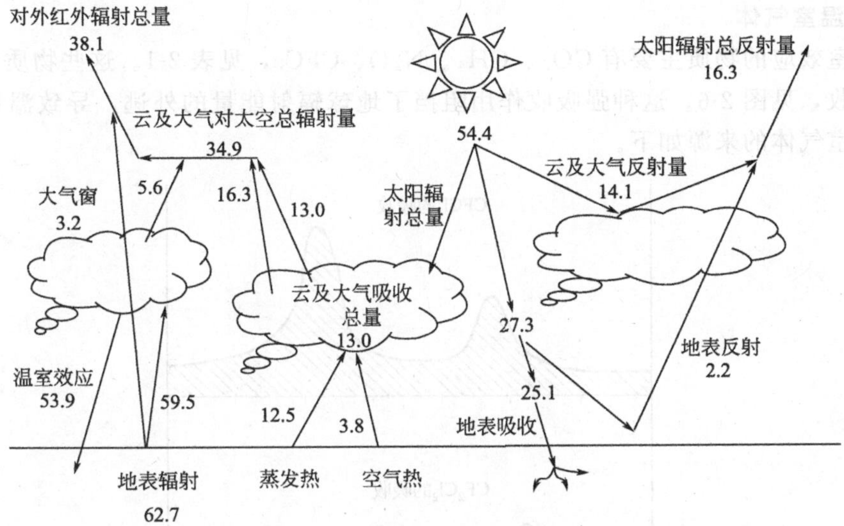

图 2-4　地球的热平衡图

四、温室效应

1. 红外吸收与分子振动

在二氧化碳和水分子中，可以发生对称伸缩振动、不对称伸缩振动和弯曲振动，见图 2-5。

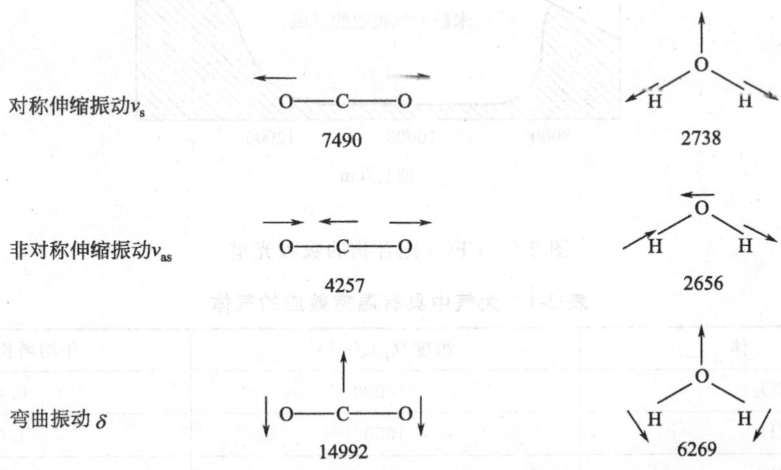

图 2-5　CO_2 和 H_2O 的分子振动

如上所述，温室效应是通过大气捕获低层热量形成的。地球的大气许可来自太阳的射线穿越，但是捕获了来自地球表面所发射的红外射线。

大气中的主要气体单原子分子不能够吸收红外线，因为它们不能满足电磁辐射的基本条件。对比之下，多原子分子有多种振动，其中一些振动能够改变瞬间偶极并对红外具有活性。水和二氧化碳的红外吸收光谱见图 2-3。可以看出，水和二氧化碳能够吸收 4000～25000nm 波长范围内的大部分的光，但是对 8000～13000nm 的光却不能吸收而使其脱逃。

2. 主要温室气体

引起温室效应的物质主要有 CO_2、CH_4、N_2O、CFCs，见表 2-1。这些物质在大气窗范围内有强吸收，见图 2-6。这种强吸收作用阻挡了地球辐射能量的外逃，导致温室效应的出现。主要温室气体的来源如下。

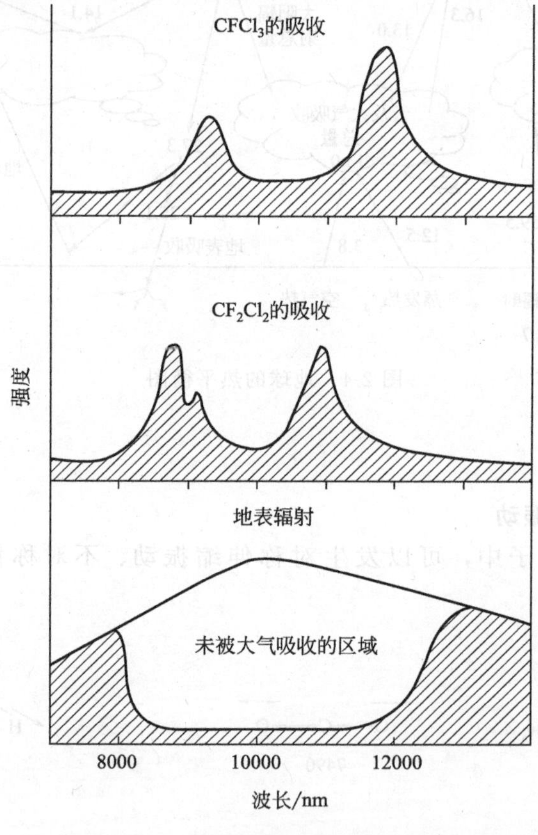

图 2-6　CFCs 化合物的吸收光谱

表 2-1　大气中具有温室效应的气体

气　　体	浓度/(μL/m³)	年均增长率/%
CO_2	340000	0.4
CH_4	1650	1.0
CO	304	0.25
$C_2H_2Cl_2$	0.13	7.0

续表

气　体	浓度/(μL/m³)	年均增长率/%
O_3	—	—
CFC-11	0.23	5.0
CFC-12	0.4	5.0
CCl_4	0.125	1.0

① 大气中 CH_4 主要来自湿地、牛群、稻田等，浓度仅次于 CO_2，其温室效应比 CO_2 强 20～30 倍。

② 我国农田土壤排放的 CO_2、CH_4、N_2O 的量分别为 260×10^6t、17.5×10^6t 和 9.6×10^6t，分别占总排放量的 8%、50%、10%。

③ CFC 也是温室气体，对温室效应的贡献率占 25%。CFC-11、CFC-12 主要吸收800～2000cm^{-1}之间的辐射；每个 CFC-12 分子产生的温室效应相当于 15000 个 CO_2 分子。

④ 新发现的温室效应最强的物质是 CF_3SF_5，1 个 CF_3SF_5 分子产生的温室效应相当于 10^5 个 CO_2。

3. 温室效应的利与弊

气候变暖，温室气体具有这样的特性：让阳光通过，但强烈吸收地面发出的长波辐射，从而使气温上升，起到了温室的作用。20 世纪全球平均气温升高了 0.5℃，如果温室气体按目前的速度增加，到 2030 年，全球平均气温将再提高 2～3℃，灾害性天气和异常天气将更加频繁。全球变暖是目前全球环境研究的一个主要议题。根据对 100 多份全球变化资料的系统分析，发现全球平均温度已升高 0.3～0.6℃。其中 11 个最暖的年份发生在 20 世纪 80 年代中期以后，因而全球变暖是一个毋庸置疑的事实。全球变暖将带来非常严重的后果，如冰川消退、海平面上升、荒漠化，还给生态系统、农业生产带来严重影响。

因此，探求全球变暖的起因成为重要的研究课题。分析表明，虽然地球演化史上曾经多次发生变暖-变冷的气候波动，但人类活动引起的大气温室效应增长可能是主要因素。

第四节　对流层大气化学反应

一、大气组成的分类

大气主要由氮、氧和几种惰性气体组成，它们约占大气总量的 99.9%以上。除气体外，大气中还悬浮着大量固体和液体颗粒。按照停留时间的长短，大气组分可分为三类：①准永久气体，如 N_2、Ar、Ne、Kr、Xe；②可变组分，如 CO_2、CH_4、H_2、N_2O、O_3、O_2；③强可变组分，如 H_2O、CO、NO、NH_3、SO_2 碳氢化合物（HC）、颗粒物、H_2S。对流层大气气体的组成及循环见表 2-2。

表 2-2　对流层大气气体的组成及循环

气　体	平均浓度/(mL/m³)	近似停留时间/a	循　环
N_2	780840	10^6	生物和微生物
O_2	209460	10	生物和微生物
Ar	9340	—	无循环

续表

气　体	平均浓度/(mL/m³)	近似停留时间/a	循　环
Ne	18	—	无循环
Kr	1.1	—	无循环
Xe	0.09	—	无循环
CH_4	1.65	7	生物活动和化学过程
CO_2	332	15	人类活动和生物活动
CO	0.05～0.2	65/d	人类活动和生物活动
H_2	0.58	10	生物活动和化学过程
N_2O	0.33	10	生物活动和化学过程
SO_2	10^{-5}～10^{-4}	40/d	人类活动和化学过程
NH_3	10^{-4}～10^{-3}	20/d	生物活动，化学过程，雨除
$NO+NO_2$	10^{-6}～10^{-2}	1/d	人类活动，化学过程，闪电
O_3	10^{2}～10^{-1}	—	化学过程
HNO_2	10^{-5}～10^{-3}	1/d	化学过程，雨除
H_2O	变化	10/d	物理化学过程
He	5.2	10	物理化学过程

大气中强可变组分主要来自人为源，其次是天然源。可变组分和强可变组分在大气中停留时间短，有可能参与平流层或对流层中的化学变化，它们在大气中的时空分布受局地源影响，在不同地区或高度，其分布往往有很大的不同。如冶炼厂、火力电厂所在地上空的大气中含烟尘、SO_2、NO_x 等强可变组分较多；在化工区周围的大气中含有较多的无机或有机物质；当这些物质在大气中达到一定浓度时，就有可能产生局部的大气污染。

二、大气组分的源、汇和气体循环

在大气中，准永久性气体浓度不变，但整个大气仍是一个动态体系；大气组分可通过大气圈与其他三个圈之间发生物理、化学或生物化学过程，不断进行着物质交换或转化，即构成所谓的“气体循环”。产生气体的过程称为气体的源，它包括大气中的化学过程、生物活动、火山喷发以及人类活动等，如燃料的燃烧。由大气中去除气体的过程如化学过程和生物活动、物理过程等就是被去除气体的汇。例如，大气中的水蒸气主要来自海水的蒸发，少量来自江河、湖泊水的蒸发以及生物圈、土壤、植物的蒸腾作用；大气中的水汽又可以遇冷凝结成雨、雪等降水回到地表，这就构成了大气中的水循环。因蒸腾作用是在不断地进行的，而降水是间歇性的，故造成了大气中水蒸气的浓度随地点而异。

大气中含有丰富的氮，但不能被植物直接利用。固氮细菌、蓝绿藻、雷电作用可将空气中的氮转变成硝酸盐，植物从土壤中吸收硝酸盐和铵盐在体内转变成各种氨基酸，然后再合成各种蛋白质。动物借食用植物而获取氮，蛋白质进入食物链。动植物死后，其中的蛋白质被生物分解成铵盐返回土壤，一部分被植物直接吸收；另一部分被细菌逐渐转化为亚硝酸盐和硝酸盐，既可被植物吸收，也可被细菌的脱氮作用转化成 N_2O，或通过反硝化作用产生 N_2；而亚硝酸盐也可通过化学去氮作用转化成 N_2 或 NO；土壤中的铵盐也可转化成 NH_3 并进入大气。由此可见，氮是通过无机、有机及微生物作用进行循环的。大气中氧等其他组分也都有各自的循环。

三、大气中重要的光吸收物质

大气中能吸收波长在 290～700nm 的光的物质主要有 NO_2、O_3、SO_2、HNO_2、RONO、RCOR、$RONO_2$、HNO_3、H_2O_2、HCHO、CH_3CHO、CH_3COCH_3 等，在对流层中，除 SO_2 外，其余都可发生光解反应。

1. NO_2 的光吸收特性

NO_2 是城市大气中最重要的光吸收分子，在低层大气中，它能吸收全部可见和紫外范围的太阳辐射。NO_2 的光解是对流层中最重要的光解反应。

$$NO_2 + h\nu(\lambda \leqslant 420nm) \longrightarrow NO + O(^3P) \tag{2-11}$$

$$O(^3P) + O_2 + M \longrightarrow O_3 \tag{2-12}$$

NO_2 光解是对流层大气中 O_3 唯一的人为来源。

$\lambda < 380nm$　　$\phi_{NO} \approx 1$

$\lambda = 405nm$　　$\phi_{NO} \approx 0.36$

$\lambda > 420nm$　　$\phi_{NO} \approx 0$

波长在 420nm 以上的光只能使 NO_2 跃迁至激发态 NO_2^*，并不能继续发生光解离。

2. O_3、HNO_2 和 H_2O_2 的光解反应

$$O_3 + h\nu(\lambda \leqslant 320nm) \longrightarrow O(^1D) + O_2 \tag{2-13}$$

$$O(^1D) + H_2O \longrightarrow 2 \cdot OH \tag{2-14}$$

式中，$O(^1D)$ 为激发态氧原子，上述反应为清洁大气中 · OH 自由基最重要的形成途径。

HNO_2 和 H_2O_2 的光解反应也都是大气中 · OH 自由基的重要来源，不过这两者相对来说在污染大气中更为重要。

$$HONO + h\nu(\lambda < 400nm) \longrightarrow \cdot OH + NO \tag{2-15}$$

$$H_2O_2 + h\nu(\lambda < 360nm) \longrightarrow \cdot OH + \cdot OH \tag{2-16}$$

3. HCHO 和 CH_3CHO 的光解反应

是对流层大气中 HO_2 · 自由基的主要来源。

4. SO_2 的光吸收

$$SO_2 + h\nu(290nm < \lambda < 360nm) \longrightarrow {}^1SO_2(\text{单重激发态})\quad\text{强吸收} \tag{2-17}$$

$$SO_2 + h\nu(340nm < \lambda < 400nm) \longrightarrow {}^3SO_2(\text{三重激发态})\quad\text{弱吸收} \tag{2-18}$$

SO_2 分子中的一个电子被激发到较高能级，分子中两个不成对电子自旋方向相同，分子处于三重态。

因为 SO—O 的键能为 5648kJ/mol，对应波长为 218nm，即只有 $\lambda < 218nm$ 的光量子才能使 SO_2 分子发生解离：

$$SO_2 + h\nu(\lambda < 218nm) \longrightarrow {}^3SO + O(^3P) \tag{2-19}$$

但这一反应在对流层不会发生，因而对流层 SO_2 的光化学反应不是光解而是与自由基的反应。

由上可见，大气中光吸收物质有两种：一是吸收光后发生解离反应，产生的自由基使大气中的化学过程异常活跃；另一种是受激后不解离，而是依赖与自由基的反应，由一次污染

物转化为二次污染物（如 SO_2）。

四、大气中的自由基

大气是一个氧化性的气氛，大气化学最基本的特征就是氧化，人们甚至把大气中发生的化学过程看做是低温燃烧的火焰，自由基则是这一火焰的助燃剂，它们是推动大气化学反应的动力。大气中常见的重要自由基有：·OH（氢氧自由基），HO_2·（氢过氧自由基），R·（烷基自由基）和 RO·（烷氧基自由基）。其中最重要的为·OH 自由基。

1.·OH 自由基的来源

在清洁大气中·OH 自由基最重要的形成途径是 O_3 的光解离：

$$O_3 + h\nu(\lambda \leqslant 320\text{nm}) \longrightarrow O(^1D) + O_2 \tag{2-20}$$

$$O(^1D) + H_2O \longrightarrow 2 \cdot OH \tag{2-21}$$

而在污染大气中，HNO_2 和 H_2O 的光解反应相对来说更为重要。

$$HONO + h\nu(\lambda < 400\text{nm}) \longrightarrow \cdot OH + NO \cdot \tag{2-22}$$

$$H_2O_2 + h\nu(\lambda < 360\text{nm}) \longrightarrow \cdot OH + \cdot OH \tag{2-23}$$

从·OH 的来源我们可以看到，其生成与太阳辐射直接有关。因此，它的大气浓度分布特点是：白天大于晚上，夏季大于冬季，热带上空为全球最高。全球平均浓度为 7×10^5 个/cm^3。

2. HO_2·自由基的主要来源

大气中甲醛的光分解：

$$HCHO + h\nu \longrightarrow H \cdot + HCO \cdot \tag{2-24}$$

$$H \cdot + O_2 + M \longrightarrow HO_2 \cdot \tag{2-25}$$

$$HCO \cdot + O_2 \longrightarrow HO_2 \cdot + CO \tag{2-26}$$

从上述反应可以看出，任何能生成 H 原子或甲酸基的反应都是对流层 HO_2·的来源。因此，乙醛也是 HO_2·的来源之一，但因其在大气中的浓度较低，不如甲醛重要。

此外，研究还发现了 HO_2·的夜间来源，与 NO_3·自由基的存在直接有关。

$$HCHO + NO_3 \cdot \longrightarrow HNO_3 + HCO \cdot \tag{2-27}$$

$$HCO \cdot + O_2 \longrightarrow HO_2 \cdot + CO \tag{2-28}$$

从这些自由基的来源和分布特征我们可以看出它们的相对重要性（白天、晚上）。

3. R·、RO·、RO_2·的来源

大气中含量最高的烷基自由基是甲基（·CH_3），它主要来自乙醇和丙酮的光解。

$$CH_3CHO + h\nu \longrightarrow \cdot CH_3 + HCO \cdot \tag{2-29}$$

$$CH_3COCH_3 + h\nu \longrightarrow \cdot CH_3 + CH_3CO \cdot \tag{2-30}$$

另外，O_2、HO·可与大气中的 HCs（碳氢化合物）发生 H 摘除反应，生成 R·。

$$RH + O_2 \longrightarrow R \cdot + HO_2 \cdot \tag{2-31}$$

$$RH + HO \cdot \longrightarrow R \cdot + H_2O \tag{2-32}$$

大气中的甲氧基 CH_3O 主要来源于甲基亚硝酸酯和甲基硝酸酯的光解。

$$CH_3ONO + h\nu \longrightarrow CH_3O \cdot + NO \cdot \tag{2-33}$$

$$CH_3NO_3 + h\nu \longrightarrow CH_3O \cdot + NO_2 \cdot \tag{2-34}$$

大气中的过氧烷基是由烷基与空气中的 O_2 结合而成的。

$$R \cdot + O_2 \longrightarrow RO_2 \cdot \tag{2-35}$$

4. 自由基的汇

(1) 自由基之间的相互转化

$$CO+OH\cdot \longrightarrow CO_2+H\cdot \tag{2-36}$$

$$H\cdot +O_2 \longrightarrow HO_2\cdot \tag{2-37}$$

$$CH_4O_2\cdot \longrightarrow CH_3\cdot +H_2O \tag{2-38}$$

$$CH_3\cdot +O_2 \longrightarrow CH_3O_2\cdot \tag{2-39}$$

$$HO_2\cdot +NO \longrightarrow NO_2+\cdot OH \tag{2-40}$$

$$HO_2\cdot +O_3 \longrightarrow 2O_2+\cdot OH \tag{2-41}$$

(2) 复合作用

$$HO_2\cdot +\cdot OH \longrightarrow H_2O+O_2 \tag{2-42}$$

$$\cdot OH+\cdot OH \longrightarrow H_2O_2 \tag{2-43}$$

除此之外，自由基还有其他的许多转化和去除途径，这里只是几个有代表性的例子。

五、对流层清洁大气的光化学过程

1. NO-NO_2-O_3的基本光化学循环

假设大气中没有其他的有机组分，只有氮氧化物（NO、NO_2）存在时，NO-NO_2-O_3之间将在阳光作用下，构成一个光稳态体系，首先是NO_2光解生成原子氧（O），原子氧能迅速与氧气分子结合生成臭氧，而臭氧一旦生成，又很快与NO反应生成NO_2。

$$NO_2+h\nu(\lambda=300\sim400nm) \longrightarrow O+NO \tag{2-44}$$

$$O+O_2+M \longrightarrow O_3+M \tag{2-45}$$

$$O_3+NO \longrightarrow O_2+NO_2 \tag{2-46}$$

一般认为大气中O_2的浓度是恒定不变的，那么在上述反应中，共涉及四个物种，即NO_2、NO、O和O_3。其动力学方程为：

$$d[O]/dt=k_1[NO_2]-k_2[O][O_2][M] \tag{2-47}$$

由于O原子极不稳定，一经生成就立即与大气中的O_2发生反应而消失，因此可采用稳态近似法来处理，即$d[O]/dt=0$，于是：

$$k_1[NO_2]=k_2[O][O_2][M] \tag{2-48}$$

由此得到的O_3稳态浓度为：

$$[O_3]=k_1[NO_2]/k_3[NO] \tag{2-49}$$

式（2-49）说明大气中若无其他反应参与，则O_3的浓度取决于$[NO_2]/[NO]$。但由此计算出的大气O_3浓度比实际测到的结果小得多，说明在大气中必定还存在着其他反应使O_3浓度增加，以及能与上述反应（2-46）竞争的反应。研究发现，这些反应主要来自含碳化合物。

2. 清洁大气中的基本化学过程

远离人为源的地区空气相对清洁，组成较为简单，浓度也比较低。但研究结果表明在这些清洁的天然大气中，仍存在着一系列化学和光化学反应，其中关键的物种是CH_4、CO、O_3和NO_x。

引发天然对流层大气中化学反应的是O_3的光分解。O_3光分解成O（^1D），再与H_2O作用生成两个·OH，由此引发一系列反应。

$$CH_4+\cdot OH \longrightarrow CH_3\cdot +H_2O \tag{2-50}$$

$$CO + \cdot OH \longrightarrow CO_2 + H\cdot \tag{2-51}$$

$H\cdot$ 和 $CH_3\cdot$ 与空气中的 O_2 结合产生了过氧自由基。

$$H\cdot + O_2 + M \longrightarrow HO_2\cdot + M \tag{2-52}$$

$$CH_3\cdot + O_2 + M \longrightarrow CH_3O_2\cdot + M \tag{2-53}$$

过氧自由基将清洁大气中的 NO 转化为 NO_2，同时又生成 $\cdot OH$ 基和 $HO_2\cdot$ 基，起到链传递的作用。

$$CH_3O_2\cdot + NO \longrightarrow NO_2 + CH_3O\cdot \tag{2-54}$$

$$HO_2\cdot + NO \longrightarrow NO_2 + \cdot OH \tag{2-55}$$

$$CH_3O\cdot + O_2 \longrightarrow HCHO + HO_2\cdot \tag{2-56}$$

主要的链终止反应是：

$$NO_2\cdot + \cdot OH \longrightarrow HNO_3 \tag{2-57}$$

$$HO_2\cdot + HO_2\cdot \longrightarrow H_2O_2 + O_2 \tag{2-58}$$

由上可见，大气中的自由基通过链反应的机制有效地将低价态的含 C 和含 N 化合物氧化到高价态，这些高价态物质最终通过降水或被地表吸收而去除。

第五节　光化学烟雾

一、光化学烟雾的出现

含有 NO_x 和 VOCs 的大气，在阳光紫外线的强烈照射下，发生一系列复杂的化学反应，产生一些氧化性很强的物质，如臭氧（O_3）、醛类（RCHO）、过氧乙酰硝酸酯(PAN)、硝酸（HNO_3）、过氧化氢（H_2O_2）等，这些氧化性产物与反应物的混合物称为光化学烟雾。

1936 年以后在美国的洛杉矶市首先出现了光化学烟雾污染。它的特征是烟雾呈浅蓝色。具有强氧化性，对人的眼睛和呼吸道有强烈刺激作用，伤害植物叶片（发白枯萎），致使橡胶开裂，并使大气能见度降低。城市大气中，这两种前体物的主要来源就是汽车尾气。继洛杉矶之后，光化学烟雾污染在世界各地不断出现。1974 年在我国的兰州西固地区也观测到了光化学烟雾污染。

二、光化学烟雾的日变化规律

对于现代化大都市，CO、NMHC（非甲烷烃）及 NO 的浓度最大值出现在早上 8：00 和下午 18：00，这两次峰值对应的时间正是人们上下班的高峰期，交通最为繁忙的阶段。随着阳光照射的逐渐增强，NO 和 NMHC 的浓度开始逐渐下降，而 NO_2 的浓度则逐渐上升并达到峰值。NO_2 浓度逐渐降低的同时，O_3 的浓度也迅速增加到最大值。

三、光化学烟雾的形成机制

从以上污染物的日变化特点可以初步推测出 NO_2 和 O_3 是在日光照射下通过大气光化学反应产生的，而不是直接由源排放的一次污染物。

为了搞清光化学烟雾中各物种浓度随时间变化的机理，人们设计了烟雾箱实验。即在一个大容器里，在人工光源（氙灯）的照射下，模拟大气化学反应。测定结果表明，当 NMHC 与 NO_x 并存时，在紫外线作用下 NO 转化 NO_2，NMHC 被氧化消耗，同时产生

O_3、PAN 等二次污染物。

利用这一简化机制可以说明光化学烟雾的日变化规律。清晨大量的碳氢化合物和NO由汽车尾气及其他源排入大气。由于夜间 NO 氧化的结果，已有少量的 NO_2存在。日出时，NO_2光解产生 O・自由基，进而与 H_2O 分子碰撞产生・OH 自由基。・OH 自由基产生后开始氧化 VOCs，并产生 HO_2・、RO_2・等自由基，这两种自由基可将大气中的 NO 有效地氧化成 NO_2，同时又生成・OH，继续氧化 HCs。如此循环往复，直到大气中的 HCs 和 NO 被消耗完。在此过程中，NO_2逐渐增多，并继续光解产生 O_3，使 O・自由基得以积累，同时 NO_2还可以与自由基反应生成 PAN、HNO_2等稳定的氧化物。当 NO_2量下降到一定程度时，O_2的生成量受到影响，而 O_2在大气中通过分解及与 HCs 和 NO 的反应也在不断消耗，当其累积速率与消耗速率达成平衡时，O_3浓度达到最大。下午，随着日光的减弱，NO_2分解受到抑制，于是反应趋于缓慢，产物的浓度相继下降。

四、光化学烟雾形成的影响因素与标志

光化学烟雾的形成需要三个条件。第一个条件是污染物的存在。常见的污染物有 HCs、NO_x、CO、VOCs 等。第二个条件是有逆温现象。地面温度低于高空温度的现象就是逆温现象。发生逆温现象时，地面上的污染物不能向上扩散，滞留在地表而容易发生光化学反应。若有风出现，则逆温现象消失。第三个条件是太阳光照射。满足以上三个条件，光化学烟雾就会发生。

光化学烟雾污染形成的氧化剂中臭氧是最重要的，占 90%以上，其次是 PAN（即过氧乙酸硝酸酯），二者可看作是光化学烟雾的标志产物。

对流层大气中臭氧的本底浓度通常在 0.02～0.06μg/mL 之间变化，平均约为 0.03 μg/mL。城市大气中 O_3浓度则高出许多，一般都在 0.1μg/mL 以上。当发生光化学烟雾时，大气中 O_3的浓度可高达 0.2～0.5μg/mL，这已是会损害人体健康和危及生物生存的危险浓度。

臭氧主要危害人体的气管和肺部，对心脏及脑组织也有一定影响。长期小剂量接触，可引发肺病、肺肿瘤和衰老等病症。臭氧可损伤植物叶片，影响其生长和降低产量，还可使织物、纸张等发脆，使橡胶老化而降低强度。

五、PAN

PAN 可由含乙基的过氧自由基与 NO_2反应产生。当 R・为 CH_3・时，产物为 PAN；当 R・为乙基、丙基或苯基时，则会形成相应的 PAN 的一系列同系物。

PAN 能强烈刺激眼睛，引起流泪和炎症；还能伤害植物，使多种植物叶子的背面呈青铜色或发生玻璃化。有研究指出它还是造成皮肤癌的可能试剂。

PAN 还是 NO 在对流层大气中的一个储库分子，可以传输和释放出 NO_2分子，不仅扩大了污染范围，还成为又一次光化学烟雾的潜在引发剂。此外，PAN 能在雨水中解离出 NO_3^- 和有机物，参与降水的酸化。

六、光化学烟雾污染的控制

许多研究结果都表明，某一地区能够生成的最大臭氧浓度和其前体物之间并非呈线性关系，也就是说，并不是氮氧化物或挥发性有机物的大气浓度增加一倍，能够生成的臭氧浓度也会相应地增加一倍。

如果氮氧化物浓度保持不变，碳氢化合物浓度的变化对臭氧的生成影响不大。但是，氮氧化物浓度的减小会导致臭氧浓度显著减小。离城市较远的郊区和乡村大气通常具有这样的大气条件特征。由于 $NMHC/NO_x$ 高，即 NO_x 相对较少，O_3 的生成主要受 NO_x 的限制，在这种情况下控制光化学烟雾，应主要采取控制氮氧化物的措施。

需要指出的是，要想对某一城市或区域的光化学烟雾污染进行有效的控制，必须针对这一地区具体的大气状况和特征，包括污染源分布和排放特征及气象条件等因素，找出污染物的变化规律，制订相应的污染控制措施。

第六节　酸沉降化学

酸沉降化学的研究开始于酸雨。20 世纪 50 年代欧洲发现了降水酸性逐渐增强的趋势，酸雨问题受到普遍重视。由于酸雨的危害较大，形成过程复杂，影响面广、持久，还可以远距离输送，因此受到了全世界的关注。各国相继大力开展对酸雨的研究，纷纷建立了酸雨的监测网站，制订长期研究计划，开展国际间合作。近几年在酸雨研究中发现酸的干沉降不能低估，引起的环境效应往往是干、湿沉降综合作用的结果。因此，过去被大量引用的“酸雨”的提法已逐渐被“酸沉降”所取代。

一、酸雨的定义及降水的酸化过程

1. 酸雨的定义

通常认为雨水的“天然”酸度为 pH＝5.6。此值来自如下考虑：影响天然降水 pH 值的因素仅为大气中存在的 CO_2，根据 CO_2 的全球大气浓度 330mL/m^3 与纯水的平衡：

$$CO_2(g)+H_2O \xrightleftharpoons{H} CO_2 \cdot H_2O$$

$$CO_2 \cdot H_2O \xrightleftharpoons{K_1} H^+ + HCO_3^-$$

$$HCO_3^- \xrightleftharpoons{K_2} H^+ + CO_3^{2-}$$

式中，H 为 CO_2 的亨利常数；按电中性原理得：

$$[H^+]=[OH^-]+[HCO_3^-]+2[CO_3{}^{2-}]$$

$$=K_W/[H^+]+K_1Hp_{CO_2}/[H^+]+HK_1K_2p_{CO_2}/[H^+]^2$$

式中，K_W 为水的离子积；p_{CO_2} 为 CO_2 在大气中的分压；K_1、K_2 分别为 $CO_2 \cdot H_2O$ 的一级和二级电离常数；计算可得洁净雨水的 pH 值为 5.6，故 pH 值小于 5.6 的雨水被认为是酸雨。

通过对降水的多年观察，近年来已经对 pH＝5.6 能否作为酸性降水的界限以及判别人为污染的界限提出了异议。唐孝炎等在《大气环境化学》中总结其主要论点如下。①高清洁大气中，除 CO_2 外还存在着各种酸、碱性气态和气溶胶物质，它们通过成云过程和降水冲刷过程进入雨水，降水酸度是其中各种酸、碱性物质综合作用的结果，其 pH 值不一定正好是 5.6。②作为对降水 pH 值有决定影响的强酸，尤其是硫酸和硝酸，并不都来自人为源。如火山爆发排放的 SO_2 和海盐中的 SO_4^{2-} 等都对雨水有贡献。③降水 pH＞5.6 的地区并不都意味着没有人为污染，有的地区空气中酸性物质污染严重，但碱性尘粒或其他碱性物质如 NH_3 含量高，降水冲刷的结果使 pH 值大于 5.6。这涉及降水的离子平衡。④H^+ 浓度不是一个守恒量，它不能表示降水受污染的程度。同一酸度的降水，其中的 SO_4^{2-}、NO_3^- 等含量可以相差很大。在偏远地区，降水 pH 值低不一定表示污染严重；城市附近的降水，有时 pH 值并不低，降水实际上已受到了污染。降水 pH 值与其中酸、碱离子的平衡有关。

综上所述，pH＝5.6 不是一个判别降水是否受到酸化和人为污染的合理界限，于是提出了降水 pH 值的背景值和降水污染与否的判别标准问题，这方面已做了不少研究。

由于世界各地区自然条件不同，如地质、气象、水文等的差异，会造成各地区降水 pH 值的不同。世界某些地区降水 pH 值的背景值为 4.79～5.00，均小于或等于 5.0（见表 2-3），因而认为将 5.0 作为酸雨 pH 值的界限更符合实际情况。

表 2-3　世界某些降水背景点的 pH 值

地点	样品数	pH 平均值
中国丽江	280	5.00
Amsterdan(印度洋)	26	4.92
Porkflot(阿拉斯加)	16	4.94
Katherine(澳大利亚)	40	4.78
Sancarlos(委内瑞拉)	14	4.81
St. Georges(大西洋白慕群岛)	67	4.79

有人认为 pH 值大于 5.6 的降水也未必没有受到酸性物质的人为干扰，因为即使有人为干扰，如果不是很强烈，由于雨水有足够的缓冲容量，不会使雨水呈酸性；而 pH 值在 5.0～5.6 之间的雨水有可能受到人为活动的影响，但没有超过天然本底硫的影响范围，或者说人为影响即使存在，也不超出天然缓冲作用的调节能力，因为雨水与天然硫平衡时的 pH 值即为 5.0。如果雨水 pH 值小于 5.0，可以确信人为影响是存在的。所以提出以 5.0 作为酸雨 pH 值的界限更为确切。

由于实际的大气是一个非常复杂的体系，所以降水的酸性是进入水中的各种物质综合作用的结果，降水的酸性取决于其中酸、碱性离子浓度的对比。许多事实说明随着矿物燃料消费的逐年增加，SO_2、NO_x 排放量亦有所增加，致使大气酸化加剧。一些背景采样点连续观测的结果表明，降水 pH 值有逐年下降的趋势。例如日本岩手县绫里 1976～1992 年的降水 pH 值从 5.20 降至 4.85。这说明全球大气气氛正在逐渐变化，这无疑与全球酸性污染物排放量逐年增加以及气溶胶酸化缓冲能力下降密切相关。大气中的 CO_2 浓度也以每年 2 mL/m^3 的速度增加，目前已近 360mL/m^3。因此，全球降水酸度背景值不是稳定不变的，从这个意义上说，将背景值下降到某一定值没有多大的实际意义。所以，现在一般仍以 pH＝5.60作为酸性降水的判断标准。

2. 降水的酸化过程

大气降水的酸度与其中的酸、碱物质的性质及相对比例有关。下面简要介绍这些物质进入降水，造成降水酸化的过程。

酸雨的形成过程包括雨除和冲刷。在自由大气里，由于存在 0.1～10μm 范围的凝结核而造成水蒸气的凝结，然后通过碰并和聚结等过程进一步生长从而形成云滴和雨滴。在云内，云滴相互碰并或与气溶胶粒子碰并，同时吸收大气气体污染物，在云内部发生化学反应，这个过程叫污染物的云内清除或雨除（in-cloud scavenging or rain out）。在雨滴下落过程中，雨滴冲刷着空气中的气体和气溶胶，雨滴内部也会发生化学反应，这个过程叫做污染物的云下清除或冲刷（below-cloud scavenging or washout）。这些过程就是降水对大气中气

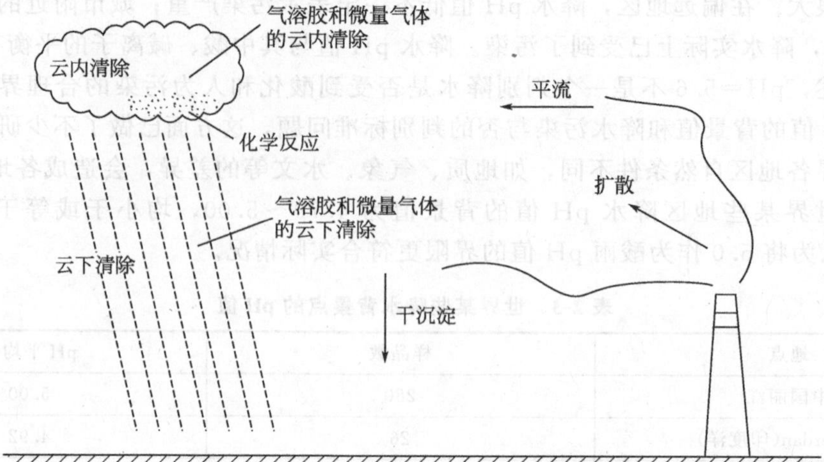

图 2-7 降水的酸化过程示意图

态物质和颗粒物的清除过程，酸化就是在这些清除过程中发生的，如图 2-7 所示。

(1) 云内清除过程（雨除）

大气中硫酸盐和硝酸盐等气溶胶可作为活性凝结核参与成云过程；此外，水蒸气过饱和时也能产生成核作用。由于水蒸气凝结在云滴上和云滴间的碰并，使云滴不断生长，与此同时，各种污染气体溶于云滴中并发生各种化学反应；当云滴成熟后即变成雨从云基下落。大气污染物的云内清除（雨除）过程包括气溶胶粒子的雨除和微量气体的雨除。

气溶胶粒子进入云滴可通过以下三种机制。①气溶胶粒子作为水蒸气的活性凝结核进入云滴。②气溶胶粒子和云滴的碰并。气溶胶粒子通过布朗运动和湍流运动与云滴碰并。粒径小于 $0.01\mu m$ 的气溶胶粒子几乎都经该机制进入云滴。③气溶胶粒子受力运动，并沿着蒸气压梯度方向移动而进入云滴。在对流层大气中，若气溶胶浓度小于 $200\sim300\mu g/m^3$，几乎全部粒子在成云过程中被清除。而作为活性凝结核进入云滴、参与雨除过程是气溶胶粒子雨除的主要机制。

污染气体的雨除对云水组成的影响与气溶胶的雨除同样重要。微量气体的雨除取决于气体分子的传质过程和在溶液中的反应性，同时还与云的类型和云滴谱有关。在污染气体的云内清除过程中，其中的一些物质被氧化，如 S(Ⅳ) 被氧化成 S(Ⅵ)；故化学氧化速率是整个过程的决速步，液相氧化反应的速率取决于氧化剂的类型和浓度，而污染气体在云滴中的溶解度取决于气相浓度和云滴的 pH 值。

(2) 云下清除过程（冲刷）

雨滴离开云基，在其下落过程中有可能继续吸收和捕获大气中的污染气体和气溶胶，这就是污染物的云下清除或降水的冲刷作用。它包括微量气体及气溶胶的云下清除。

微量气体的云下清除：云下清除过程与气体分子同液相的交换速率、气体在水中的溶解度和液相氧化速率以及雨滴在大气中的停留时间等因素有关。雨滴进入大气后会产生污染气体从气相向液相的传质过程或从液相向气相的传质过程，传质系数随雨滴粒径的增大而减小。至于雨滴内的化学反应，由于雨滴在大气中的停留时间较短，所以一些快反应如离子反应、强氧化剂 H_2O_2、O_3、自由基·OH、HO_2·及金属离子 Mn^{2+}、Fe^{3+} 等对 S(Ⅳ) 的氧化反应，才会对雨滴的化学组成产生影响，而大多数的慢反应对雨滴的影响较小。在污染气体的云下清除过程中，气液间传质速率和液相反应速率共同决定污染气体在液相的反应速率。气液传质速率控制了大雨滴中的液相反应速率，化学反应速率则控制了小雨滴的液相反应速率。在水中溶解度极大或在溶液中仅参加快速离子反应使溶解度增大的微量气体，它们的云下清除是不可逆的，其去除率与已进入液相的浓度无关，仅与气相浓度有关。不可逆清除的气体在液相中的总浓度随雨滴降落距离的增加而线性增加。气体不可逆清除的清除率与其他气体无关，仅与自身的特性和雨滴谱及降水量有关；清除率随降水量的增加而增加。研究表明，雨水对 SO_2 气体的清除系数与雨强、气相 SO_2 浓度、NH_3 浓度、氧化剂浓度以及雨水的 pH 初值有关，它们之间有复杂的相互作用。

气溶胶的云下清除：雨滴在下落过程中捕获气溶胶粒子。气溶胶被捕获后，其中的可溶部分如 SO_4^{2-}、NO_3^-、NH_4^+、Ca^{2+}、Mg^{2+}、Mn^{2+}、Fe^{2+}、H^+ 及 OH^- 等会释放出来，从而影响雨滴的化学组成和酸度。刘帅仁等研究了大气气溶胶在云下雨水酸化过程中的作用。结果表明：①气溶胶对雨水酸度有影响，若气溶胶的 pH 值低于雨水的 pH 值，则气溶胶起酸化作用；反之，则起碱化作用；气溶胶的酸化作用强于碱化作用。②在一般浓度（每立方厘米含 10^3 个）下，酸性气溶胶是雨水 H^+ 的重要来源，碱性气溶胶可消耗雨水中的 H^+；气溶胶对雨水 $SO_4{}^{2-}$ 的贡献较小。③酸性气溶胶对雨水的酸化作用随 SO_2 浓度的增大而减弱，而碱性气溶胶对雨水的碱化作用随 SO_2 浓度的增大而增强；云内清除过程是雨水 SO_4^{2-} 的重要来源，云下气溶胶清除过程对 SO_4^{2-} 的贡献较小。④HNO_3 对雨水 H^+ 的贡献比同浓度的 SO_2 要大几倍，气溶胶对雨水 $NO_3{}^-$ 的贡献相当于 1ng/mL HNO_3 的贡献，随着 HNO_3 浓度的增大，气溶胶的相对贡献迅速减小。⑤气溶胶是雨水中 NH_4^+ 的重要来源，相当于 NH_3 浓度 5～8ng/mL 时对雨 NH_4^+ 的贡献。

云内清除和云下清除过程受大气污染程度和许多环境参数的影响。云内清除和云下清除对酸雨形成的相对重要性在不同地理区域、不同源排放和不同气象条件等情况下是不同的。观测结果表明，在我国一些重污染地区，云下清除过程是很重要的。如对重庆和北京地区云下清除过程的数值模拟结果表明，重庆雨水中的 H^+ 来源以云下 SO_2 氧化为主，气溶胶起碱化作用；北京雨水中的 H^+ 来源以云内清除过程为主，云下气体 NH_3 和气溶胶起碱化作用。可见，北京地区 NH_3 浓度高是雨水不酸的首要原因，同时也使雨水中 SO_4^{2-} 浓度偏高。

由此可见，酸雨的形成过程是非常复杂的。酸雨前体物 SO_2、NO_x 除局地源排放外，还可以由远距离输送而来。如我国南方酸性降水存在着局地冲刷和中长距离传输双重形成机制。南方大气 SO_2 浓度较高的某些山区城市降水酸化的主要来源为局地冲刷。南方春季常出现大面积酸性降水，其主要来源是污染物中长距离传输。污染物在一定气象条件下可传输数百公里或更远，使下风区域云水和降水酸化。南方多数的酸性降水为局地冲刷和传输双重来源。当云水酸度相同时，各个城市酸性气体的浓度和气溶胶的缓冲能力决定了局地冲刷和传输对降水中酸的相对贡献。

如前所述，导致降水呈酸性的主要物质是硫酸，其次是硝酸，还有有机酸等其他酸类。现以 SO_2、NO_x 造成降水酸化为例，概述酸雨的形成过程。

① 由源排放的气态 SO_2、NO_x 经气相反应生成 H_2SO_4、HNO_3 或硫酸盐、硝酸盐气溶胶。

② 云形成时，含 SO_4^{2-} 和 NO_3^- 的气溶胶粒子以凝结核的形式进入降水。

③ 云滴吸收了 SO_2、NO_x 气体，在水相氧化形成 SO_4^{2-}、NO_3^-。

④ 云滴成为雨滴，降落时清除了含有 SO_4^{2-}、NO_3^- 的气溶胶。

⑤ 雨滴下降时吸收 SO_2、NO_x，再在水相中转化成 SO_4^{2-}、NO_3^-。

途径②、③为雨除，④、⑤为冲刷（云下清除过程）；在雨除和冲刷过程中同时进行着 SO_2、NO_x 的吸收及其液相氧化；H_2O_2、O_3 及 $\cdot OH$、$HO_2\cdot$ 对 SO_2、NO_x 的液相氧化起了重要作用。

大气中的其他气态物质和 NH_3、H_2O_2、O_3 和碳氢化合物等也会被清除进入降水，其中的一些物质（如碳氢化合物）可发生氧化转化，从而对降水起酸化作用。因此，酸雨的形成是酸化的化学过程与清除的物理过程交织在一起的。

二、酸雨污染造成的危害

1. 对土壤的影响

酸性物质不仅通过降雨湿性沉降，也可通过干性沉降于土壤。一方面土壤中的钙、镁、钾等养分被淋溶，导致土壤日益酸化、贫瘠化，影响植物的生长；另一方面酸化的土壤影响微生物的活性。

2. 对植物的影响

受到酸雨侵蚀的叶子，其叶绿素含量降低，由于光合作用受阻，使农作物产量降低，也可使森林生长速度降低。

3. 对水体的影响

酸雨可使湖泊、河流等地表水酸化，污染饮用水源。当水体 pH<5 时，鱼类的生长繁殖即会受到严重影响；流域土壤和湖、河底泥中的有毒金属，如铝等则会溶解在水中，毒害鱼类。水质变酸还会引起水生生态结构上的变化；酸化后的湖泊与河流中，鱼类会减少甚至绝迹。

4. 对材料的影响

酸雨加速了许多用于建筑结构、桥梁、水坝、工业装备、供水管网及通信电缆等材料的腐蚀，还能严重损害古迹、历史建筑以及其他重要文化设施。

5. 对人体健康的影响

酸雨不仅可造成很大的经济损失，还可危害人体的健康，这种危害可以是间接的，也可以是直接的。1974 年中国科学院生态环境研究中心等单位的科学家首先在北京西郊进行了酸雨监测；而后 20 世纪 70 年代末期又奔走于重庆、贵阳、武汉、南京和北京等城市，进行监测，并在长江以南测到酸雨。引起了国人的重视。

三、我国酸雨的污染现状及控制对策

1. 污染现状

我国对酸雨的研究始于 20 世纪 70 年代末期。当时在北京、上海、南京、重庆和贵阳等城市开展了局部研究，发现这些地区在不同程度上存在着酸雨问题，西南地区则很严重。1982～1984 年我国开展了酸雨的调查，为了弄清降水酸度及化学组成的时空分布情况，1985～1986 年在全国范围内布设了 189 个监测站，523 个降水采样点，对降水数据进行了全面系统的分析。结果表明，降水年平均 pH 值小于 5.6 的地区主要分布在秦岭淮河以南，在此以北仅有个别地区；降水年平均 pH 值小于 5.0 的地区主要在西南、华东及东南沿海一带。即我国酸雨由北向南逐渐加重，长江以南酸雨已是比较普遍的问题。酸雨情况最严重的西南地区，如重庆、贵阳两市的雨水酸度的月平均值几乎全在 pH=5 以下。我国酸雨的主要致酸物质是硫酸盐，降水中 SO_4^{2-} 的含量普遍都很高。

(1) 浙江酸雨现状

浙江省基本被酸雨覆盖，酸雨率已达 83.7%；降水 pH（酸度）年均值都在 5.6 以下，其中 96.9%的地区降水 pH（酸度）<5.0，特别是临安、开化、安吉等地小于 4.5；降水主要酸性污染物是 SO_4^{2-}，但是 NO_3^- 的影响正在逐年加重。

某地收集到酸雨样品，还不能算是酸雨区，因为一年可有数十场雨，某场雨可能是酸雨，某场雨可能不是酸雨，所以要看年均值。目前我国定义酸雨区的科学标准尚在讨论之中，但一般认为：年均降水 pH 值高于 5.65，酸雨率是 0～20%，为非酸雨区；pH 值为 5.30～5.60，酸雨率是 10%～40%，为轻酸雨区；pH 值为 5.00～5.30，酸雨率是 30%～60%，为中度酸雨区；pH 值为 4.70～5.00，酸雨率是 50%～80%，为较重酸雨区；pH 值小于 4.70，酸雨率是 70%～100%，为重酸雨区。这就是所谓的五级标准。其实，北京、西宁、兰州、乌鲁木齐等城市也收集到了几场酸雨，但年均 pH 值和酸雨率都在非酸雨区标准内，故为非酸雨区。

(2) 云贵川酸雨状况

20 世纪 80 年代初期，云南某地区乡镇企业起步，片面追求利润，实施土法炼硫，工艺陈旧落后，硫的回收率仅达 30%。其余的硫都燃烧成为 SO_2，迷漫于大气之中，加之云南气候湿润，有人曾收集检测到每场降雨都是酸雨，pH 值可达到 3.2～3.8。其后果是生产区周围，树上看不到昆虫，地下找不到蚂蚁，天上望不到飞鸟，1500 亩（1 亩 $=\frac{1}{15}hm^2$）良田寸草不长，一片荒凉的灾区景象。1993 年采取了一系列治理措施，问题已得到缓解。

贵州北部、西北部某些县也有土法炼硫。它们以硫铁矿为原料制备硫黄，硫的回收率仅为 45%，一半以上的硫，以 SO_x 的形式排向大气，污染环境。说来惊人！一个年产千吨的土硫黄厂，方圆 400m 之内，寸草不生，空气中 SO_2 的浓度超过 $6mg/m^3$；在 $2km^2$ 内，SO_2 浓度超过 $0.5mg/m^3$。

川南和川东某些县市土法炼硫更为严重。据统计有 8～100t 容量规模的土磺炉两千余座，每生产 1t 硫黄向大气排放 SO_2 和 H_2S 气体为 1.8～2.2t，后者在大气中也能逐渐转化为 SO_2。经监测，土法炼硫炉群生产区空气中 SO_2 浓度为 $2.2～1087mg/m^3$；平均浓度 $78～185mg/m^3$，超标百倍以上；生活区空气中 SO_2 浓度为 $1.1～27.5mg/m^3$，平均浓度 $8.8 mg/m^3$，竟超标 27 倍。令人无法忍受。

生产区域周围，酸雨率 100%。年降水 pH 值为 3.0～5.2，均值为 3.75，属于重酸雨区。四川省土法炼硫的生态危害范围为 $350～400km^2$。损失粮食 5500 万斤（1 斤 = 0.5kg）。区内，番茄亩产 7200 斤；海椒 3700 斤/亩。而在区外，番茄亩产 11350 斤；海椒 4850 斤/亩。四川名胜峨眉山，风景秀丽，全靠山深林秀。但近十年来，酸雨像空中死神缠住了此佛教圣地，冷杉林成片死亡。金顶海拔 3077m，降水 pH 值平均为 4.34；酸雨率达到 85.7%，硫酸根占阳离子总量的 60%。其他风景区千佛顶、太子坪、七里坡、雷洞坪也处于酸雨严重污染区。峨眉主要风景林是冷杉，死亡率超过 30%的中度受害面积达到 $7.00km^2$；死亡率超过 50%严重受害面积达到 $2.28km^2$。七里坡接引殿一带，有 4%的树木枯死；金顶附近 600 余亩，几乎全部死绝，秃秃光光，景观全非。猴子也跑到其他山沟里去了。

(3) 重庆酸雨状况

1994 年重庆及其郊区下了数场黑雨，色如墨汁，且有强酸性。人们发生了恐慌，纷纷询问地区环保部门，欲知其详。经化学分析，黑色物质是煤屑，原来在锅炉内化石燃料未能燃烧充分，析出一些细的炭粒，通过烟囱排向高空。酸性物质主要是硫酸根，来自煤中燃烧的杂质硫。结论是：黑雨就是强酸性雨；也可以说是酸雨发展到某种极端情况的产物。

无独有偶，1991 年我国喜马拉雅山区，下了数场“黑雪”。它来自于中东战争，伊拉克军队从科威特撤退时，放了一把大火，主要油气井火光冲天，喷出浓浓的黑烟，夹杂着令人窒息的 SO_x 和 NO_x 气味，直向高空，随风飘向东方，遇到喜马拉雅高山，难以跨越，随雪落下，成为有酸性的黑雪。人迹罕至的世界屋脊也未能逃脱“空中死神”的灾难。

(4) 福建酸雨状况

福建省三明市是一座新兴的工业城市，它位于该省内陆山区中部市区，面积约 $1000km^2$，四面环山，东西长，南北短。奇怪的是，多年来城市东部降水呈明显酸性，近几年来降水 pH 值最低达到 4.0；而相差仅 4km 的城市西部，降水呈明显碱性，降水 pH 值最高可达 9.89；同一场雨，城东酸雨，城西碱雨，现象前所未闻。通过长期监测和污染源调查，科学家做出了合理的解释。该城市坐落于高山环绕的盆地中，降水的酸碱性主要为当地环境污染源所制约。城市东部，是上风向，降水主要受高空大气层积雨层影响，该层含有城市燃煤所排放的酸雨前兆体 SO_x 和 NO_x，故呈酸性，落地就是酸雨。而城西部，是在下风向，特别时三明钢铁厂，三明化工厂所排出的碱性粉尘和颗粒物，遇到高山，堆集在西部低空，在降雨过程中被冲刷，各种酸碱物质中和，碱性大于酸性，降水明显呈碱性。此种碱雨还不同于新疆的碱雨，前者仅含碱性颗粒物；而后者是在酸雨的基础上增加碱性粉尘和颗粒物形成的，阴、阳离子含量甚高，污染更为严重，治理也就更难。

2. 控制对策

我国自“六五”期间就开始组织了大规模的酸雨研究，国家和地方政府也采取了多项措施控制和缓解酸雨的方案。1995 年，全国人大常委会通过了新修订的《大气污染防治法》，其中专门规定在全国范围内划定酸雨控制区和二氧化硫控制区，即所谓的“两控区”。

在划定的“两控区”内，SO_2 的年平均浓度超过国家环境质量的二级标准，日平均浓度超过国家三级标准，这两级标准分别是保护居民和生态环境不受损害的基本要求和不受急性危害的最低浓度。而酸雨控制区是指降水的监测结果 pH 值低于 4.5，而且硫的沉降超过酸沉降的临界负荷，同时考虑地区经济发展状况。这里，酸沉降的临界负荷是指生态系统能承受的最大酸性物质的沉降量。根据这样的划分方法，“两控区”的总面积为 109 万平方千米，其中酸雨控制区主要是长江以南、四川和云南以东地区，面积大约 80 万平方千米，二氧化硫控制区面积约为 $29km^2$。

酸雨和二氧化硫污染之间存在密切的相互联系，应协同控制。降低局地大气中的二氧化硫污染水平、改善局地空气环境质量，需要重点控制对二氧化硫污染贡献大的局地污染源，而控制酸雨不仅要控制局地源，而且要从区域的角度控制对酸雨和超临界负荷区的形成有贡献的所有二氧化硫和氮氧化物排放源。特别是随着全国燃油和燃煤电厂的持续增加，我国二氧化硫排放源的平均排放高度将不断增加，地区间的相互影响将越来越大，区域协调控制的作用会显得越来越重要。

3. 主要措施

(1) 针对煤炭使用的技术措施

我国50%以上的初级能源来自燃煤，因此煤炭的使用贡献了我国二氧化硫排放量的约50%。因此，煤炭消费过程中的二氧化硫排放控制是一项十分关键的工作。通过限制高流煤的开采，从煤炭的源头上即开始实施控制。两控区内将停止建设煤层含硫分大于3%的煤矿；对于新建、改建、扩建含硫分大于1.5%的煤矿，将配套建设相应规模的洗选设施；对于已经建成的硫分大于2%的煤矿，补建煤炭洗选设施。同时提倡使用低硫煤，和在煤炭使用前通过选煤、洗煤等技术脱硫；大力推广使用型煤、汽化煤、水煤浆，集中供暖供热及高效燃煤新技术等有利于提高煤炭的利用效率。在煤炭燃烧的末端，采取各种切实有效而又经济可行的技术从烟道中脱硫，也是减少二氧化硫向大气排放的有效途径。

(2) 以调整能源结构和提高能源利用效率为主的能源措施

在酸雨污染严重的地区，应考虑制订能源结构调整计划，以洁净的燃气、燃油或水电、核电逐步改变一次性能源过分依赖煤炭的状况，尤其是减少民用燃煤量。同时，加强工艺改造和技术革新，努力通过能源的使用效率，对我国许多地区和企业将是具有很大潜力的重要措施。

(3) 推行有效的管理制度和经济政策

① 实行区域二氧化硫排放总量控制和重点污染源排污许可证制度。

国家根据不同阶段两控区总量和基本控制单元的二氧化硫排放总量向各省、直辖市、自治区下达总量控制指标，各级环境保护部门根据两控区污染控制目标确定二氧化硫污染源的允许排放总量，并利用排污许可证方式将排污总量分配到各个排污单位，对污染源排放二氧化硫实施监督管理。

② 试行二氧化硫排污交易政策。

在排污许可证制度的基础上，政府环境保护行政主管部门在保证排污总量不超过规定的控制目标的前提下，可试行排污企业间买卖政府分配的允许二氧化硫排放总量指标，以使排污企业结合自己的实际情况，选择实现污染物总量控制指标和排污许可限值费用最小的方法，将排污总量降到最低水平。

③ 推行有利于污染控制的二氧化硫排污收费政策。

提高二氧化硫排污收费标准，使其逐步达到等于或高于治理成本，真正使污染控制成本成为产品总成本的组成部分，形成谁污染谁就会在经济上受损失的机制，促使排污企业积极增加投入，主动治理污染。

另外，在我国两控区的控制规划实施过程中，还应大力加强环境监测。环境监测是两控区酸雨和二氧化硫污染状况及规划实施控制效果科学评估的重要手段，监测结果还可为进一步的控制工作提供指导建议。

第七节 城市空气质量问题

一、我国城市空气质量的现状和发展趋势

我国正处在社会经济高速发展的时期，在发展过程中面临各种各样的环境问题的挑战。伴随着城市化和工业化的进程，交通运输带来的污染已成为我国日益突出的严重的环境问题。在国家的经济发展过程中，交通运输条件是主要的基础设施，而支持这基础设施的则是

交通工业尤其是汽车工业的发展。

1. 城市汽车数量发展状况

我国目前汽车工业的发展水平还比较低，汽车的社会保有量还比较小，20 世纪 90 年代以前汽车保有量增长缓慢。同时，中国汽车保有量的地区分布极不平衡。由于城市化进程迅速，人口和机动车向城市和经济快速发展地区高度密集，机动车主要分布在北京、上海和广州等一些大城市，如北京市的汽车保有量占全国的 1/10 左右，且发展很快。

另外，我国机动车的发展还伴随着与发达国家不同的两个特征，即路况条件差和车况条件差。我国目前汽车单车污染物排放水平较高，轿车的排放量是日本的 10～20 倍，卡车的排放量是美国的 1～8 倍。虽然汽车工业发展造成的大气环境问题目前还没有在我国普遍发生，但从发达国家的经验教训和我国可持续发展的总体战略的角度来看，这一问题的严重性应该得到充分的认识。

2. 机动车排放的污染物

机动车排放的污染物包括一次污染物，以及直接排放的物质经大气化学过程形成的二次污染物如细粒子和 O_3。主要污染物有：一氧化碳、氮氧化物、挥发性有机物、铅、颗粒物、二氧化碳等。

这些一次污染物进入大气后还会发生种种大气化学反应，生成如细粒子和大气氧化剂等对人体健康和生态系统危害更大的二次污染。全球范围内，机动车是造成大气氧化性增强和能见度下降的重要因素。特别需要指出的是目前我国机动车使用的燃油与以往及国外有很大差别，为了提高无铅汽油的抗爆性能，汽油中烯烃和芳香烃的含量有较大提高，这些物质可以通过蒸发和燃烧排气等方式进入大气，由于其光化学活性较强，所以更易引起光化学污染。

3. 机动车排放对我国大气环境的影响

机动车的快速发展，对我国大气环境质量有着本质性的影响，主要体现在我国大气污染的类型和特征将随之发生变化，污染所影响的区域也在不断扩展。

我国是一个传统的煤烟型污染国家，SO_2 和大气气溶胶一直是我国大气最主要的污染物。近年来随着我国经济的快速发展以及部分城市和地区能源结构的转换，在一些大中城市机动车排放污染和燃油、燃气排放污染呈迅速加重的趋势。广州市和北京市分别于 1985 年和 1993 年氮氧化物的浓度超过了二氧化硫，显示这两个城市大气污染的特征和性质已经发生了明显的变化。伴随着这种变化，一些城市和地区出现了以 NO_x。污染为代表的光化学污染，形成了复合型大气污染，如何在控制 SO_2 和大气气溶胶污染的同时研究和控制 NO_x 污染和光化学污染是近期内必须考虑和认真处理的问题。

我国于 20 世纪 70 年代末在兰州西固石油化工区发现光化学烟雾。经研究，兰州光化学烟雾主要由石油工业排放的 VOC 和电厂排放的 NO_x 引起，其造成的影响也仅限于兰州西固的局部地区。1986 年夏在北京发现光化学烟雾迹象。北京市的光化学烟雾则主要由汽车尾气和燃煤排放所致。近 30 年来，随着经济的发展和机动车保有量的快速增长，上海、广州、深圳等城市频繁观测到光化学烟雾污染的现象，机动车排放已成为大气污染的主要来源，而 NO_x 则成为主要污染物。

我国城市光化学烟雾在今后一定时期内还将呈恶化之势，由机动车排放的污染物经过大气化学过程导致大气氧化能力不断增强，大气细粒子污染严重。细粒子和大气氧化剂的输送使得污染从城近郊区蔓延到区域，形成区域性的大气氧化剂污染。

二、城市空气质量日报

近年来，我国越来越多的城市开始提供空气质量日报服务。如北京环境保护监测中心每日提供二氧化硫、一氧化碳、臭氧、二氧化氮和可吸入颗粒物的污染指数，并指明首要污染物和空气污染程度。表 2-4 和表 2-5 分别给出了这些污染指数所对应的污染物浓度限值和相应的空气质量状况，可供参考。

空气污染分指数的计算方法：当某种污染物分指数的浓度 $C_n < C_i < C_{n+1}$ 时：

表 2-4　污染指数及污染物浓度限值

污染指数 API	污染物浓度/(mg/m^3)				
	总悬浮颗粒	二氧化硫	氮氧化物	一氧化碳	臭氧
500	1.000	2.620	0.940		
400	0.875	2.100	0.750	>20	1000
300	0.625	1.600	0.565	10	800
200	0.500	0.250	0.150	6	200
100	0.300	0.150	0.100	40	160
50	0.120	0.050	0.050	2	120

① 其污染分指数 I_i 为：

$$I_i = \frac{(C_i - C_n)}{C_{n+1} - C_n} \cdot (I_{n+1} - I_n) + I_n$$

② 各种污染参数的污染分指数都计算出以后。取最大者为该区域或城市的空气污染指数 API。即：

$$API = \max(I_1, I_2, \cdots, I_n)$$

例题：二氧化硫在空气中的浓度为 $1.8mg/m^3$，计算其污染指数。

解：

$$I_i = \frac{(C_i - C_n)}{C_{n+1} - C_n} \cdot (I_{n+1} - I_n) + I_n$$

$$I_i = \frac{(1.8 - 1.6)}{2.1 - 1.6} \times (400 - 300) + 300 = 340$$

答：二氧化硫在空气中产生的污染指数为 340。

表 2-5　空气污染指数及空气质量级别

API	空气质量级别	空气质量状况	对健康的级别	对应空气质量的使用范围
0～50	Ⅰ	优	可正常活动	自然保护区、风景名胜区和其他需要特殊保护的地区
51～100	Ⅱ	良	可正常活动	城镇规划中确定的居住区、商业交通居民混合区、文化区、一般工业区、农村地区
101～200	Ⅲ	轻度污染	长期接触，易感人群症状有轻度加剧，健康人群出现刺激症状	—
201～300	Ⅳ	中度污染	一定时间接触后，心脏病和肺病患者症状显著加剧，运动耐受力降低，健康人群普遍出现不适症状	—
>300	Ⅴ	重度污染	健康人群明显出现强烈症状，运动耐受力降低，提前出现某些疾病	—

思考题

1. 从地球表面垂直方向按温度变化规律划分，大气层可分为哪几个层？
2. 一般情况下，在对流大气层中每升高 100m 温度降低多少？
3. 简述“大气窗”的形成和功能。
4. 试述酸雨的形成机理及其危害。
5. 空气中二氧化硫的浓度为 2.43，试计算其污染指数。

第三章　水环境化学

本章要点：水资源和利用；地球上水的分布；水循环；天然水的组成；水体污染与自净；水体中的污染物；污染物进入水体后的运动过程；水体污染对人体健康的影响；有机污染程度的指标；水体中某些有机污染物的降解；废水的一级处理；废水的二级处理；废水的三级处理；城市废水处理的典型流程。

第一节　天然水概况

一、水资源和利用

地球上水的总量是固定的，约 13.86 亿立方千米，但可利用的淡水只占水总量的 0.3%。虽然淡水资源有限，但如果时空分布得当，并保持恰当水质，还是可以满足全球目前和将来的淡水需要的。遗憾的是，地球上淡水资源的时空分布极其不均匀，加上水污染日益严重以及工农业和生活用水量的增加，许多国家和地区出现了水资源严重短缺的局面。水的短缺和污染不仅影响了生物生存，而且直接或间接地给人类生存带来了威胁和危害，同时也造成了重大的经济损失。当前主要的缺水类型有：资源型缺水、工程型缺水和水质型缺水。由于缺水，危害了农作物生长，并影响着工业生产，威胁着人体健康和生态以及国家安全。

二、地球上水的分布

地球表面约有 70%以上被水所覆盖，所以地球素有“水的行星”之称。地球上的水分布在海洋、湖泊、沼泽、河流、冰川、雪地，以及大气、生物体、土壤和地层之中。水的总量约为 13.86 亿立方千米（见表 3-1 和表 3-2），其中海水占 96.5%，淡水为 0.35 亿立方千米，占总水量的 2.35%。由于开发困难或技术、经济的限制，到目前为止，海水、深层地下淡水、冰雪固态淡水、盐湖水等很少被直接利用。比较容易开发利用的，与人类生活和生产关系密切的淡水储量为 400 多万立方千米，仅占淡水的 11%，总水量的 0.3%。

我国地表水径流总量约 2.8 万亿立方米，地下水资源约 8000 亿立方米，冰川年平均融水量约 500 亿立方米，近海海水约 500 万立方千米。我国目前可供利用的水资源量每年约有 11000 亿立方米，平均每人占有地表水资源约 $2700m^3$，居世界第 88 位，仅为世界人均占有量的 1/4。每亩土地占有地表水 $1755m^3$，只相当于世界平均水平的 1/2。总的来说，我国淡水资源并不丰富，处于缺水状态，而且水资源的时空分布非常不均衡，东南多，西北少，耕地面积只占全国 33%的长江流域和长江以南地区，水资源占全国的 70%。我国许多地方缺水或严重缺水，水污染比较严重。另一方面，由于人口剧增，工农业生产迅速发展，我国和西方国家一样，也面临水质下降、水源不足的威胁。因此，控制水体污染，保护水资源已成为刻不容缓的任务。

表 3-1 自然环境中的水量分布

水的类型	水量/km^3	比例/%
海水	1338000000	96.5
地下水	23716000	1.71
冰雪水	24064100	1.74
湖泊水	176400	0.013
沼泽水	11470	0.0008
河水	2120	0.0002
大气水	12900	0.001
生物水	1120	0.0001
总计	1385984610	

表 3-2 全世界淡水储量

水的类型	水量/km^3	比例/%
地下水	10846500	30.96
冰雪水	24064100	68.70
湖泊水	91000	0.260
沼泽水	11470	0.033
河水	2210	0.006
大气水	12900	0.037
生物水	1120	0.0032
总计	35029210	

三、水循环

自然界的水不仅受地球引力作用沿着地壳倾斜方向流动，而且由于水在太阳能和地球表面热能的作用下发生形态变化，蒸发的水分随着气流运行而转移，遇冷凝结成云或以降水形式到达地面，到达地表的水又重新蒸发、凝结、降落，这个周而复始的过程，称为水循环。仅在局部地区（陆地或海洋）完成的水的循环过程称为小循环。自海洋蒸发的部分水汽，随气流转移至陆地上空，并以降水形式到达地面后，经过江河湖泊或渗入地下，再归入海洋，这种海洋和陆地之间水的往复运动过程，称为水的大循环。环境中水的循环是大、小循环交织在一起的，并在全球范围内和地球上各个地区内不停地进行着。影响水循环的作用有三个基本因素：

① 水的物理性质决定了水的循环成为可能；

② 太阳辐射是水循环的原动力；

③ 循环路线的结构和性质，特别是地表循环途径的结构和性质，如地质地貌、土壤和生物等的类型和性质，它不但影响降水的分布和输送，而且还影响下渗及输水道的特性。

这些因素相互作用决定了天然水的循环方向和强度，造成了自然界错综复杂的水文现象。蒸发、降水、径流是水循环的基本要素。河川径流是人类用水的基本来源，它是地球上水循环较活跃的部分，世界河床蓄水量大约 16d 就可更换一次，因此它有很强的自净能力。

四、天然水的组成

1. 水体的定义

在水污染化学中，水体指河流、湖泊、沼泽、水库、地下水、冰川、海洋等储水体的总称。水体的组成不仅包括水，而且也包括其中的悬浮物质、胶体物质、溶解物质、底泥和水生生物，所以水体是个完整的生态系统，或是被水覆盖地段的自然综合体。区分水和水体是重要的。例如，重金属（如 Pb、Cu、Cr、Zn、Cd 等）污染物通过沉淀、吸附、螯合等途径，很容易从水相转移到底泥中，所以水中重金属的含量一般都不高。如果光从水来看，似乎没有受到污染，但整个水体可能受到严重污染。

2. 天然水的组成

在水的自然循环中，地表径流部分水量虽然不大，但是它在自然界所起的作用十分巨大，在风化（包括物理、化学、生物风化）过程和水迁移过程中，输送了土壤中的矿物质和有机质，使天然水具有各种组分。天然水体中除水以外，还有其他各种物质；根据它们在水中存在的状态不同，可将这些物质分为三类，即悬浮物质、胶体物质、溶解物质（见表 3-3）。在此只讨论天然水体中溶解物质的种类及其存在形态。

表 3-3 天然水体的组成

分类	主要物质
悬浮物质	细菌、病毒、藻类及原生物、泥沙、黏土等颗粒物
胶体物质	硅、铝、铁的水合氧化物胶体物质，黏土矿物胶体物质，腐殖质等有机高分子化合物
溶解物质	氧、二氧化碳、硫化氢、氮等溶解气体，钙、镁、钠、铁、锰等离子的卤化物，碳酸盐、硫酸盐等盐类，其他可溶性有机物

溶解在天然水中的物质大致可分为：主要离子、溶解气体、营养物质、微量元素和有机物质。溶解物质在水中的含量，除与物质和性质有关外，还与气候条件、水文特征、岩石与土壤的组成等因素有关。表 3-4 和表 3-5 分别是海水或河水中部分元素的含量及理论上可能存在的形态。

（1）主要离子

天然水体中的主要阳离子有 Ca^{2+}、Mg^{2+}、Na^{+}、K^{+} 等，主要阴离子有 Cl^{-}、SO_4^{2-}、HCO_3^{-}、CO_3^{2-} 等，这八种离子可占水中溶解固体总量的 95%～99% 以上。海水中又以 Na^{+} 与 Cl^{-} 含量占优势，河水中以 Ca^{2+}、HCO_3^{-} 含量占优势。陆地水中下列成分的含量顺序一般为 $HCO_3^{-}>SO_4^{2-}>Cl^{-}$，$Ca^{2+}>Na^{+}>Mg^{2+}$，而海水中相应的含量顺序为 $Cl^{-}>SO_4^{2-}>HCO_3^{-}$，$Na^{+}>Mg^{2+}>Ca^{2+}$。地下水受局部环境地质条件限制，其优势离子变化较大。

（2）微量元素

微量元素是指含量在 ng/mL 级的元素。天然水体中微量元素的种类很多，如 As、Cd、Hg、Ni、Pb、Sb、Sn、Zn、Mn、Cu 等。微量元素多数属于重金属元素，它们在天然水中的含量及存在形式见表 3-4 和表 3-5。

表 3-4 海水中部分元素的含量及可能存在的形态

元　素	平均含量/(μg/L)	可能存在形态
Na	10.77×10^{6}	Na^{+}
K	3.99×10^{5}	K^{+}
Mg	1.29×10^{6}	Mg^{2+}
Ca	4.12×10^{5}	Ca^{+}

续表

元　素	平均含量/(μg/L)	可能存在形态
Sr	7.9×10^3	Sr^{2+}
Cl	19.35×10^6	Cl^-
S	9.05×10^5	SO_4^{2-}、HSO_4^-
C	2.8×10^4	HCO_3^-、CO_3^{2-}
Br	6.73×10^4	Br^-
F	1.3×10^3	F^-
B	4.4×10^3	$B(OH)_3$
N	6.40×10^2	NH_4^+、NO_3^-
P	60	$HPO_4{}^{2-}$、$MgPO_4^-$
Si	2.20×10^3	$Si(OH)_4$、胶体
Fe	2	胶体、$Fe(OH)_2^+$
Mn	0.2	Mn^{2+}、$MnCl^+$
Cu	0.25	$Cu(OH)^+$、$CuCO_3$
Cd	0.11	胶体、$CdCl_2$
Hg	0.03	$HgCl_4^{2-}$、有机汞
Ni	0.56	Ni^{2+}、$NiCO_3$
Pd	0.03	$PdCO_3$、胶体
As	3.7	$HAsO_4^{2-}$、有机砷
Sb	0.24	$Sb(OH)_6^-$
Sn	0.004	$H_3SnO_4^-$、有机锡
Zn	4.9	Zn^{2+}、$Zn(OH)^+$
U	3.2	$UO_2(CO_3)_3^{4-}$

表 3-5　河水中部分元素的含量及其可能存在的形态

元　素	中位/(μg/L)	含量范围/(μg/L)	可能存在形态
As	0.5	0.2～230	As(Ⅲ、Ⅴ)、砷的甲基化物
B	15	7～500	$B(OH)_3$
Br	14	0.05～55	Br^-
C	1.1×10^4	$6\times10^3\sim1.9\times10^4$	HCO_3^-
Ca	1.5×10^4	$2\times10^3\sim1.2\times10^5$	Ca^{2+}
Cd	0.1	0.01～3	有机配合物、螯合物
Cl	7.0×10^3	$1\times10^3\sim3.5\times10^4$	Cl^-
Cu	3	0.2～30	有机配合物、螯合物
F	100	$50\sim2.7\times10^3$	F^-
Fe	500	$10\sim1.4\times10^3$	胶体
Hg	0.1	$1.0\times10^{-4}\sim2.8$	有机配合物、螯合物
K	2.2×10^3	$500\sim1.0\times10^4$	K^+
Mg	4.0×10^3	$400\sim6.0\times10^3$	Mg^{2+}
Mn	8	0.02～130	
N	50	$2\sim1.8\times10^3$	NO_3^-
Na	6.0×10^3	$700\sim2.5\times10^4$	Na^+
Ni	0.5	0.02～27	

续表

元　素	中位/(μg/L)	含量范围/(μg/L)	可能存在形态
P	20	1～300	$H_2PO_4^-$
Pb	3	0.06～120	
S	3.7×10^3	$200\sim4.0\times10^4$	SO_4^{2-}
Sb	0.2	0.01～5	Sb(Ⅴ)
Si	7.0×10^3	$500\sim1.2\times10^4$	
Sn	9.0×10^{-3}	0.004～0.09	Sn(Ⅳ)、锡的甲基化物
Sr	70	$3\sim1.0\times10^3$	Sr^{2+}
Zn	15	0.2～100	Zn^{2+}有机配合物、螯合物

(3) 营养物质

营养物质是指与生物生长有关的元素，包括氮、磷、硅等非金属元素以及某些微量元素(如 Mn、Fe、Cu 等)。这些元素的含量一般在 mg/mL～ng/mL 之间，它们存在的形态与水体酸碱性、氧化还原性有关。

氮、磷是水生生物生长和繁殖所必需的营养元素，但如果水中氮、磷含量过高，会发生“富营养化”现象。铁是常见的一种矿物元素；地下水中的铁主要以 Fe(Ⅱ) 价态存在，含量可达每升几十毫克；在地表水中由于溶解氧充足，铁常以 Fe(Ⅲ) 价态存在。天然水中硅主要来自硅酸盐和铝硅酸盐的水解。

(4) 有机物质

天然水体中有机物的种类繁多。它们是水生植物光合作用的产物和水生动物在不同阶段分解产物的混合物。通常将水体中的有机物分为非腐殖质和腐殖质。非腐殖质包括碳水化合物、脂肪、蛋白质、维生素及其低分子量有机物等。水体中大部分有机物是褐色或黑色无定形腐殖质，它们的分子量范围为几百至几万。通常根据腐殖质在酸、碱中的溶解情况，可将腐殖质分为胡敏素（腐黑物）、腐殖酸（HA），后者又可分为富里酸（黄腐酸、FA）和胡敏酸（腐黑物）等，如图 3-1 所示。

腐殖质中除含有大量苯环外，分子中还含有羟基—OH、羧基—COOH、羰基 $>C=O$

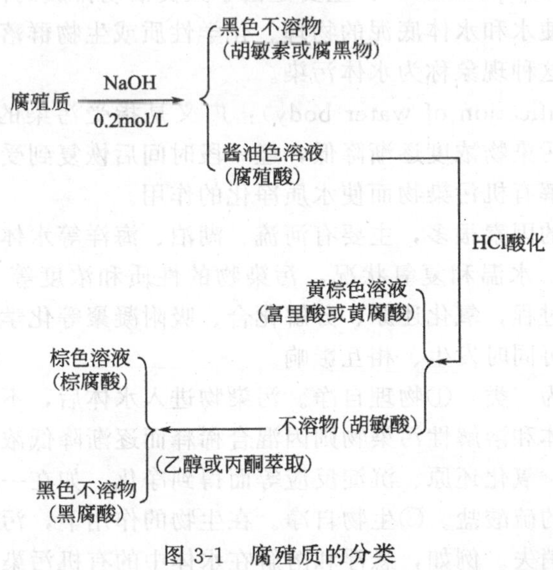

图 3-1　腐殖质的分类

等活性基团，因而腐殖质具有弱酸性、离子交换性、配位化合及氧化还原等化学活性。它能与水体中的金属离子形成稳定的水溶性或不溶性化合物，还能与有机物相互作用。腐殖质对水体中重金属等污染物的迁移转化具有较大的影响。

(5) 溶解气体

一般情况下，天然水中存在的气体有氧气、二氧化碳、硫化氢、氮气和甲烷等。它们来自大气中各种气体的溶解、水生动植物的活动、化学反应等。海水中的气体还可来自海底火山爆发。表 3-6 列出了海水中主要溶解气体的含量范围。

表 3-6 海水中主要溶解气体的含量范围

气体	O_2	N_2	CO_2	H_2S	Ar
含量范围/(mg/L)	0～8.5	8.4～14.5	34～56	0～22	0.2～0.4

溶解于水中的气体以氧（O_2）和二氧化碳（CO_2）比较重要，它们能影响水生生物的生存和繁殖及水中物质的溶解、化合等化学和生化行为。水中氧的输入，一是富集空气中的氧，当水中含氧量小于其饱和含量时才能使水进行富集氧；二是水生植物光合作用逸出的氧。水中氧的输出部分包括有机物的氧化、有机体的呼吸和生物残骸的发酵腐烂作用等。天然水中氧的含量一般为 0～14mg/L。水中二氧化碳来源于有机体的氧化分解、水生动植物的新陈代谢作用及空气中二氧化碳的溶解；消耗于碳酸盐类溶解和水生植物的光合作用。天然水除氧、二氧化碳以外，在通气不良的条件下，有时还会有硫化氢气体存在。水体中硫化氢气体来自厌氧条件下，含硫有机物的分解及硫酸盐的还原，而大量硫化氢则是火山喷发的产物。由于 H_2S 易于氧化，所以只能在缺氧条件下存在，同时空气中硫化氢分压很低，水中硫化氢容易逸出，所以地表水中硫化氢含量甚微或为零，否则就表明已遭受人为污染。

第二节 水体污染

一、水体污染与自净

水体污染（water body pollution）：主要是由于人类活动排放的污染物进入河流、湖泊、海洋或地下水等水体，使水和水体底泥的物理、化学性质或生物群落组成发生变化，从而降低了水体的使用价值，这种现象称为水体污染。

水体自净（self-purification of water body）：广义是指受污染的水体由于物理、化学、生物等方面的作用，使污染物浓度逐渐降低，经一段时间后恢复到受污染前的状态；狭义是指水体中微生物氧化分解有机污染物而使水质净化的作用。

影响水体净化过程的因素很多，主要有河流、湖泊、海洋等水体的地形和水文条件，水中微生物的种类和数量，水温和复氧状况，污染物的性质和浓度等。水体自净机理包括沉淀、稀释、混合等物理过程，氧化还原、分解化合、吸附凝聚等化学和物理化学过程以及生物化学过程。各种过程可同时发生、相互影响。

水体自净作用可分为三类。①物理自净。污染物进入水体后，不溶性固体逐渐沉至水底形成污泥；悬浮物、胶体和溶解性污染物则因混合稀释而逐渐降低浓度。②化学自净。污染物进入水体后，经络合、氧化还原、沉淀反应等而得到净化。如在一定条件下水中难溶性硫化物可以氧化为易溶性的硫酸盐。③生物自净。在生物的作用下，污染物的数量减少，浓度下降，毒性减轻，直至消失。例如，悬浮和溶解在水体中的有机污染物，在需氧微生物的作用下，氧化分解为简单、稳定的无机物，如二氧化碳、水、硝酸盐和磷酸盐等，使水体得到净化。一般来说，物理和生物化学过程在水体自净中占主要地位。对有机物来说，生物自净作用是最重要的。水体自净作用可以在同一介质中进行，也可在不同介质之间进行。例如，河水自净过程大致如下：当污水进入河流之后，首先是混合稀释、扩散，以及反应生成的沉淀物质和吸附有污染物的固体沉入水底，使水中污染物浓度下降；水的最终净化主要靠微生物的作用。微生物把污染物质作为营养源，通过生物化学过程，把复杂化合物变成简单化合物，最终产物是二氧化碳、水等无机物。此外，各类水生生物摄取较多的固体食物或其他生物，包括细菌、植物，这在河水自净中也起着重要作用。藻类和其他绿色植物的光合作用，也有助于水的净化。

水体自净作用是有限的，当人类直接或间接排放的污染物大量进入水体，而超过它的自净作用时，就会造成水体污染。原则上，进入水体的污染物最终都能被净化，但由于环境差异，污染物的性质及污染程度不同，净化的难易和净化的速度也不同。了解污染物的性质与含量以及它们在水体中的存在形式、化学行为，对于研究水体的自净能力、采取措施防止和克服污染所造成的危害具有重大的意义。

二、水体中的污染物

水体污染主要是由人为污染造成的。污染物的种类繁多，包括无机和有机有毒物质、耗氧有机物、石油类、放射性物质、热污染以及病原微生物等。下面对几类主要的水体污染物分别加以说明。

1. 病原体污染物

生活污水、畜禽饲养场污水以及制革、洗毛、屠宰业和医院等排出的废水，常含有各种病原体，如病毒、病菌、寄生虫。水体受到病原体的污染会传播疾病，如血吸虫病、霍乱、伤寒、痢疾、病毒性肝炎等。历史上流行的瘟疫，有的就是水媒型传染病。如 1848 年和 1854 年英国两次霍乱流行，死亡万余人；1892 年德国汉堡霍乱流行，死亡 750 余人，均是水污染引起的。

受病原体污染后的水体，微生物激增，其中许多是致病菌、病虫卵和病毒，它们往往与其他细菌和大肠杆菌共存，所以通常规定用细菌总数和大肠杆菌指数及菌值数作为病原体污染的直接指标。病原体污染的特点是：①数量大；②分布广；③存活时间较长，④繁殖速度快；⑤易产生抗药性，很难绝灭；⑥传统的二级生化污水处理及加氯消毒后，某些病原微生物、病毒仍能大量存活。常见的混凝、沉淀、过滤、消毒处理能够去除水中 99% 以上的病毒，如出水浊度大于 0.5 度时，仍会伴随病毒的穿透。病原体污染物可通过多种途径进入水体，一旦条件适合，就会诱发人体疾病。

2. 耗氧污染物

在生活污水、食品加工和造纸等工业废水中，含有碳水化合物、蛋白质、油脂、木质素等有机物质。这些物质以悬浮或溶解状态存在于污水中，可通过微生物的生物化学作用而分解。在其分解过程中需要消耗氧气，因而被称为耗氧污染物。这种污染物可造成水中溶解氧减少，影响鱼类和其他水生生物的生长。水中溶解氧耗尽后，有机物进行厌氧分解，产生硫化氢、氨和硫醇等难闻气味，使水质进一步恶化。水体中有机物成分非常复杂，耗氧有机物浓度常用单位体积水中耗氧物质生化分解过程中所消耗的氧量表示，即以生化需氧量（BOD）表示。一般用 20℃时，五天生化需氧量（BOD_5）表示。

3. 植物营养物

植物营养物主要指氮、磷等能刺激藻类及水草生长、干扰水质净化，使BOD_5升高的物质。水体中营养物质过量所造成的“富营养化”对于湖泊及流动缓慢的水体所造成的危害已成为水源保护的严重问题。

富营养化（eutrophication）是指在人类活动的影响下，生物所需的氮、磷等营养物质大量进入湖泊、河口、海湾等缓流水体，引起藻类及其他浮游生物迅速繁殖，水体溶解氧量下降，水质恶化，鱼类及其他生物大量死亡的现象。在自然条件下，湖泊也会从贫营养状态过渡到富营养状态，沉积物不断增多，先变为沼泽，后变为陆地。这种自然过程非常缓慢，常需几千年甚至上万年。而人为排放含营养物质的工业废水和生活污水所引起的水体富营养化现象，可以在短期内出现。

植物营养物质的来源广、数量大，有生活污水（有机质、洗涤剂）、农业（化肥、农家肥）、工业废水、垃圾等。每人每天带进污水中的氮约50g。生活污水中的磷主要来源于洗涤废水，而施入农田的化肥有50%～80%流入江河、湖海和地下水体中。天然水体中磷和氮（特别是磷）的含量在一定程度上是浮游生物生长的控制因素。当大量氮、磷植物营养物质排入水体后，促使某些生物（如藻类）急剧繁殖生长，生长周期变短。藻类及其他浮游生物死亡后被需氧生物分解，不断消耗水中的溶解氧，或被厌氧微生物所分解，不断产生硫化氢等气体，使水质恶化，造成鱼类和其他水生生物的大量死亡。藻类及其他浮游生物残体在腐烂过程中，又把生物所需的氮、磷等营养物质释放到水中，供新一代的藻类等生物利用。因此，水体富营养化后，即使切断外界营养物质的来源，也很难自净和恢复到正常水平。水体富营养化严重时，湖泊可被某些繁生植物及其残骸淤塞，成为沼泽甚至干地。局部海区可变成“死海”，或出现“赤潮”现象。

常用氮、磷含量，生产率（O_2）及α-叶绿素作为水体富营养化程度的指标。防治富营养化，必须控制进入水体的氮、磷含量（如肥料）。

4. 有毒污染物

有毒污染物指的是进入生物体后累积到一定数量能使体液和组织发生生化和生理功能的变化，引起暂时或持久的病理状态，甚至危及生命的物质。如重金属和难分解的有机污染物等。污染物的毒性与摄入机体内的数量有密切关系。同一污染物的毒性也与它的存在形态有密切关系。价态或形态不同，其毒性可以有很大的差异。如Cr（Ⅵ）的毒性比Cr(Ⅲ）大；As(Ⅲ）的毒性比As(Ⅴ）大；甲基汞的毒性比无机汞大得多。另外，污染物的毒性还与若干综合效应有密切关系。从传统毒理学来看，有毒污染物对生物的综合效应有三种。

① 相加作用，即两种以上毒物共存时，其总效果大致是各成分效果之和。

② 协同作用，即两种以上毒物共存时，一种成分能促进另一种成分毒性急剧增加。如铜、锌共存时，其毒性为它们单独存在时的8倍。

③ 拮抗作用，两种以上的毒物共存时，其毒性可以抵消一部分或大部分。如锌可以抑制镉的毒性；又如在一定条件下硒对汞能产生拮抗作用。总之，除考虑有毒污染物的含量外，还须考虑它的存在形态和综合效应，这样才能全面深入地了解污染物对水质及人体健康的影响。

有毒污染物主要有以下几类。

① 重金属。如汞、镉、铬、铅、钒、钴、钡等，其中汞、镉、铅危害较大；砷、硒和铍的毒性也较大。重金属在自然界中一般不易消失，它们能通过食物链而被富集；这类物质

除直接作用于人体引起疾病外，某些金属还可能促进慢性病的发展。

② 无机阴离子。主要是NO_2^-、F^-、CN^-。NO_2^-是致癌物质。剧毒物质氰化物主要来自工业废水排放。

③ 有机农药、多氯联苯。目前世界上有机农药大约有6000种，常用的大约有200多种。农药喷在农田中，经淋溶等作用进入水体，产生污染作用。有机农药可分为有机磷农药和有机氯农药。有机磷农药的毒性虽大，但一般容易降解，累积性不强，因而对生态系统的影响不明显；而绝大多数的有机氯农药，毒性大，几乎不降解，累积性甚高，对生态系统有显著影响。多氯联苯（PCB）是联苯分子中一部分氢或全部氢被氯取代后所形成的各种异构体混合物的总称。

多氯联苯有剧毒，脂溶性大，易被生物吸收，化学性质十分稳定，难以和酸、碱、氧化剂等作用，有高度耐热性，在1000～1400℃高温下才能完全分解，因而在水体和生物中很难降解。

④ 致癌物质。致癌物质大体分为三类：稠环芳香烃（PAHs），如3,4-苯并芘等；杂环化合物，如黄曲霉素等；芳香胺类，如甲、乙苯胺，联苯胺等。

⑤ 一般有机物质。如酚类化合物就有2000多种，最简单的是苯酚，均为高毒性物质；腈类化合物也有毒性，其中丙烯腈的环境影响最为引人注目。

5. 石油类污染物

石油污染是水体污染的重要类型之一，特别在河口、近海水域更为突出。排入海洋的石油估计每年有高达数百万吨至上千万吨，约占世界石油总产量的5‰。石油污染物主要来自工业排放，清洗石油运输船只的船舱、机件及发生意外事故、海上采油等均可造成石油污染。而油船事故属于爆炸性的集中污染源，危害是毁灭性的。

石油是烷烃、烯烃和芳香烃的混合物，进入水体后的危害是多方面的。如在水上形成油膜，能阻碍水体的复氧作用，油类黏附在鱼鳃上，可使鱼窒息；黏附在藻类、浮游生物上，可使它们死亡。油类会抑制水鸟产卵和孵化，严重时使鸟类大量死亡。石油污染还能使水产品质量降低。

6. 放射性污染物

放射性污染是放射性物质进入水体后造成的。放射性污染物主要来源于核动力工厂排出的冷却水，向海洋投弃的放射性废物，核爆炸降落到水体的散落物，核动力船舶事故泄漏的核燃料；开采、提炼和使用放射性物质时，如果处理不当，也会造成放射性污染。水体中的放射性污染物可以附着在生物体表面，也可以进入生物体蓄积起来，还可通过食物链对人产生内照射。

水中主要的天然放射性元素有^{40}K、^{238}U、^{286}Ra、^{210}Po、^{14}C等。目前，在世界上的任何海区几乎都能测出^{90}Sr、^{137}Cs。

7. 酸、碱、盐无机污染物

各种酸、碱、盐等无机物进入水体（酸、碱中和生成盐，它们与水体中某些矿物相互作用产生某些盐类），使淡水资源的矿化度提高，影响各种用水水质。盐污染主要来自生活污水和工矿废水以及某些工业废渣。另外，由于酸雨规模日益扩大，造成土壤酸化、地下水矿化度升高。

水体中无机盐的增加能提高水的渗透压，对淡水生物、植物生长产生不良影响。在盐碱化地区，地面水、地下水中的盐将对土壤质量产生更大影响。

8. 热污染

热污染是一种能量污染，它是工矿企业向水体排放高温废水造成的。一些热电厂及各种工业过程中的冷却水，若不采取措施，直接排放到水体中，均可使水温升高，水中化学反应、生化反应的速度随之加快，使某些有毒物质（如氰化物、重金属离子等）的毒性提高，溶解氧减少，影响鱼类的生存和繁殖，加速某些细菌的繁殖，助长水草丛生，厌气发酵，恶臭。

鱼类生长都有一个最佳的水温区间。水温过高或过低都不适合鱼类生长，甚至会导致死亡。不同鱼类对水温的适应性也是不同的。如热带鱼适于15～32℃，温带鱼适于10～22℃，寒带鱼适于2～10℃的范围。又如鳟鱼虽在24℃的水中生活，但其繁殖温度则要低于14℃。一般水生生物能够生活的水温上限是33～35℃。

除了上述八类污染物以外，洗涤剂等表面活性剂对水环境的主要危害在于使水产生泡沫，阻止了空气与水接触而降低溶解氧，同时由于有机物的生化降解耗用水中溶解氧而导致水体缺氧。高浓度表面活性剂对微生物有明显毒性。

水体污染的例子很多，如京杭大运河（杭州段）两岸有许多工厂，每天均有大量废水排入运河，使水体中固体悬浮物、有机物、重金属（Zn、Cd、Pb、Cu等）及酚、氰化物等含量大大超过地面水标准，有的超过几十倍，使水体处于厌氧的还原状态，乌黑发臭，鱼虾绝迹，不能用于生活、农业等；水体自净能力差，若不治理，并控制污染源，水体污染还会进一步加剧。

水环境中的污染物，总体上可划分为无机污染物和有机污染物两大类。在水环境化学中较为重要的，研究得较多的污染物是重金属和有机物。我国水污染化学研究始于20世纪70年代，从重金属、耗氧有机物、DDT、六六六等农药污染开始，目前研究的重点已转向有机污染物，特别是难降解有机物，因其在环境中的存留期长，容易沿食物链（网）传递积累（富集），威胁生物生长和人体健康，因而日益受到人们重视。本章着重介绍重金属和有机污染物在水体中迁移转化的环境化学行为。

三、污染物进入水体后的运动过程

污染物进入水体后立即发生各种运动。下面以海洋为例作一简介，其他水体的情况，可以类推。污染物排入水体后的运动过程如图3-2所示。

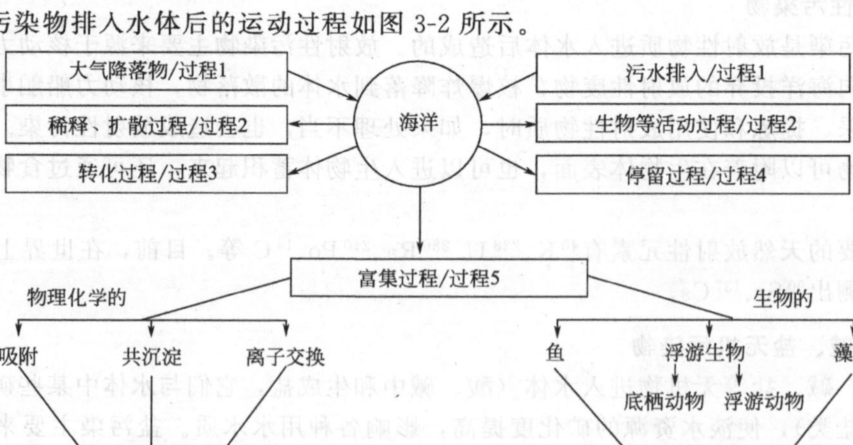

图3-2 污染物排入水体后的运动过程示意图

图中过程 1、2 是污染物在海水中通过稀释、扩散、生物活动等引起污染物空间位置的移动。影响该过程的主要因素是海区的水文条件，其中包括潮汐、海流、海水平流和涡流等多种变量。

图中过程 3 是污染物在海洋中的转化过程，包括化学、物理化学、水化学和生物化学的作用。通过转化过程，污染物的浓度，尤其是形态将发生变化，从而影响到污染物在海洋中的分布、迁移、停留、富集以及它们的毒性、底质的二次污染等。这是水污染化学中重要的研究课题。

图中过程 4 是污染物在海水中的停留过程。该过程涉及污染物危害持续时间的长短问题，它取决于海水的交换和更新以及水中胶体、悬浮物的沉积速率。污染物在海水中停留的时间 τ_i 可用下式计算：

$$\tau_i = A_i/(\mathrm{d}A_i/\mathrm{d}t)$$

式中，A_i 为排入水体的污染物 i 的总量，$\mathrm{d}A_i/\mathrm{d}t$ 为污染物 i 在海洋中的沉积速率。一般情况下，污染物在海水中的活性越大，停留时间就越短。

图中过程 5 为污染物在海洋中的富集过程，它主要取决于吸附等物理化学的富集沉降以及食物链的选择性吸收，其结果是污染物脱离海水，使后者得到净化，同时将在不同程度上有害于生物，并将增加底质中污染物的积累，有可能引起海水的二次污染。

四、水体污染对人体健康的影响

水体污染的危害是多方面的，这里简单介绍一下水体污染对人体健康的影响。

1. 引起急性和慢性中毒

水体受有毒有害化学物质污染后，通过饮水或食物链便可能造成中毒。著名的水俣病、痛痛病是由水体污染引起的。

2. 致癌作用

某些有致癌作用的化学物质如砷、铬、镍、铍、苯胺、苯并（a）芘和其他多环芳烃、卤代烃污染水体后，可被悬浮物、底泥吸附，也可在水生生物体内积累，长期饮用含有这类物质的水，或食用体内蓄积了这类物质的生物（如鱼类）就可能诱发癌症。

3. 发生以水为媒介的传染病

人畜粪便等生物污染物污染水体，可能引起细菌性肠道传染病如伤寒、痢疾、肠炎、霍乱等；肠道内常见病毒如脊髓灰质类病毒、柯萨奇病毒、传染性肝炎病毒等，皆可通过水体污染引起相应的传染病。1989 年上海的“甲肝事件”，就是由水体污染引起的。在发展中国家，每年约有 6000 万人死于腹泻，其中大部分是儿童。

4. 间接影响

水体污染后，常可引起水的感官性状恶化，如某些污染物在一定浓度下，对人的健康虽无直接危害，但可使水发生异臭、异色，呈现泡沫和油膜等，妨碍水体的正常利用。铜、锌、镍等物质在一定浓度下能抑制微生物的生长和繁殖，从而影响水中有机物的分解和生物氧化，使水体自净能力下降，影响水体的卫生状况。水体污染既可严重危害生态系统，还可造成严重的经济损失。

第三节 有机物污染的水环境化学

一、有机污染程度的指标

水体中有机污染物的种类繁多、组成复杂，现代分析技术难以分别测定它们的含量。因此，只能利用它们共同的特点，用一些指标间接反映水体中有机物的污染程度。常见的指标有：溶解氧、生化需氧量、化学需氧量、总有机碳和总需氧量。

1. 溶解氧（dissolved oxygen，简称 DO）

溶解氧即在一定温度和压力下，溶解于水中的氧的含量，是水质的重要指标之一。水中溶解氧含量受到两种作用的影响：一是耗氧作用，包括耗氧有机物降解的耗氧、生物呼吸耗氧等，使 DO 下降；另一种是复氧作用，主要有空气中氧的溶解、水生植物的光合作用等，使 DO 增加。这两种作用的相互消长，使水中溶解氧含量呈现时空变化。此外，DO 随水温升高而降低，还随水深增加而减小。常温下，水体中 DO 为 8～14mg/L；在水藻繁生的水中，DO 可能处于饱和状态；如果水体中的有机污染量较多，耗氧作用大于复氧作用，水中 DO 减少；有机物污染严重时，DO 为零。在缺氧的水体中，水生动植物生长将受到抑制，甚至死亡。例如，当 DO<4mg/L 时，鱼类将死亡。因此测定水体中的溶解氧含量，可评价水体污染程度及自净状况。测定水中 DO 的方法有碘量法、叠氮化钠修正法、$KMnO_4$ 修正法和膜电极法，其中最常用的是碘量法。

2. 生化需氧量（biochemical oxygen demand，BOD）

水体中微生物分解有机物的过程中消耗水中的溶解氧的量称为生化需氧量，通常用 BOD 表示，其单位为 mg（O_2）/L。BOD 反映了水体中可被微生物分解的有机物的总量。有机物的微生物氧化分解分为两个阶段进行。第一阶段主要是有机物被转化为无机的 CO_2、H_2O 和氨；第二阶段是氨被转化为 NO_2^-、NO_3^-。第二阶段的环境影响较小，所以生化需氧量一般是指第一阶段有机物经微生物氧化分解所需的氧量。微生物分解有机物的速度和程度与温度、时间有关。如在 20℃时，通常生活污水中的有机物需要 20d 左右才能基本完成第一阶段的生化氧化，但经过 5d 也可完成第一阶段转化的 70%左右。为缩短测定时间，同时使 BOD 值有可比性，因而采用在 20℃条件下，培养 5d 测定生化需氧量作为标准方法，称为五日生化需氧量，以 BOD_5 表示。BOD 基本上能反映出有机物在自然状况下氧化分解所消耗的氧量，较确切地说明了需氧有机污染物对环境的影响。但 BOD 的测定时间长，对毒性大的废水因微生物活动受到抑制，而难以准确测定。若要尽快知道水中有机物的污染状况，可测定化学需氧量。

3. 化学需氧量（chemical oxygen demand，COD）

水体中能被氧化的物质在规定条件下进行化学氧化过程中所消耗氧化剂的量，以每升水样消耗氧的毫克数表示，通常称为 COD，其单位为 mg（O_2）/L。水体的 COD 值越高，表示有机物污染越严重。水中各种有机物进行化学氧化反应的难易程度是不同的，因此化学需氧量只表示在规定条件下，水中可被氧化物质的需氧量的总和。目前测定化学需氧量的常用方法有 $KMnO_4$ 法和 $K_2Cr_2O_7$ 法，前者氧化性相对较弱，适合于测定较清洁的水样，后者则用于污染严重的水样和工业废水。同一水样用上述两种方法测定的结果是不同的。因此，在报告化学需氧量的测定结果时要注明测定方法。

同生化需氧量相比较，COD的测定不受水质条件的限制，测定时间短，但COD不能良好地表示出微生物所能氧化的有机物的量。化学氧化剂不能氧化某些需氧有机物，但能氧化无机还原性物质（硫化物、亚铁等）。所以，作为需氧有机物污染的评价指标，化学需氧量不如生化需氧量合适。但在条件不具备或受水质限制不能做BOD测定时，可用COD代替。此外，在水质相对稳定的条件下，化学需氧量同生化需氧量之间有比较密切的相关性。一般来说，重铬酸钾法 COD＞BOD_5＞高锰酸钾法 COD。

4. 总有机碳（TOC）和总需氧量（TOD）

总有机碳（total organic carbon，简称TOC）是水中几乎全部有机物的含碳量。总需氧量（total oxygen demand，简称TOD）是水中几乎全部可被氧化的物质（基本上是有机物）变成稳定氧化物时所需的氧量。由于BOD测定费时，为实现快速反映有机污染程度的目的，而采用TOC与TOD测定法，一次测定只需3min左右，可以连续自动测定。它们都可用化学燃烧法测定，前者测定结果以碳表示，后者则以氧表示需氧有机物的含量。它们是评价水中需氧有机污染物的一种指标。但是，总有机碳和总需氧量的测定绝不是水中有机物的完全氧化，测定时的氧化条件与自然界的氧化条件相差很远，对总需氧量有影响的无机物质未必是自然界的耗氧物质，并且测定器的标准化问题还未完全解决，所以不能把它们当作评价水体需氧有机污染物的万能指标。由于测定时耗氧过程不同，而且各种水体中有机物成分不同，生化过程差别也较大，所以各种水质之间，TOC或TOD与BOD_5不存在固定的相关性。在水质条件基本相同的条件下，水体BOD_5与TOC或TOD之间有一定的相关性。如日本多摩川河水BOD与TOC的相关系数为0.6，其回归方程为：

$$BOD_5 = 1.72TOC - 1.9 \text{（TOC对} BOD_5 \text{回归）}$$

$$BOD_5 = 1.04TOC + 4.2 \text{（} BOD_5 \text{对TOC回归）}$$

与BOD_5之间的相关系数为0.86，其回归方程是：

$$TOD = 1.34BOD + 4.7 \text{（TOD对BOD回归）}$$

$$TOD = 1.15BOD + 10.4 \text{（BOD对TOD回归）}$$

DO、COD、BOD_5通常可作为水质评价指标，地面水分级标准见表3-7。

表3-7　地面水分级标准　　单位：mg/L

项　目	Ⅰ类水	Ⅱ类水	Ⅲ类水	Ⅳ类水	Ⅴ类水
DO　≥	饱和率90%	6	5	3	2
BOD_5　≤	3以下	3	4	6	10
COD_{Mn}　≤	2	4	6	8	10

5. 生物污染物指标

生物污染物主要是指废水中的致病微生物和其他的有机体。废水中，特别是生活废水中常含有许多微生物。它们大部分是无害的，但其中也含有对人体及其他生物有害的病原菌。这种污染物主要来自医院、屠宰厂、生物研究所及生活污水等。

生物污染物指标主要有细菌总数、大肠菌数、病原菌和病毒等。细菌总数是指1mL水中所含有各种细菌的总数。在水质分析中，是把一定量的水接种于琼脂培养基中，在37℃条件下培养24h后，数出生长的细菌菌落数，然后计算出每毫升水中所含的细菌数。大肠菌数是指1L水中所含大肠菌的个数。由于大肠菌在外部环境中的生存条件与肠道传染病的细菌、寄生虫卵相似，而且大肠菌的数量多，比较容易检验，所以把大肠菌数作为生物污染指标。

6. 感官污染物指标

废水中的异色、浑浊、泡沫、不良气味等会引起人们感官上不快的污染物称为感官污染物，特别是在游览水域，这种感官污染物的影响更为严重。感官污染物的常用指标有色泽、色度、臭和味等。色泽是指废水的颜色种类，通常用文字描述，如：废水呈深蓝色、棕黄色、浅绿色、暗红色等。色度是指废水所呈现的颜色深浅程度。色度有两种表示方法：一是采用铂钴标准比色法，规定在1L水中含有氯铂酸钾（K_2PtCl_6）2.491mg及氯化钴（$CoCl_2 \cdot 6H_2O$）2.00mg时，也就是在1L水中含铂1mg及钴0.5mg时所产生的颜色深浅为1度；二是采用稀释倍数法，即将废水稀释，用把废水稀释到接近无色时所需的稀释倍数表示色度。臭和味是判断水质优劣的感官指标之一。洁净的水是没有气味的，受到污染后会产生各种臭味。常见的水臭味有霉烂臭味 粪便臭味、汽油臭味、臭蛋味、氯气味等。臭味的表示方法现行的是用文字描述臭的种类，用强、弱等字样表示臭的强度。比较准确的定量方法是臭阈法，即用无臭水将待测水样稀释到接近无臭程度的稀释倍数表示臭的强度，倍数越大说明臭味越大。

二、水体中某些有机污染物的降解

有机物在水环境介质中的降解是环境污染物自然净化的主要过程，它主要通过水解、氧化、光解、生物化学分解等途径来实现。水体中有些物质如碳水化合物、脂肪、蛋白质等比较容易降解；有机氯农药、多氯联苯、多环芳烃等难降解。下面分别简单介绍有机农药、合成洗涤剂、石油、多环芳烃和多氯联苯等的降解情况。

1. 有机农药的降解

目前世界上的有机农药有1000多种，常用的大约有200多种。有机农药按用途可分为杀虫剂、杀菌剂、除草剂、选种剂等；按化学成分，农药则可分为有机氯农药、有机磷农药、有机汞农药、氨基甲酸酯类农药等。有机氯农药品种较多，大多数用作杀虫剂，如DDT、六六六、艾氏剂等；特点是化学性质稳定、不易分解、毒性释放较缓慢、残留时间长、微溶于水而溶于脂肪、蓄积性很强，水生生物对其的富集系数可高达几十万倍。有机氯农药目前已经限制使用，我国于1983年开始停止生产。有机磷农药大多数也用作杀虫剂，如对硫磷、敌百虫、敌敌畏等，其特点是毒性大，但易分解，蓄积作用微弱，因而对生态系统的影响不明显，有取代有机氯农药的趋势。氨基甲酸酯农药，如杀虫剂西维因、除草剂灭草灵、芽根灵等，这类农药对动物的毒性低，残留时间短，易于分解。有机汞农药多是杀菌剂，如赛力散、西力生等，但由于汞污染，现已减少使用。

农药进入水体后可与各类物质接触，发生一系列的物理、化学和生化反应。它们的行为可归纳为以下几个方面：被水中颗粒物吸附，被生物吸附并积累，发生降解反应，使农药含量逐渐降低。下面简单讨论几种常见有机农药DDT、六六六、有机磷和氨基甲酸酯的降解。

(1) DDT的降解

DDT可以通过光化学、催化和生化反应降解。

DDT的光化学降解：在紫外线的作用下，DDT可经碳碳键均裂过程而生成氯苯自由基，后者相互结合生成二氯联苯。这类光化反应中可能还伴生有三氯联苯、四氯联苯、3,6-二氯苯并呋喃等，并可按自由基历程生成多氯联苯。

DDT的催化降解：金属对可以通过催化还原降解DDT。DDT的脱氯反应可能有两种方式：即脱去一个氯原子变成DDD，然后再脱氯得DDMS和DDEt；或是同时脱去三个氯原子得到DDEt。DDD的毒性与DDT相当，而DDEt的毒性不及DDT的千分之一，所以它是DDT还原降解最理想的产物。

DDT的生化降解：关于DDT的生化降解研究得较多，有人曾用各种微生物在缺氧和有氧条件下做培养试验。在不同情况下检出的降解产物包括DDD、DDE、DDMV、DDMS、DDNV、DDA、DDM、DBH、DBP、Kelthane和DDCN等。DDT的降解途径目前还没有搞清楚。Rhead（1975）提出了DDT降解的可能途径如图3-3所示。

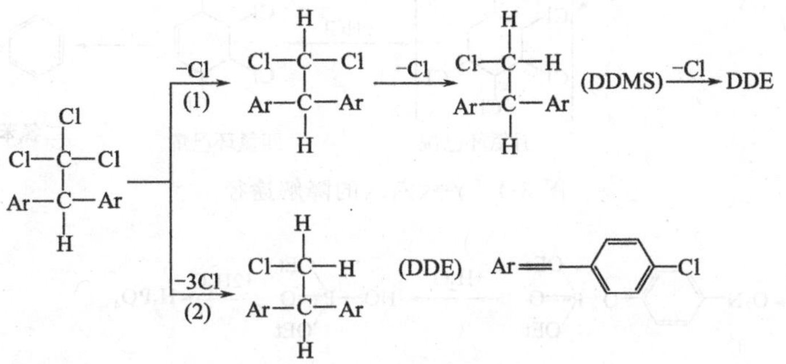

图3-3 DDT降解的可能途径

不同微生物对DDT的降解并不完全相同，然而DDE及DDD是一般的降解产物，其中DDE要比DDD稳定得多。DDCN是下水道污泥中厌氧降解的一个主要产物，据推测DDCN经化学过程形成的可能性较大，而污泥中的生物作用可能有助于保证维持适宜的还原条件，从而间接影响DDT、DDCN的转变。

（2）六六六的解降

六六六较DDT容易降解，其降解同样是通过光化学、化学及生化反应等途径。

六六六的光化降解：一般情况下六六六不易直接吸收光子而发生光化反应，但当水体中存在某些物质（如苯胺、芳酚等芳香化合物，X 、$S_2O_3^{2-}$等无机阴离子）时，这些物质的分子或离子能在光的激发下发生光化反应，产生某种还原性活性中间体，如水合电子（e_{aq}^-），而e_{aq}^-可以进一步与六六六分子反应，使六六六像直接接受光能一样，发生化学键的断裂而降解。有人根据动力学稳定态处理方法得出六六六的光化学反应为一级反应，并指出光化学反应的速率常数k与水体中给电子物质的浓度成正比；k与水体中其他能结合电子物质的浓度成反比，这些结合电子的物质（如Cu^{2+}、Zn^{2+}、溶解氧、H_3O^+）等的存在能降低六六六的光化学反应速率。所以水体中有溶解氧存在时，六六六的光化学降解速率极慢。

六六六的化学降解：有人认为六六六化学降解的产物为四氯环己烯、五氯环己烯，降解时均需脱除HCl，因此在碱性介质中六六六易降解，而在酸性介质中六六六比较稳定。

六六六的生化降解：在六六六的降解中，微生物对其降解起着决定性作用。γ-六六六可经脱氯化氢形成γ-五氯环己烯，也可以在厌氧条件下转化为α，β，δ-六六六异构体，在水中完全迅速地降解。实验结果表明，在微生物存在下，六六六浸水2个月后，四种异构体基本消失，见图3-4。

（3）有机磷农药的降解

有机磷农药既能直接水解，也能被微生物降解。如甲基对硫磷、乙基对硫磷、杀螟松等有机农药均可在微生物的作用下发生氧化、还原、水解等过程，以完成降解。图3-5为乙基对硫磷的生化降解途径。乙基对硫磷的氧化过程即为硫代磷酸酯的脱硫氧化，还原过程即为

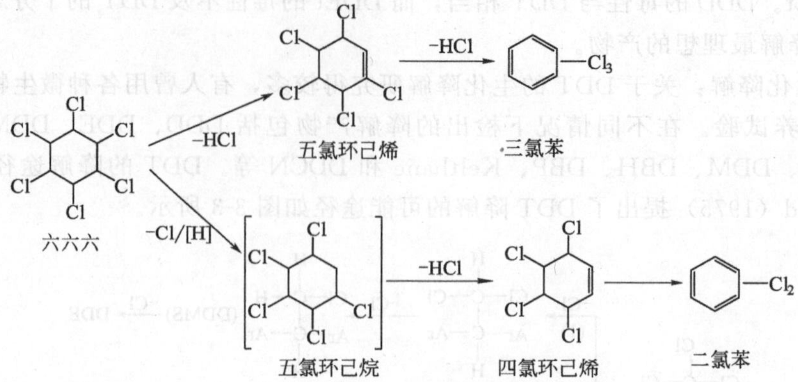

图 3-4　γ-六六六的降解途径

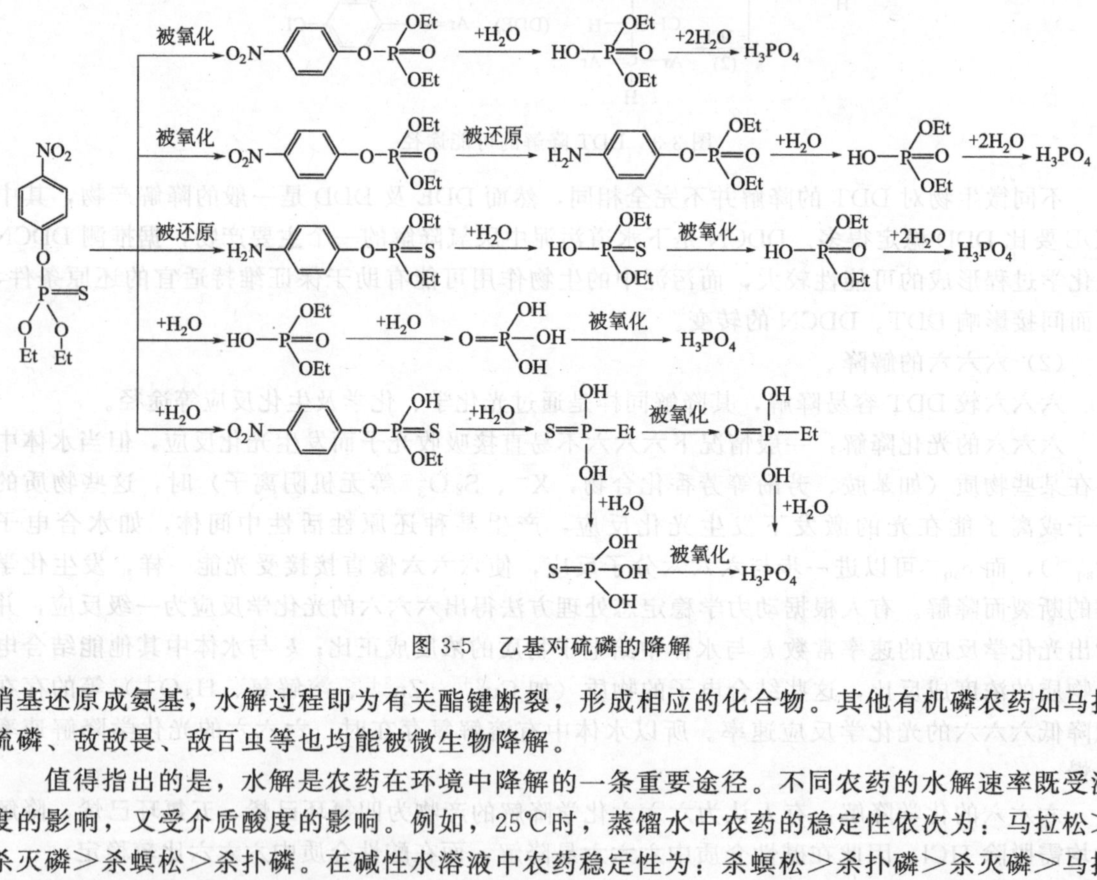

图 3-5　乙基对硫磷的降解

硝基还原成氨基，水解过程即为有关酯键断裂，形成相应的化合物。其他有机磷农药如马拉硫磷、敌敌畏、敌百虫等也均能被微生物降解。

值得指出的是，水解是农药在环境中降解的一条重要途径。不同农药的水解速率既受温度的影响，又受介质酸度的影响。例如，25℃时，蒸馏水中农药的稳定性依次为：马拉松＞杀灭磷＞杀螟松＞杀扑磷。在碱性水溶液中农药稳定性为：杀螟松＞杀扑磷＞杀灭磷＞马拉松。又如敌敌畏和甲基对硫磷均属于水解不稳定的化合物，在地面水条件（25℃、pH＝7）下，它们的 $t_{1/2}$ 分别为 43.9h 和 83.5h；在地下水条件（15℃、pH＝7）下，它们的 $t_{1/2}$ 分别为 133.3h 和 3013h；这两种农药在酸性介质中也具有一定的稳定性，pH＝3.04、25℃时，其 $t_{1/2}$ 分别为 533h 和 2772h。

由此可见，每一种农药都有自己的降解过程，即使是同种农药，在不同条件下或由不同微生物降解时，降解过程也不会完全相同。

2. 多氯联苯（PCBs）和多环芳烃（PAHs）

多氯联苯（PCBs）和多环芳烃（PAHs）的污染在全球已较为普遍。许多天然水体及土壤样品中均存在 PCBs 和 PAHs 的污染。PCBs 和 PAHs 有剧毒，脂溶性大，易被生物吸收，化学性质十分稳定；一些 PAHs 又是“三致”物质，因而对人体健康构成威胁。研究 PAHs 在水体中的环境化学行为和生物效应，已越来越受到人们的重视。

文献报道，一氯联苯生物降解产生氯苯甲酸，二氯联苯降解的产物是二氯苯甲酸。氯苯甲酸还会继续分解，但其速率缓慢，滞缓的周期取决于所用细菌接种体。

R. E. Ballay 等在 1980 年发现，河水中一氯联苯的三种异构体都会迅速地被生物降解，在低浓度下它们的降解速率大致相同，但在浓度较高时，生物降解比较迟缓；然而，2,2′,4,4′-四氯联苯在培育 98d 以后，仍没有发现有降解现象。李敏学等人研究了第二松花江中 PCBs 与有机氯农药的迁移和分布。结果表明，在自然条件下，江水中 PCBs 基本上不被光和微生物降解，由于 PCBs 的蒸气压很低，加上第二松花江冰封季节近 5 个月，因此江水中 PCBs 的挥发量也极小。因 PCBs 的黏度和密度均较大，且难溶于水，易被悬浮颗粒物吸附而沉入江底，故在底泥中 PCBs 的含量比在江水中高出一个数量级。第二松花江水中的 PCBs 的检出率虽不高，但在鱼体中富集较多，主要是 PCBs 在水环境中难以挥发和降解，同时由于 PCBs 的脂溶性好，可在水生动物体内富集，再通过食物链的传递在鱼体内达到高度富集，富集系数达 10^4。

刘红果等研究了多氯联苯的微生物降解，以 PCBs 为唯一碳源，通过选择性富集培养，从变电站变压器油污染的土样中分离出 25 株 PCBs 降解菌；底质生长试验结果表明，它们可以利用多种 PCBs 作为唯一碳源；用气相色谱法和氯离子测定法检测了八株菌对各种 PCBs 的降解能力，降解率一般为 20％～50％，高的可达 60％～100％；降解能力随 PCBs 中含氯量的增高而降低，不同异构体降解的难易程度有很大差别。天然水中 PAHs 的浓度较低，一般为0.001～10μg/L。PAHs 具有脂溶性而难溶于水，故进入水体的 PAHs 主要被吸附在悬浮物、水生生物或沉积物上，最终沉入底泥。这是 PAHs 在水体中存在的主要形式。

孔令仁（1986 年）研究了苯并（a）蒽（BaA）在水体中的光化学降解，认为水体中有溶解氧时，基态氧受紫外辐射使其激发，激发态氧容易与基态或被激发了的 BaA 反应生成 BaA 的氧化物，即苯并（a）蒽，7,12-醌；这里，BaA 与溶解氧发生了氧化反应，而光对反应起了催化作用。王连生等（1991 年）系统地研究了常见的 17 种多环芳烃在甲醇-水、乙腈-水溶液中的光化学降解。结果表明，大多数 PAHs 的光解速率常数（k）与它们的极谱氧化半波电位有关；半波电位低，光化学降解速率快；黑暗条件下 PAHs 不能降解。还测得苯的光解产物为邻苯二甲酸，并发现溶解氧除去后溶液中的萘不能光解，表明大部分 PAHs 光解为光氧化反应。

综上所述，有机污染物的降解过程是复杂的。在实际水体中，某种有机污染物往往可以有几种降解途径同时进行，或因环境条件不同而以某种降解途径为主。有机物的最终分解要靠微生物作用来完成。

第四节 废水处理方法

处理废水的方法很多，其分类方法也不同。这里仅介绍两种。第一种是按废水中污染物从废水中除去的方式分，可分为三类：①分离处理，通过各种方法使污染物从废水中分离出来，一般不改变污染物的化学本性；②转化处理，通过化学或生物化学的方法，使废水中的污染物转化为无害的物质，或是转化为易于分离的物质，然后再分离；③稀释处理，这类方法既不改变污染物的化学特性也不把污染物分离，而是通过稀释混合降低污染物的浓度，但污染物的总量不变。这是一种消极的方法。

另一种分类方法是按废水处理的程度，或说按处理的阶段来分类。一般按处理的程度不同可把废水处理分为三级：一级处理、二级处理和三级处理。一级处理也叫初级处理，该过程只能除去废水中的大颗粒的悬浮物及漂浮物，很难达到排放标准。二级处理一般可以除去细小的或呈胶体态的悬浮物及有机物，一般能达到排放标准。三级处理也称高级处理，进一步除去废水中的胶体及溶解态的污染物，一般可达到回用的目的。下面按这种分类方法把处理废水常用的方法概述一下。

一、废水的一级处理

1. 重力分离方法

依靠重力的作用，使污染物分离，又分为沉降分离和浮上分离。沉降法用于除去水中密度比水大的污染物，上浮法用于除去水中密度比水小的漂浮物。

2. 阻力截留法

这种方法是利用筛网等与悬浮固体之间几何尺寸的差异截留固体悬浮物。包括格栅、筛网和粒状介质截留法。

3. 稀释法

稀释法即用没有污染物的或污染物含量低的水与污染物含量高的水相互混合而降低污染物浓度的方法。

4. 中和法

中和法是利用酸碱中和的原理来消除废水中酸或碱污染物的方法。

废水的一级处理方法较简单，多数情况下达不到排放标准，需要进一步处理。

二、废水的二级处理

1. 气浮法

气浮法是利用废水中的污染物的疏水性，或是添加某种药剂使废水中的污染物变得疏水，然后向废水中通入气泡，疏水的污染物就会吸附到气泡上，而随气泡浮到水面上而形成泡沫层，把泡沫层与水分离即可把污染物与水分离。

2. 混凝法

混凝法是向废水中投加电解质或混凝剂或通过机械搅拌，使废水中呈胶体状态存在的污染物互相凝聚，形成大而重的絮凝体，然后再用重力沉降的方法分离。

3. 萃取法

萃取法是利用分配定律的原理，用一种与水不互溶，而对废水中某些污染物溶解度大的有机溶剂，从废水中分离除去污染物的方法。

4. 氧化还原法

氧化还原法是向废水中投加氧化剂或还原剂，将有害的污染物氧化或还原为无害或害处较小的物质的方法。

5. 电解法

电解法是指用电解的基本原理，将废水中的污染物通过电解过程在阴、阳极上分别发生氧化或还原反应转化为无害物质，以实现废水净化目的的方法。

6. 生物法

生物法是利用水中的微生物来氧化分解污染物的方法，又可分为好氧生物法和厌氧生物法。好氧生物法是在水中有溶解氧存在的条件下，利用好氧微生物和兼性微生物分解污染物的方法。厌氧生物法是在无溶解氧的条件下，利用厌氧微生物和兼性微生物分解废水中污染物的方法。生物法是目前应用较广的二级废水处理方法，特别是对于城市废水，几乎都是用生物法处理。

7. 吹脱法

这种方法用来除去废水中的气态或挥发性污染物。使空气与废水充分接触，使溶解在废水中的气体或挥发性污染物扩散到空气中而除去。

8. 汽提法

这种方法也用于除去废水中的挥发性污染物，是利用蒸汽直接加热废水至沸腾，挥发性污染物随水蒸气一起逸出而除去的方法。污染物浓度高时，可把蒸汽冷凝后回收污染物。

三、废水的三级处理

1. 吸附法

让废水与固体吸附剂接触，使分子或离子状态的污染物吸附于吸附剂上，然后分离水与吸附剂即可把污染物与水分离。一般吸附剂再生后可以循环使用。

2. 膜分离法

按作用原理的不同，膜分离法可分为超过滤、电渗析和反渗透三种方法。

超过滤也称为精密过滤，它也是利用过滤介质除去废水中污染物的方法，与一般过滤不同的是超过滤所用的介质孔径很小，一般为 1～0.1μm。这种方法可以除去水中的胶体物质或大分子的污染物。电渗析是使废水通过由阴离子交换膜和阳离子交换膜交替排列组成的通道，在直流电场的作用下，离子能有选择性地透过不同的膜，某些通道中的污染物被浓缩，另一些通道中的废水则得到净化。反渗透法是用半透膜把废水与清水分开，在废水表面施加压力，使水分子透过半透膜，而污染物不能透过，从而分离或浓缩污染物的方法。

3. 磁过滤法

磁过滤法是依靠磁场的作用，用高梯度磁过滤器截留磁性的污染物，或投加磁种，使非磁性的污染物吸附于磁种上，然后再分离的方法。

4. 离子交换法

离子交换法是将废水与固体离子交换剂接触，离子态污染物能与离子交换剂上的同号离子互相交换，使废水中的有害离子分离出来的方法。

四、城市废水处理的典型流程

对于城市废水有相对典型的处理流程，一级二级处理典型流程图如图 3-6 所示，三级处理典型流程图如图 3-7 所示。

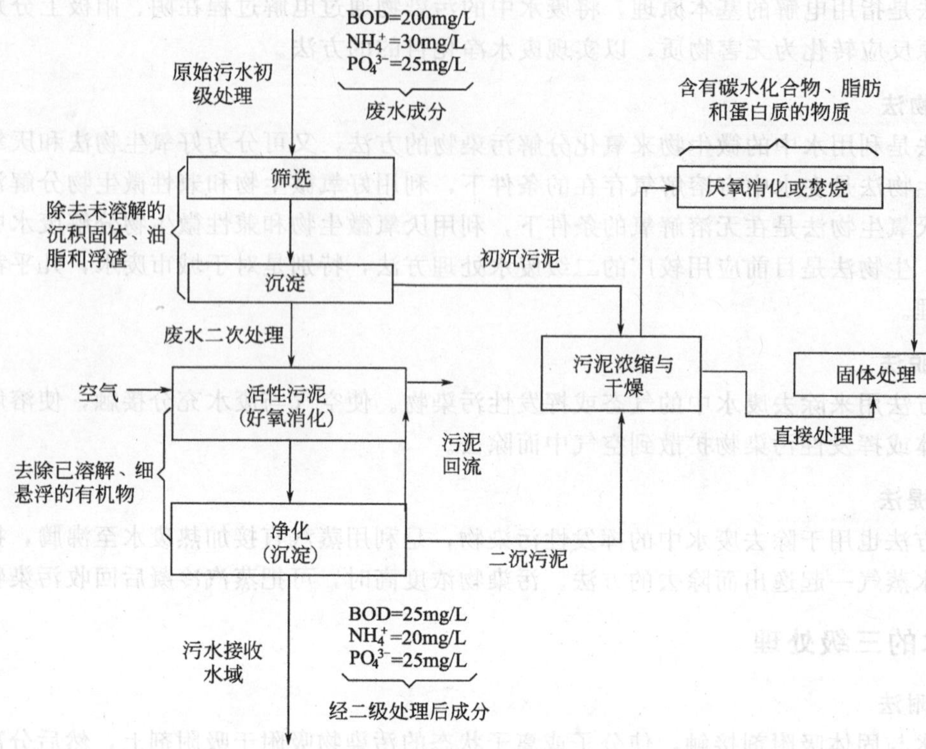

图 3-6 城市废水一级二级处理典型流程图

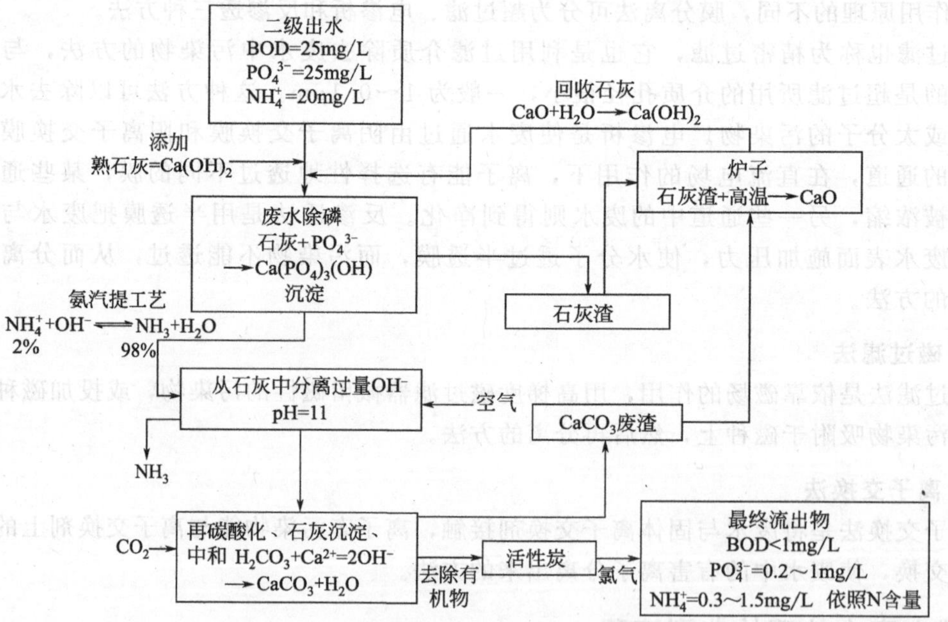

图 3-7 城市废水三级处理典型流程图

思考题

1. 举例说明天然水中物质的基本类型。
2. 什么是水体自净？水体自净的方式有哪几种？举例说明河水自净过程。
3. 水体污染物可分为哪几类？
4. 重金属污染的特点是什么？
5. 胶体微粒吸附对水体中的污染物有何影响？
6. 水体中的有机物是怎样迁移转化的？

第四章　土壤环境化学

本章要点：土壤形成的实质；土壤形成的自然因素；土壤形成的人为作用；土壤矿物质；土壤有机质；土壤溶液；土壤空气；土壤的物理性质；土壤的胶体性质；土壤的配合和螯合作用；土壤的氧化还原性质；土壤的酸碱性；土壤的生物学性质；土壤的自净作用；土壤污染源；土壤的主要污染物；控制和消除土壤污染源；土壤改良与修复；发展绿色食品，促进人类健康。

土壤是环境的要素之一，它介于生物界与非生物界之间，是一切生物赖以生存的基础。随着现代工农业生产的发展，化肥、农药的大量施用，工矿废水不断侵袭农田，城市工业废物和其他人工合成物质不断进入土壤，导致严重的污染事故不断发生。污染物进入土壤后，可能对地表水、地下水等造成次生污染，还会影响植物生长发育及土壤内部生物群的变化与物质的转化；污染物可通过土壤植物系统，经由食物链进入人体，危害人类健康。因此，防治土壤污染是环境科学的重要研究课题之一，而了解污染物在土壤中的存在及迁移转化，是采取防治措施的重要依据。

第一节　土壤形成

一、土壤形成的实质

土壤形成的实质是地质大循环和生物小循环的矛盾统一。

1. 地质大循环

植物营养物质由大陆流到海洋，海洋又变为大陆后，这些物质又由新的大陆流向新的海洋。营养物质的这种循环过程称为植物营养物质的地质大循环。

风化作用下大块岩石不断崩解，由大变小，由粗变细，最后成为疏松多孔的散碎体。在此过程中，还产生了一些溶解于水的矿物质（如硅酸盐内含有的一些钾、钠、钙、镁、磷、硫等营养元素）。这些物质被降水不断地淋洗，并随着地表径流从高处流向低处，经过河流最后流到海洋中去。流入海洋后的这些物质，与流入海洋的岩石和泥沙等在浓缩、沉淀、堆积的过程中，经胶结和硬化的成岩作用，形成沉积岩。在地壳上升的运动中，沉积岩由海洋的底部上升形成大陆；沉积岩暴露于地表之后，又重新进行风化和淋溶，重复地进行着这种作用。

2. 生物小循环

通过植物（包括所有参与这一过程的生物）的反复吸收利用和积累营养物质的这一过程，称为营养物质的生物小循环。

由于风化作用的结果，产生了母质。母质具有松散性、多孔性、透气性、透水性和保水性。这就给植物的生长提供了水分、空气、养分等条件，也就是提供了植物在母质上生长的可能性。而最初在原始幼年土壤上生活的，只是一些需要较少养分的低等植物，它们从原始幼年土中吸收矿质养分、水分以及获取其他生活物质和条件，来建造自身的有机体。这样就使得地质大循环过程中的一些可溶性养分得到了保存。当这些植物死亡之后，经微生物的分解作用，有机残体中的营养元素又变成无机物质，一部分又重新进入地质大循环的过程中，另一部分可被植物重新吸收利用。

3. 地质大循环与生物小循环的矛盾统一

土壤形成过程是建立在地质大循环（营养元素的释放和淋溶过程）与生物小循环（营养元素被生物吸收积累和释放过程）的基础上的。

生物小循环是构成地质大循环中地表物质运动的一部分。地质大循环使营养元素不断向下淋失，而生物小循环却从地质大循环中不断地积累生物所必需的营养元素。按照生物界的发展过程，原始生物对营养元素的要求是种类少、数量低，所以积累在土壤中的营养物质也少。随生物种类的进化，通过生物的选择性吸收，土壤中的养分积累也越来越多。生物对土壤的影响是逐渐扩大和加深的。

在地质大循环和生物小循环这一对矛盾相互作用、不断发展的过程中，土壤肥力也得了到不断发展。

二、土壤形成的自然因素

自然成土因素：影响土壤形成过程的因素，在没有受人类经济活动影响的成土过程中的，称为自然成土因素。属于自然成土因素的有：生物、气候、母质（或岩石）、地形和时间。

$$S=B+C+P+Tg+Ti(+M)$$

1. 母质

母质是岩石风化的产物，是自然土壤形成的物质基础。它对土壤的物理性质和化学性质的影响极为明显。如花岗岩中的长石、云母易风化，并富含钾素；而石英则不易风化，经常呈砂粒残留在土壤中。因此在花岗岩母质上发育的土壤，往往砂粒比例适中。由于页岩是由富含黏土的物质经过硬化后形成的，所以在其风化产物上形成的土壤，质地较黏重。石英砂岩主要是由石英颗粒组成的。因此在其风化产物上形成的土壤，往往砂性较强，养分较少，并含有较多的石砾。

2. 生物

植物着生于母质后，就开始了土壤的形成作用。植物和其死体所产生的物理作用和化学作用，不断地改善着土壤的肥力状况。其中主要的是高等绿色植物通过选择吸收养分，合成有机质并在死亡后积累在土壤中。土壤微生物及小动物分解有机质，释放出养分，同时还合成稳定的腐殖质物质。这样，一方面增加了有机质，另一方面也改造了土壤的物理性质，形成各种土壤结构。由于生物作用的结果，使土壤的肥力状况不断地得到发展。

3. 气候

气候对土壤形成的作用十分复杂。它直接影响在土壤形成过程中起重要作用的水分和热量条件，同时在很大程度上决定着各种植被类型的分布，从而影响土壤矿物和土壤有机质的分解和合成。

温度直接影响着土壤形成过程的强度和方向。在寒冷地带，土壤中的化学作用比较弱，植物生长也较缓慢，有机质形成量小，土壤微生物活动不旺盛，因而土壤中养分的转化也很缓慢。反之，在热带地区，土壤中的矿物质除石英外大部分被分解，植物生长迅速，有机质形成量大，微生物活动旺盛，生物小循环较寒冷地区快。

降水量对土壤形成的影响也极为显著，在干旱气候条件下，盐类不断积累，使土壤发生盐渍化现象。在潮湿气候条件下，盐基离子遭到降水不断地淋洗，使土壤胶体呈不饱和状态。上述这些现象在我国不同类型土壤的形成过程中，都强烈地表现出来。

4. 地形

地形在土壤形成过程中所起的作用是多方面的。首先地形能影响热量的重新分配，不同坡度和方位的斜坡，接受太阳的热量不同。南坡最多，土温高；北坡则相反，土温低。在不同方位的坡向上，由于温度和湿度的差异，植物的分布也是不同的，因而在某些地区，土壤类型在不同的坡向上的分布也会有所不同。

地形还能影响土壤水分、养分和机械组成的分配状况。在分水岭和斜坡地区，水分及其夹带的养分（包括盐类），以及土壤细粒，经常以地表径流或土壤径流的方式向下坡及低地移动，引起坡地和分水岭的土壤不同。坡地上部的土壤经常保持良好的排水状况，土层较薄，质地较粗，养分（及盐基）较少。下部及低平地区，因水分集中，土壤含水量较大。

地形的影响还能通过海拔绝对高度的变化表现出来。随海拔的增加，气候变冷湿，土壤的水热条件和植被都因此而发生相应的改变，所以山区土壤的分布和海拔高度的变化有很密切的关系。

5. 时间

土壤的形成过程随着时间的进展不断加深。任何一个成土因素对土壤的影响，随时间而加深。土壤形成过程的程度是以时间为转移的，随时间持续不同，土壤中物质的淋溶与聚积的程度也不同，因此土壤形成过程必然受到当地地质年龄的影响。在其他土壤形成条件相同的情况下，发育年龄不同的土壤，其肥力状况是不同的。如菜园土因培育的时间长短不同，其熟化程度也不同。

6. 五个因素的关系

五个成土因素中的每一个因素，不仅可以直接影响土壤的形成过程，而且可以间接影响其他因素而起作用。如气候因素，一方面它对土壤形成过程有直接影响，另一方面又影响着植被的组成、有机质的数量以及微生物的分解等方面。植被也能影响近地表的大气层，造成特殊的小气候，并通过这种气候影响土壤的水热状况。

三、土壤形成的人为作用

人类的生产活动直接影响土壤的肥力状况，而且也会对自然成土因素有所影响。长期以来，不少自然土壤早为人类开垦利用，为提高土壤生产力，人类积极地控制自然因素，使之向对生产有利的方向发展。如通过精耕细作、合理施肥、灌溉排水等各种土壤改良措施，来改善土壤的肥力性状。

第二节　土壤的组成

土壤是陆地地表具有肥力并能生长植物的疏松层，它是在地球地面岩石风化过程和母质成土过程的综合作用下形成的。具有肥力是土壤异于其他物质最本质的特征。土壤仅是岩石圈上薄薄的一层，大约 2m 左右，它能提供植物生长所必需的物质和能量。

土壤由固、液、气三相物质组成。固相包括土壤矿物质和有机质，占土壤总质量的

90%～95%，占体积的50%左右。液相指土壤水分及所含的可溶物，也称土壤溶液，占土壤体积的20%～30%。气相指土壤空气，占土壤体积的20%～30%。土壤中还有数量众多的细菌和微生物。因此，土壤是一个以固相为主的不均质多相体系，三相物质互相联系、制约，构成一个有机整体，如图4-1所示。土壤中与土壤污染化学行为关系密切的组分主要是矿物质、有机质和微生物。

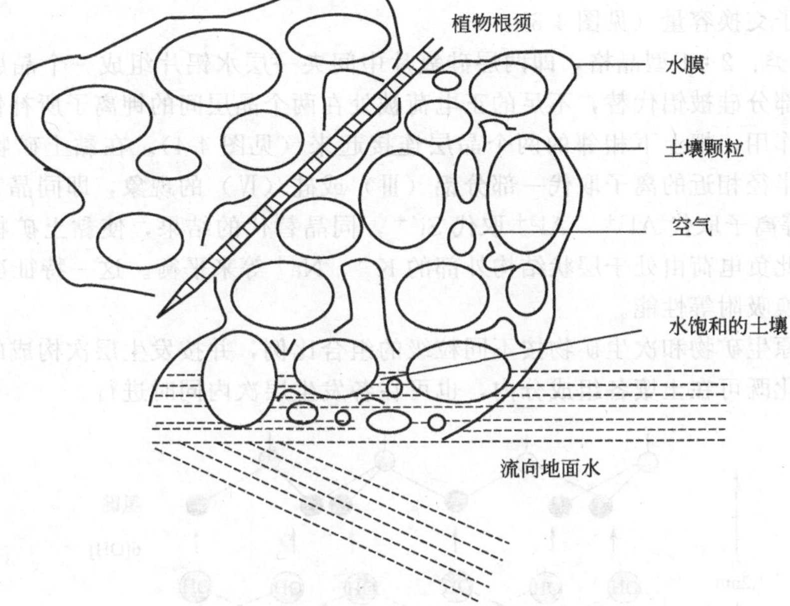

图4-1 土壤中固相、液相、气相结构示意图

一、土壤矿物质

土壤矿物质按其成因可分为原生矿物和次生矿物。原生矿物是指那些在风化过程中未改变化学组成的原始成岩矿物，主要有石英、长石类、角闪石类、云母类等。原生矿物的粒径较大，如砂粒的粒径为0.02～2mm，粉砂粒为0.002～0.02mm。它具有坚实而稳定的晶格，不透水，不具有物理化学吸收性能，不膨胀。

次生矿物是岩石经历化学风化形成的新矿物，其粒径较小，大部分以黏粒和胶体（粒径小于0.002mm）分散状态存在。许多次生矿物具有活动的晶格、强的吸附和离子交换能力，吸水后膨胀，有明显的胶体特征。次生矿物是构成土壤的最主要的组成部分，对土壤中无机污染物的行为和归宿影响很大。

次生矿物有晶态和非晶态之分。非晶态次生矿物主要呈胶膜状态，它裹于土粒表面，如水合氧化铁、铝及硅等；也有呈粒状凝胶成为极细的土粒，如水铝类石等；后者是一种无固定组成的硅铝氧化物，并有较高的阳离子和阴离子代换量，特别是无定形氧化物具有巨大的比表面积和较高的化学活性。晶态次生矿物主要是铝硅酸盐类黏土矿物，如高岭石、蒙脱石、伊利石、蛭石等；它们由硅氧四面体（一个硅原子与四个氧原子组成，形成一个三角锥形的晶格单元）和铝氧四面体（一个铝原子与六个氧原子或氢氧原子组成，形成具有八个面的晶格单元）的层片组成。

黏土矿物通常分为1∶1和2∶1两种类型。根据构成晶层时硅氧四面体（硅氧片）与铝氧八面体（水铝片）的数目和排列方式不同，黏土矿物可分为三大类。

① 高岭石类：由一硅氧片与一水铝片组成一个晶层，属 1∶1 型二层黏土矿物（见图 4-2）。晶层的一面是氧原子，另一面是氢氧原子组，晶层之间通过氢键相连接。晶层间的距离很小，仅 0.72nm，故内部空隙不大，水分子和其他离子都难以进入层间。

② 蒙脱石类：由两层硅氧片中间夹一层水铝片组成一个晶层，属于 2∶1 型三层黏土矿物。晶层表面都是氧原子，没有氢氧原子组，晶层间没有氢键结合力，只有松弛的联系；晶层间的距离为 0.96～2.14nm。水分子或其他交换性阳离子可以进入层间。因此，蒙脱石具有较高的阳离子交换容量（见图 4-3）。

③ 伊利石类：2∶1 型晶格，即两层硅氧片中间夹一层水铝片组成一个晶层。但伊利石类晶格中有一部分硅被铝代替，不足的正电荷被处在两个晶层间的钾离子所补偿；这些钾离子似乎起桥梁作用，把上下相邻的两个晶层连接起来（见图 4-4）。在黏土矿物的形成过程中，常常发生半径相近的离子取代一部分铝（Ⅲ）或硅（Ⅳ）的现象，即同晶替代作用。如 Mg^{2+}、Fe^{3+} 等离子取代 Al^{3+}，Al^{3+} 取代 Si^{4+}，同晶替代的结果，使黏土矿物微粒具有过剩的负电荷。此负电荷由处于层状结构外部的 K^+、Na^+ 等来平衡。这一特征决定了黏土矿物具有离子交换吸附等性能。

土壤是由原生矿物和次生矿物按不同粒级的组合比例，并按发生层次构成的。土体内物质的迁移、转化既可在土壤各组成分内，也可在各发生层次内同时进行。

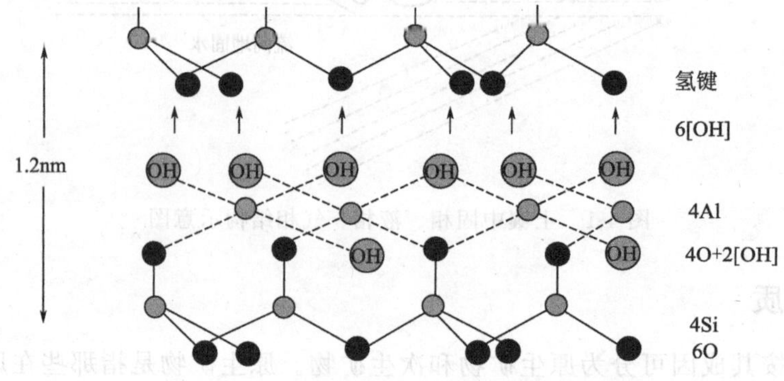

图 4-2　1∶1 型黏土矿物（高岭石）结构示意图

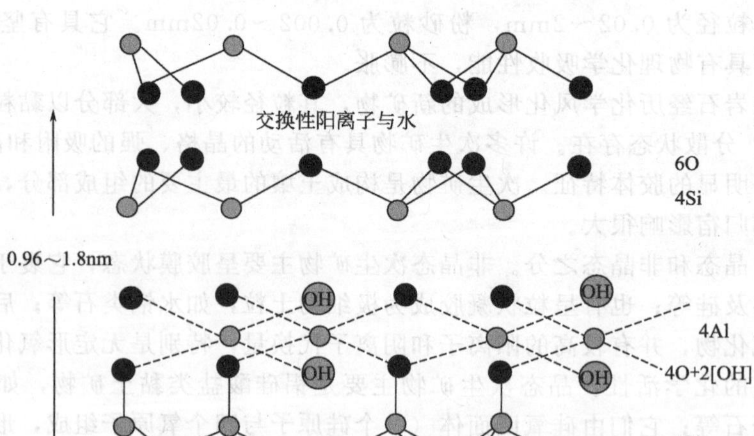

图 4-3　2∶1 型黏土矿物（蒙脱石）结构示意图

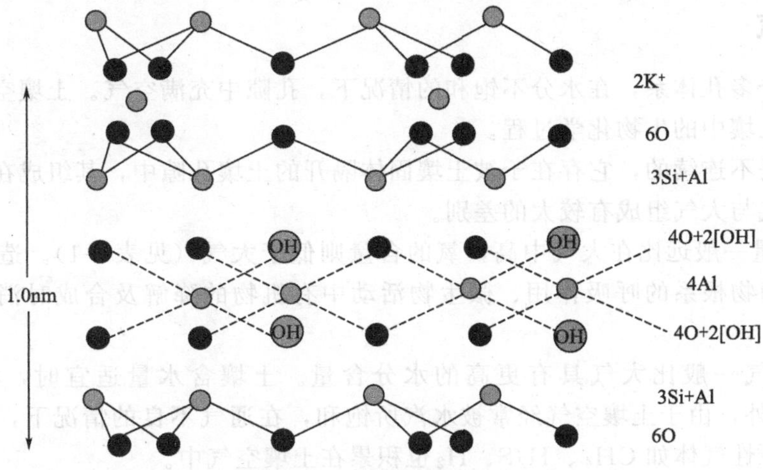

图 4-4　伊利石结构示意图

二、土壤有机质

土壤有机质包括生物死亡的残骸、施用的有机肥料、微生物活动生成的有机物等。土壤有机质的化学组分有腐殖质、糖类、木质素、有机氮、脂肪、蜡质、有机磷等，其中腐殖质占有机质总量的70%～90%。和土壤矿物质相比，有机质含量不高，只占土壤总量的5%左右。但有机质对土壤的一系列物理化学性质有很大的影响，对土壤肥力有重大作用。

土壤有机质不但含有丰富的营养元素，而且在自身缓慢的分解过程中，把生成的CO_2释放到空气中，成为光合作用的物质来源；与此同时，产生的有机酸可以促进矿物养分的溶出，为作物生长提供丰富的养分。土壤有机质，尤其是胡敏酸具有芳香族多元酚官能团，能增强植物呼吸，提高细胞膜的渗透性，促进根系的生长。有机质中的维生素、生长素、抗生素等对植物起促生长、抗病害的作用。有机质还能促进土壤良好结构的形成，增加土壤疏松性、通气性、透水性和保水性。腐殖质有巨大的比表面积，可强烈吸附土壤中的可溶性养分，保持土壤肥力；具有两性胶体性质的有机物可缓冲土壤溶液的pH值。有机物可作为土壤微生物的营养物，而微生物活动又增加土壤养分，促进作物生长。

土壤有机质和微生物是土壤中最活跃的组成部分。有机质的合成与分解、微生物的代谢和转化活动不仅具有肥力意义，从环境角度看，腐殖质对土壤中有机、无机污染物的吸附、络合或螯合作用，微生物对有机污染物的代谢、降解活动等具有重要意义。

三、土壤溶液

土壤溶液占土壤总体积的20%～30%，含有Na^+、K^+、Mg^{2+}、Ca^{2+}、Cl^-、NO_3^-、SO^{2+}、HCO_3^-等无机离子，还含有机物。土壤水分是土壤三相（固、液、气）中的要素。它把土壤、大气中的植物养分溶解成营养溶液，输送到植物根部，最大限度地提供给植物体。因此，土壤水分是植物吸收养料的主要媒介。

土壤水分主要来源于降雨、雪和灌溉。在地下水位接近于地面（2～3m）的情况下，地下水也是上层土壤水分的重要来源。土粒表面的吸附力和微细孔隙的毛细管力可把进入土壤的水分保持住。土壤固体保持水分的牢固程度，在很大程度上决定了土壤中水分的运动和植物对水分的利用。当水分进入土壤后，即和其他组成物质发生作用，其中的一些可溶性物质如盐类和空气将溶解在水里。这种溶有盐类和空气的土壤水即为土壤溶液。

四、土壤空气

土壤是一个多孔体系，在水分不饱和的情况下，孔隙中充满空气。土壤空气主要来自大气，其次来自土壤中的生物化学过程。

土壤空气是不连续的，它存在于被土壤固体隔开的土壤孔隙中，其组成在不同处是有差异的。土壤空气与大气组成有较大的差别。

① CO_2含量一般远比在大气中高，氧的含量则低于大气（见表4-1)。造成这种差别的原因是土壤中植物根系的呼吸作用、微生物活动中有机物的降解及合成时消耗其中的O_2，放出CO_2。

② 土壤空气一般比大气具有更高的水分含量。土壤含水量适宜时，相对湿度接近100%。除此之外，由于土壤空气经常被水汽所饱和，在通气不良的情况下，厌氧细菌活动产生的少量还原性气体如CH_4、H_2S、H_2也积累在土壤空气中。

表4-1 土壤空气与大气组成 单位：%

气体	氧	二氧化碳	氮
土壤空气	18.00～20.03	0.15～0.65	78.8～80.24
近地大气	20.95	0.03	78.09

土壤空气的含量和组成在很大程度上取决于土水关系。作为气体混合物的土壤空气，只进入未被水分占据的那些土壤孔隙。细孔隙比例大的土壤，往往通气条件较差。在这类土壤中，水分占优势，土壤空气的含量和组成不适于植物的最佳生长。在土壤孔隙里储存的水分和空气，它们的相对含量经常随自然条件的改变而变化。

第三节 土壤的基本性质

土壤与植物的生命活动紧密相连，具有储存、转化太阳能和生物能的功能。土壤是农业生态系统中连接生物与非生物界、有机界与无机界的重要枢纽。土壤也是组成环境的各个部分相互作用的地方。土壤是一个复杂的物质体系，其中生存着大大小小有生命的有机体，还有各种无机和有机物，在这些物质的各相界面上进行着多种多样物理的、化学的、生物化学的变化。

环境中的有机、无机污染物可以通过各种途径进入土壤-植物系统。土壤本身是一个活的过滤器，对污染物产生过滤、稀释等物理效应，土壤的吸附作用对重金属等污染物的迁移转化有较大的影响，土壤微生物和植物生命活动产生的化学、生物化学反应对污染物也有显著的净化、代谢作用。土壤的性质是多方面的，下面简单介绍与污染物迁移转化有关的一些基本性质。

一、土壤的物理性质

土壤为固、液、气三相共存的多相体系。土壤固体的分散程度越高，土粒直径越小，其总表面积就越大。比表面积大的物质具有较高的表面能，表现出胶结、吸附及其他不同的物理及物理化学性质。例如，细颗粒具有特别大的胶结性和吸附能力。一般土壤内部空间并没有全部被填满，它总是按一定方式排列，其间有许多孔隙。土壤孔隙系统包括大量形状及大小各异的粒间孔隙，它们之间被比孔隙本身直径还小的通道互相联结。土壤孔隙可分为大孔隙（非毛管孔隙）和小孔隙（毛管孔隙）。土壤团聚体间的孔隙以非毛管孔隙为主，而团聚体内部的孔隙以毛管孔隙为主。毛管孔隙能吸持水分，不易通气、透水，非毛管孔隙不能吸持水分，易通气、透水。

因土壤疏松多孔，一些污染物挥发或呈气体状态，另一些污染物溶解于水中或被吸附于固体颗粒上。它们在土壤孔隙中随空气和水分的运动而挥发、扩散、稀释和浓集，以致迁出土体。这一过程与土壤温度、含水量及土壤孔隙的大小、数量和分布情况有关。

二、土壤的胶体性质

土壤胶体可分为无机胶体、有机胶体及有机无机复合胶体。

1. 有机胶体

土壤有机胶体主要是腐殖质。腐殖质胶体是非晶态的无定形物质，有巨大的比表面积，其范围为 $350\sim900m^2/g$；由于胶体表面羧基或酚羟基中 H^+ 的解离，使腐殖质带负电荷，其负电量平均为 200mmol/100g，高于层状硅酸盐胶体，其阳离子交换量可达 150～300mmol/100g，甚至可高达 400～900mmol/100g。

腐殖质可分为胡敏酸、富里酸和胡敏素等。胡敏素与土壤矿物质结合紧密，一般认为它对土壤吸附性能的影响不明显。胡敏酸和富里酸是含氮羟酸，它是土壤胶体吸附过程中最活跃的分散性物质，特点是功能团多，带负电量大，故其阳离子吸附量均很高。

胡敏酸和富里酸的主要区别是后者的移动性强、酸度高，有大量的含氧功能团。此外，富里酸有较大的阳离子交换容量。在相同条件下，富里酸和胡敏酸的阳离子交换量分别为 200～670mmol/100g 和 180～500mmol/100g。因此，富里酸对重金属等阳离子有很高的螯合和吸附能力，其螯合物一般是水溶性的。富里酸吸附重金属离子以后呈溶胶状态，易随土壤溶液运动，可被植物吸收，也可流出土体，进入其他环境介质中。胡敏酸除与一价金属离子（如 K^+、Na^+）形成易溶物外，与其他金属离子均形成难溶的絮凝态物质，使土壤保持有机碳和营养元素，同时也吸持了有毒的重金属离子，缓解其对植物的毒害。由此可见，胡敏酸含量高的腐殖质可大大提高土壤对重金属的容纳量。在研究土壤环境容量时，应考虑腐殖质中胡敏酸和富里酸（H/F）的相对比例。

2. 无机胶体

无机胶体包括次生黏土矿物、铁铝水合氧化物、含水氧化硅两性胶体。次生黏土矿物主要有蒙脱石、伊利石、高岭石，均是粒径小于 5nm 的层状铝硅酸盐，对土壤中分子态、离子态污染物有很强的吸附能力。其原因是：①黏土矿物颗粒微细、具有很大的表面积，其中以蒙脱石类表面积最大（$600\sim800m^2/g$），它不仅有外表面，而且有巨大的内表面，伊利石次之（$100\sim200m^2/g$），高岭石最小（$7\sim30m^2/g$），巨大的表面积伴随产生巨大的表面能，因此能够吸附进入土壤中的气、液态污染物；②黏土矿物带负电荷，阳离子交换量大，对土壤中离子态污染物有较强的交换固定能力，蒙脱石和高岭石的阳离子交换容量分别为 80～120mmol/100g 和 3～15mmol/100g。负电荷部分来源于晶层间同晶代换作用，部分来源于胶体等电点时晶格表面羟基解离出 H^+ 后产生的可变负电荷。据研究，蒙脱石类永久负电荷占总负电荷量的 95%，伊利石占 60%，高岭石占 25%。高岭石吸附的金属阳离子位于晶格表面离子交换点上，易被解吸；而蒙脱石、伊利石吸附的盐基离子部分位于晶格内部，不易解吸。

3. 有机无机复合胶体

它是无机胶体和有机胶体结合而成的一种胶体，其性质介于上述两种胶体之间。土壤胶体大多是有机-无机复合胶体。

如上所述，土壤胶体有很大的比表面积，并带有电荷，因此具有从土壤溶液中吸附和交换离子的特殊能力。土壤胶体对金属离子的吸附作用主要有表面分子吸附和阳离子交换吸附，其吸附交换量与胶体的比表面积及所带电荷量的大小有关。土壤胶体对有机污染物的吸附作用主要有表面吸附作用和分配作用。因此，土壤对污染物的吸附作用与其种类、组成、交换吸附容量、污染物本身的性质及介质的 pH 值有关。

分子表面吸附是指细小的胶体颗粒具有巨大的表面积，处于表面的分子常因受力不均匀产生剩余力，此力产生的能量称为自由能或表面能。自然界中的物体均有降低其自由能，保持其分散系统稳定性的趋势，一般情况下它通过吸附分子态物质，消耗自由能来达到。这是土壤胶体对分子态物质或螯合物吸附的重要机理之一。

阳离子交换吸附是指土壤胶体对离子态物质的吸附和保持作用，实际是胶体分散系统中扩散层的阳离子与土壤溶液中的阳离子相互交换达到平衡的过程。它有以下三个特点。

① 阳离子吸附过程是一种可逆反应的动态平衡。进入土壤的金属离子浓度愈高、价态愈高，则愈易被胶体吸附。当进入胶体表面的重金属离子过量时，土壤胶体吸附能力减弱，重金属有可能被解吸出来。

② 阳离子交换量是等量进行的。

③ 各种阳离子被胶体物质吸附的亲和力大小各不相同。对同价离子，离子半径愈小，愈易被吸附，胶体吸附金属离子的能力还常常受到土壤环境条件的影响。

不同胶体物质所能交换的阳离子量不同。当土壤溶液为中性时，吸附阳离子的最大量为该土壤的阳离子交换量（CEC），又称土壤阳离子最大吸附容量，单位为 mmol/100g 土。我国土壤的阳离子交换量（CEC）从南到北，依次由低到高：东北黑土为 24.44～34.34；华北褐土约为 16.40；长江流域黄褐土约为 13.23；南方红黄壤仅为 4.77 和 4.09。Levi-Minzi 等的试验表明，决定土壤重金属吸附量的因素首先是土壤交换量，其次是腐殖质含量，而黏土的作用不明显。不同黏土矿物对金属离子的吸附亲和力顺序是不一样的。例如：

蒙脱石对二价金属离子的吸附顺序为：$Ca^{2+}>Pb^{2+}>Cu^{2+}>Mg^{2+}>Cd^{2+}>Zn^{2+}$；

高岭石为：$Pb^{2+}>Ca^{2+}>Cu^{2+}>Mg^{2+}>Zn^{2+}>Cd^{2+}$；

伊利石为：$Pb^{2+}>Cu^{2+}>Zn^{2+}>Ca^{2+}>Cd^{2+}$。

土壤有机质对重金属离子等的吸附作用和螯合作用可同时发生。当重金属离子浓度高时以交换吸附为主，低浓度时以螯合作用为主。Hasler 指出土壤有机质结合金属离子能力的顺序为：$Pb>Cu>Ni>Co>Zn>Mn>Mg>Ba>Ca>Hg>Cd$。Jonasson 认为胡敏酸、富里酸吸附金属的顺序为 $Hg^{2+}>Cu^{2+}>Pb^{2+}>Zn^{2+}>Ni^{2+}>Co^{2+}$。土壤有机质一般只占固相部分的 5%，其负电量平均占土壤总负电量的 21%（5%～42%）；所以有机胶体对金属离子的吸附总贡献小于无机胶体，但土壤有机胶体对污染物，特别是有机污染物的迁移转化及生物效应有重要的影响。

土壤胶体对重金属及农药的吸附，对于控制它们在土壤-植物系统中的迁移起着重要作用。如土壤中重金属元素的活性在很大程度上取决于土壤的吸附作用。土壤中的黏土矿物和腐殖质对重金属有很强的吸附能力，能降低重金属的活性。

三、土壤的配合和螯合作用

土壤中的有机、无机配体能与金属离子发生配合或螯合作用，从而影响金属离子迁移转化等行为。

土壤中的有机配体主要是腐殖质，其表面含有多种含氧、含氮等多配位原子的官能团，能与金属离子形成配合或螯合物。不同配位基与金属离子亲和力的大小顺序为：$-NH_2 >$ $-OH > -COOH > C{=}O$。腐殖质官能团中羧基和酚羟基分别约占功能团的50%和30%，成为腐殖质-金属配合物的主要配位基团。我国土壤中，胡敏酸的羟基含量（340～480mmol/100g）低于富里酸（700～800mmol/100g），前者羧基与羟基的比值为0.7～1.9，后者为2.7～5.6；富里酸的配合、还原能力高于胡敏酸。

土壤中常见的无机配体有Cl^-、SO_4^{2-}、HCO_3^-、OH^-等。它们与金属离子生成各种配合物，如$Cu(OH)^+$、$Cu(OH)O_2$、$CuCl^+$、$CuCl_3^-$等。金属配合物或螯合物的稳定性与配位体或螯合剂、金属离子种类及环境条件有关。土壤有机质对金属离子的配合或螯合能力的顺序为：$Pb^{2+}>Cu^{2+}>Ni^{2+}>Zn^{2+}>Hg^{2+}>Cd^{2+}$。土壤介质pH值对螯合物的稳定性有较大的影响。pH值低时，H^+与金属离子争夺螯合剂，螯合物的稳定性较差；pH值较高时，金属离子可形成氢氧化物、磷酸盐或碳酸盐等不溶性无机化合物。

配合或螯合作用对金属离子迁移的影响取决于所形成螯合物的可溶性。若形成的螯合物易溶于水，则有利于金属的迁移；反之，有利于金属在土壤中滞留，降低其活性。一般来说，胡敏酸与金属离子形成的螯合物是难溶的，富里酸与金属离子形成的螯合物是易溶的。重金属与腐殖质生成可溶性稳定螯合物，能够有效阻止重金属作为难溶盐而沉淀。例如，腐殖质与Fe、Al、Ti、V等形成的螯合物易溶于中性、弱酸或弱碱性土壤溶液中，使之以螯合物的形式迁移，当缺乏腐殖质时它们便析出沉淀。

重金属离子与无机、有机配体的配合反应以及影响因素，在水污染化学中已作了详细论述，土壤中的情况与之类似。例如，在土壤表层的溶液中，汞主要以$Hg(OH)O_2$、$HgClO_2$的形式存在；在氯离子浓度高的盐碱土中，则以$HgCl_4^{2-}$为主。Cd^{2+}、Zn^{2+}、Pb^{2+}则可生成$MClO_2$、MCl_3^-、MCl_4^-配离子。盐碱土的pH值高，重金属可发生水解，生成羟基配离子，此时即发生羟基配合作用与氯配合作用相竞争，或形成各种复杂的配离子，如HgOHCl、CdOHCl等。重金属与羟基及氯离子的配合作用，可提高难溶重金属化合物的溶解度，并减弱土壤胶体对重金属的吸附，从而影响土壤中重金属的迁移转化。

四、土壤的氧化还原性质

土壤中存在着多种有机、无机氧化还原物质，并始终进行着氧化还原反应。土壤中氧化还原作用的强弱同样可用氧化还原电位（E）表示。土壤中的氧化剂主要有游离氧、高价金属离子，还原剂有低价金属离子、土壤有机质及在厌氧条件下的分解产物。土壤中氧化还原反应除了纯化学反应，还有生物反应。如果土壤中游离氧占优势，则以氧化作用为主，如果以有机物为主，则以还原作用为主。土壤E受通气状况及人为措施的影响。旱地E较高，以氧化作用为主，在土壤深处，E较低；水田的E可降至负值，以还原作用为主。例如，土壤中Hg^{2+}可被有机质、微生物等还原为Hg^+、再发生歧化反应则还原为HgO。旱地土壤的E为8.5～11.8V，Hg以$HgCl_2$和$Hg(OH)_2$形态存在；水田土壤E为−5.1V，产生的H_2S与Hg^{2+}形成HgS沉淀。此外，土壤E还与pH有关；pH降低，E增高。

土壤E可影响有机物和无机物的存在形态，从而影响它们在土壤中的迁移转化和对作物的毒害程度，这对那些变价元素尤为重要。例如，在E很低的还原条件下，Cd^{2+}形成难溶的CdS沉淀，难以被植物吸收，毒性降低。但水田落干后，CdS被氧化为可溶性$CdSO_4$，易被植物吸收，若Cd^{2+}含量较高，则会影响植物生长。因此，可通过调节土壤的E及pH

值，降低污染物的毒性。

五、土壤的酸碱性

土壤的酸碱性取决于土壤溶液中 H^+ 的浓度，而 H^+ 主要来自土壤中交换性铝的水解：

$$Al^{3+} + H_2O \longrightarrow Al(OH)^{2+} + H^+$$

$$Al(OH)^{2+} + H_2O \longrightarrow Al(OH)_2^+ + H^+$$

$$Al(OH)_2^+ + H_2O \longrightarrow Al(OH)_3 + H^+$$

土壤中的 CO_2 溶解于水生成碳酸也能离解出 H^+。土壤中有机酸、无机酸和其他盐碱物质的含量及吸附在胶体表面上正离子的种类对土壤酸碱度均有影响。后者是由胶体吸附的 H^+（或 Al^{3+}）在被土壤盐溶液中的阳离子所代换时才表现出来的，所以又称代换性酸或潜在酸。例如，将土壤与中性盐溶液如 KCl 相互作用，则 H^+ 被代换入溶液，等量的 K^+ 则被胶体吸附。

$$胶体—H + KCl \longrightarrow 胶体—K + HCl$$

从胶体中代换出的 H^+ 愈多，土壤的代换性酸度愈大。

土壤正常的 pH 值在 5～8 之间。酸性土壤的 pH 值可能小于 4，碱性土壤的 pH 值高达 11。一般在湿润地区，淋溶作用强，土壤呈酸性或强酸性，$pH<6.5$。在干旱地区，淋溶作用弱，土壤大多呈碱性或强碱性，$pH>7.5$。在半旱半湿地区，土壤的 pH 值介于两者之间。酸雨能使土壤酸化，这已成为世界性环境问题。

pH 值是土壤的重要指标之一。土壤酸碱度直接或间接影响污染物在土壤中的迁移转化：①影响重金属等离子的溶解度；②影响污染物氧化还原体系的电位；③影响土壤胶体对重金属离子等的吸附，如硅酸胶体吸附金属离子的最佳 pH 值范围为 Co^{2+}，5～7；Cr^{3+}，3.5～7；④土壤酸碱度对土壤中重金属的活性有明显的影响。例如，镉在酸性土壤中溶解度大，对植物的毒性较大；在碱性土壤中则溶解度减小，毒性降低。

六、土壤的生物学性质

土壤中存在着由土壤动物、原生动物和微生物组成的生物群体。土壤微生物是土壤生物的主体，它的种类繁多，数量巨大。特别是在土壤表层中，每克土壤含有以亿和十亿计的细菌、真菌、放线菌和酵母等微生物。它们能产生各种专性酶，因而在土壤有机质的分解转化过程中起主要作用。土壤微生物和其他生物对有机污染物有强的自净能力，即生物降解作用。土壤这种自身更新能力和去毒作用为土壤生态系统的物质循环创造了决定性的有利条件，也给土壤肥力的保持提供了必要的保证。

污染物可被生物吸收并累积在体内，植物根系对污染物的吸收是植物污染的主要途径。

七、土壤的自净作用

土壤的上述性质对污染物的迁移转化有很大的影响。如土壤胶体能吸附各种污染物并降低其活性，微生物对有机污染物有特殊的降解作用，使得土壤具有优越的自身更新能力，而无须借助外力。土壤的这种自身更新能力，称为土壤的自净作用。污染物进入土壤后，其自净过程大致如下：污染物在土壤内经扩散、稀释、挥发等物理过程降低其浓度；经生物和化学降解为无毒或低毒物质，或通过化学沉淀、配合或螯合作用、氧化还原作用转化为不溶性化合物，或是被土壤胶体牢固吸附，难以被植物吸收，而暂时退出生物小循环，脱离食物链或被排至土体之外的大气或水体中。

土壤自净能力与土壤的物质组成和其他特性及污染物的种类、性质有关。不同土壤的自净能力是不同的，同一土壤对不同污染物的净化能力也是有差异的。总的来说，土壤的自净速度比较缓慢，污染物进入土壤后较难净化。

第四节　土壤污染

人类活动产生的污染物进入土壤并积累到一定程度，引起土壤质量恶化的现象即为土壤污染。土壤与水体和大气环境有诸多不同，它在位置上较水体和大气相对稳定，污染物易于集聚，故有人认为土壤是污染物的“汇”。

污染物可通过各种途径进入土壤。若进入土壤的污染物的量在土壤自净能力范围内，仍可维持正常生态循环。土壤污染与净化是两个相互对立又同时存在的过程。如果人类活动产生的污染物进入土壤的数量与速度超过净化速度，造成污染物在土壤中持续累积，表现出不良的生态效应和环境效应，则最终会导致土壤正常功能的失调，土壤质量下降，影响作物的生长发育，作物的产量和质量下降，即发生了土壤污染。土壤污染可从以下两个方面来判别：①地下水是否受到污染；②作物生长是否受到影响。

土壤受到污染后，不仅会影响植物生长，同时会影响土壤内部生物群的变化与物质的转化，即产生不良的生态效应。土壤污染物会随地表径流而进入河、湖，当这种径流中的污染物浓度较高时，会污染地表水。例如，土壤中过多的 N、P，一些有机磷农药和部分有机氯农药、酚和氰的淋溶迁移常造成地表水污染。因此，污染物进入土壤后有可能对地表水、地下水造成次生污染。土壤污染物还可通过土壤-植物系统，经由食物链最终影响人类的健康。如日本的“痛痛病”就是土壤污染间接危害人类健康的一个典型例子。

一、土壤污染源

土壤污染源可分为人为污染源和自然污染源。

1. 人为污染源

土壤污染物主要是工业和城市的废水和固体废物、农药和化肥、牲畜排泄物、生物残体及大气沉降物等。污水灌溉或污泥作为肥料使用，常使土壤受到重金属、无机盐、有机物和病原体的污染。工业及城市固体废弃物任意堆放，引起其中有害物的淋溶、释放，也可导致土壤及地下水的污染。现代农业大量使用农药和化肥，也可造成土壤污染。例如，六六六、DDT 等有机氯杀虫剂能在土壤中长期残留，并在生物体内富集；氮、磷等化学肥料，凡未被植物吸收利用和未被根层土壤吸附固定的养分，都在根层以下积累，或转入地下水，成为潜在的环境污染物。禽畜饲养场的厩肥和屠宰场的废物，其性质近似人的粪尿，利用这些废物作肥料，如果不进行适当处理，其中的寄生虫、病原菌和病毒等可引起土壤和水体污染。大气中的二氧化硫、氮氧化物及颗粒物通过干沉降或湿沉降到达地面，可引起土壤酸化。

2. 自然污染源

在某些矿床或元素及化合物的富集中心周围，由于矿物的自然分解与风化，往往形成自然扩散带，使附近土壤中某些元素的含量超出一般土壤的含量。

土壤污染按性质不同可分为化学污染源、物理污染源和生物污染源，其污染源十分复杂。土壤的化学污染最为普遍、严重和复杂。

二、土壤的主要污染物

土壤污染物种类繁多，总体可分以下几类。

1. 无机污染物

包括对动、植物有危害作用的元素及其无机化合物，如镉、汞、铜、铅、锌、镍、砷等重金属；硝酸盐、硫酸盐、氟化物、可溶性碳酸盐等化合物也是常见的土壤无机污染物；过量使用氮肥或磷肥也会造成土壤污染。

(1) 汞

土壤中汞的背景值为 0.01～0.15μg/g。除来源于母岩以外，汞主要来自污染源，如含汞农药的施用、污水灌溉等，故各地土壤中汞含量差异较大。来自污染源的汞首先进入土壤表层。土壤胶体及有机质对汞的吸附作用相当强，汞在土壤中的移动性较弱，往往积累于表层，而在剖面中呈不均匀分布。土壤中的汞不易随水流失，但易挥发至大气中，许多因素可以影响汞的挥发。土壤中的汞按其化学形态可分为金属汞、无机汞和有机汞，在正常的 E 和 pH 范围内，土壤中汞以零价汞的形式存在。在一定条件下，各种形态的汞可以相互转化。进入土壤的一些无机汞可分解而生成金属汞，当土壤在还原条件下时，有机汞可降解为金属汞。一般情况下，土壤中都能发生 $Hg_2O \longrightarrow Hg + HgO$ 反应，新生成的汞可以挥发。在通气良好的土壤中，汞可以任何形态稳定存在。在厌氧条件下，部分汞可转化为可溶性甲基汞或气态二甲基汞。

阳离子态汞易被土壤吸附，许多汞盐如磷酸汞、碳酸汞和硫化汞的溶解度亦很低。在还原条件下，Hg^{2+} 与 H_2S 生成极难溶的 HgS；金属汞也可被硫酸还原细菌变成硫化汞；所有这些都可阻止汞在土壤中的移动。当氧气充足时，硫化汞又可慢慢氧化成亚硫酸盐和硫酸盐。以阴离子形式存在的汞，如 $HgCl_3^-$、$HgCl_4^{2-}$ 也可被带正电荷的氧化铁、氢氧化铁或黏土矿物的边缘所吸附。分子态的汞，如 $HgCl_2$，也可以被吸附在 Fe、Mn 的氢氧化物上。$Hg(OH)_2$ 溶解度小，可以被土壤强烈地保留。由于汞化合物和土壤组分间强烈的相互作用，除了还原成金属汞以蒸气挥发外，其他形态的汞在土壤中的迁移很缓慢。在土壤中的汞主要以气相的形式在孔隙中扩散。总体而言，汞比其他有毒金属容易迁移。当汞被土壤有机质螯合时，亦会发生一定的水平和垂直移动。

汞是危害植物生长的元素。土壤中含汞量过高，它不但能在植物体内积累，还会对植物产生毒害。通常有机汞和无机汞化合物以及蒸气汞都会引起植物中毒。例如，汞对水稻的生长发育产生危害。中国科学院植物研究所水稻的水培实验表明，采用含汞量为 0.074μg/mL 的培养液处理水稻，产量开始下降，秕谷率增加；以 0.74μg/mL 浓度处理时，水稻根部已开始受害，并随着试验浓度的增加，根部更加扭曲，呈褐色，有锈斑；当介质含汞量为 7.4μg/mL 时，水稻叶子发黄，分蘖受抑制，植株高度变矮，根系发育不良。此外，随着浓度的增加，植物各部分的含汞量上升。介质浓度为 22.2μg/mL 时，水稻严重受害，水培水稻受害的致死浓度为 36.5μg/mL。但是，在作物的土培实验中，即使土壤含汞量达 18.5μg/g，水稻和小麦产量也未受到影响。可见，汞对植物的有效性和环境条件密切相关。不同植物对汞的敏感程度有差别。例如，大豆、向日葵、玫瑰等对汞蒸气特别敏感；纸皮桦、橡树、常青藤、芦苇等对汞蒸气抗性力较强；桃树、西红柿等对汞蒸气的敏感性属中等。

汞进入植物主要有两条途径：一是通过根系吸收土壤中的汞离子，在某些情况下，也可吸收甲基汞或金属汞；其次是喷施叶面的汞剂、飘尘或雨水中的汞以及在日夜温差作用下土壤所释放的汞蒸气，由叶片进入植物体或通过根系吸收。由叶片进入到植物体的汞，可被运转到植株其他各部位，而被植物根系吸收的汞，常与根中蛋白质发生反应而沉积于根上，很少向地上部分转移。

植物吸收汞的数量不仅取决于土壤含汞量，还取决于其有效性。汞对植物的有效性和土壤氧化还原条件、酸碱度、有机质含量等有密切关系。不同植物吸收积累汞的能力是有差异的，同种植物的各器官对汞的吸收能力也不一样。植物对汞的吸收与土壤中汞的存在形态有关。

土壤中不同形态的汞对作物生长发育的影响存在差异。土壤中无机汞和有机汞对水稻生长发育影响的盆栽实验表明，当汞浓度相同时，汞化合物对水稻生长和发育的危害为：醋酸苯汞＞$HgCl_2$＞HgO＞HgS。HgS不易被水稻吸收。即使是同一种汞化合物，当土壤环境条件变化时，可以不同的形态存在，对作物的有效性也就不一样。

（2）镉

地壳中镉的丰度为5μg/g，我国部分地区镉的背景值为0.15～0.20μg/g。

土壤中的镉污染主要来自矿山、冶炼、污灌及污泥的施用。镉还可伴随磷矿渣和过磷酸钙的使用而进入土壤。在风力作用下，工业废气中镉扩散并沉降至土壤中。交通繁忙的路边土壤常发现有镉污染。

土壤中镉一般可分为可给态、代换态和难溶态。可给态镉主要以离子态或络合态的形式存在，易被植物所吸收；被黏土或腐殖质交换吸附的为代换态镉；难溶态镉包括以沉淀或难溶性螯合物形式存在的镉，不易被植物吸收。

土壤中的镉可被胶体吸附。被吸附的镉一般在0～15cm的土壤表层内累积，15cm以下含量显著减少。大多数土壤对镉的吸附率在80%～90%。土壤对镉的吸附同pH值呈正相关；被吸附的镉可被水所溶出而迁移，pH值越低，镉的溶出率越大。如pH=4时，镉的溶出率超过50%；pH=7.5时，镉很难溶出。

土壤中镉的迁移与土壤的种类、性质、pH值等因素有关，还直接受氧化还原条件的影响。水稻田是氧化还原电位很低的特殊土壤，当水田灌满水时，由于水的遮蔽效应形成了还原性环境，有机物厌氧分解产生硫化氢；当施用硫酸铵肥料时，硫还原细菌的作用使硫酸根还原产生大量的硫化氢。在淹水条件下，镉主要以CdS的形式存在，抑制了Cd^{2+}的迁移，难以被植物所吸收。当排水时造成氧化淋溶环境，S^{2-}氧化成SO_4^{2-}，引起pH值降低，镉溶解在土壤中，易被植物吸收。土壤中PO_4^{3-}等离子均能影响镉的迁移转化；如Cd^{2+}和PO_4^{3-}形成难溶的$Cd_3(PO_4)_2$，不易被植物所吸收。因此，土壤的镉污染，可通过施用石灰和磷肥，调节土壤pH值至5.0以上，以抑制镉害。

在旱地土壤里，镉以$CdCO_3$、$Cd_3(PO_4)_2$及$Cd(OH)_2$的形式存在，而其中又以$CdCO_3$为主，尤其是在pH＞7的石灰性土壤中，形成$CdCO_3$的反应为：

$$Cd^{2+}+CO_2+H_2O \rightleftharpoons CdCO_3+2H^+ \quad \lg K=-6.07$$

可导出土壤中Cd^{2+}为：$-\lg[Cd^{2+}]=-6.07+2pH-\lg[CO_2]$

如土壤空气中，CO_2的分压为0.0003atm，则：$-\lg[Cd^{2+}]=2pH-9.57$

可见旱地土壤中Cd^{2+}的浓度与pH值呈负相关。

镉是危害植物生长的有毒元素。镉对作物的危害，在较低浓度时，虽在外观上无明显的症状，但通过食物链可危及人类健康。当土壤镉浓度达到一定含量时，不仅能在植物体内残留，而且也会对植物的生长发育产生明显的危害。水稻盆栽实验表明：土壤含镉量为10μg/g时，对水稻产生不利影响；含镉量为300μg/g时，水稻生长受到显著影响；土壤含镉量为500μg/g

时，严重影响水稻生长发育。镉对植物的生物效应与其在土壤中的存在形态有关。

植物对镉的吸收与累积取决于土壤中镉的含量和形态、镉在土壤中的活性及植物的种类。许多植物均能从土壤中摄取镉，并在体内累积到一定数量。植物吸收镉的量不仅与土壤的含镉量有关，还受其化学形态的影响。例如，水稻对三种无机镉化合物吸收累积的顺序为：$CdCl_2$>$CdSO_4$>CdS。不同种类的植物对镉的吸收存在着明显的差异；同种植物的不同品种之间，对镉的吸收累积也会有较大的差异。谷类作物如小麦、玉米、水稻、燕麦和粟子都可通过根系吸收镉，其吸收量依次是玉米>小麦>水稻>大豆。同一作物，镉在体内各部位的分布也是不均匀的，其含量一般为：根>茎>叶>籽实。植物在不同的生长阶段对镉的吸收量也不一样，其中以生长期吸收量最大。由此可见，影响植物吸收镉的因素很多。

镉可通过土壤-植物系统等途径，经由食物链进入人体，危害人类健康。因此，环境的镉污染是人们极为关注的问题。

（3）铅

地壳中铅的丰度为12.5μg/g，土壤中铅的平均背景值为15～20μg/g。土壤的铅污染主要由汽油燃烧和冶炼烟尘的沉降、降水及矿山、冶炼废水污灌引起。因此，城市和矿山、冶炼厂附近的土壤含铅量比较高。汽车尾气造成的铅污染主要集中在大城市和公路两侧。距公路越近，交通量越大，土壤铅污染越严重。如一公路旁土壤含铅量为809.6μg/g，距公路91m处则含铅量为32.5μg/g。

进入土壤的Pb^{2+}容易被有机质和黏土矿物所吸附。不同土壤对铅的吸附能力如下：黑土（771.6μg/g）>褐土（770.9μg/g）>红壤土（425.0μg/g）；腐殖质对铅的吸附能力明显高于黏土矿物。铅也和配位体形成稳定的金属配合物和螯合物。土壤中铅主要以$Pb(OH)_2$、$PbCO_3$、$PbSO_4$的固体形式存在。而在土壤溶液中可溶性铅的含量很低，故土壤中铅的迁移能力较弱，生物有效性较低。当土壤pH值降低时，部分被吸附的铅可以释放出来，使铅的迁移能力提高，生物有效性增强。

植物对铅的吸收与累积取决于土壤中铅的浓度、土壤条件及植物的种类与部位，还有叶片的大小和形状。铅进入植物体的途径，一是被植物根部吸收，二是被叶面所吸收。被植物吸收和输送到地面上部的铅，取决于植物种类和环境条件，但吸收的铅主要集中在根部。土壤条件不同，植物对铅的吸收也不尽相同；在酸性土壤中，植物对铅的吸收累积大于在碱性土壤中。土壤中其他元素可以与铅发生竞争而被植物吸收。例如，在石灰性土壤中，钙与铅竞争而被植物根系吸收。一般有钙存在时，由于钙与铅的竞争作用，铅被吸收在酶化学结构不重要的位置上，即使植物体内铅的浓度较高，也没有明显的毒性。又如，当土壤中同时存在铅和镉时，镉可能会降低作物中铅的含量，而铅会增加作物体中镉的含量。因此，影响植物体对铅吸收累积的因素是复杂的。

铅不是植物生长发育的必需元素。铅进入植物的过程主要是非代谢性地被动进入植物根内。铅在环境中比较稳定，一定浓度的铅对作物生长不会产生危害。作物受铅的毒害依其对铅的敏感程度而异，通常认为铅对植物是有害的。如大豆对铅的危害比较敏感。土壤中高浓度的铅能抑制水稻生长，主要表现在叶片的叶绿素含量降低，影响光合作用，延缓生长，推迟成熟而导致减产。一般情况下，土壤含铅量增高会引起作物产量下降；在严重污染地区，能使植物的覆盖面大大减小；在另一些情况下，生长在严重污染地区的植物，往往具有耐高浓度铅的能力。作物吸收铅与土壤含铅量之间的关系目前还没有一致的结论。

（4）铬

地壳中铬的丰度为200μg/g，铬的土壤平均背景值为100μg/g。土壤中铬以四种形态存在，即三价铬离子 Cr^{3+}、CrO_2^- 及六价阴离子 CrO_4^{2-} 和 $Cr_2O_7^{2-}$，其中三价铬稳定。土壤中可溶性铬只占总铬量的 0.01%～0.4%。铬的迁移转化与土壤的pH值、氧化还原电位、有机质含量等因素有关。

三价铬进入土壤后，90%以上迅速被土壤吸附固定，以铬和铁氢氧化物的混合物的形式存在或被封闭在铁的氧化物中，故土壤中的三价铬难以迁移。土壤溶液中，三价铬的溶解度取决于pH值。当pH值大于4时，三价铬溶解度降低；当pH=5.5时，全部沉淀；在碱性溶液中形成铬的多羟基化合物。此外，在pH值较低时，铬能形成有机配合物，迁移能力增强。

土壤胶体对三价铬的强烈吸附作用与pH值呈正相关。Cr^{3+} 甚至可以交换黏土矿物晶格中的 Al^{3+}，黏土矿物吸附三价铬的能力约为六价铬的30～300倍。六价铬进入土壤后大部分游离在土壤溶液中，仅有8.5%～36.2%被土壤胶体吸附固定。不同类型的土壤或黏土矿物对六价铬的吸附能力有明显的差异；吸附能力大致如下：红壤＞黄棕壤＞黑土＞黄壤；高岭石＞伊利石＞蛭石＞蒙脱石。土壤中的有机质越多，负电性越强，对六价铬阴离子的吸附力就越弱。

土壤中铬的迁移转化受氧化还原条件的影响较大。在土壤常见的pH值和 E 范围内，Cr(Ⅵ) 可被有机质等迅速还原为 Cr(Ⅲ)。在不同水稻田中，Cr(Ⅵ) 的还原率与有机碳含量呈显著的正相关。当砖红壤中有机碳含量为1.56%或1.33%时，Cr(Ⅵ) 的还原率分别为89.6%和77.2%；一般情况下，土壤中有机碳增加1%，Cr(Ⅵ) 的还原率约增加30%。有机质对 Cr(Ⅵ) 的还原作用与土壤pH值呈负相关。当土壤有机质含量极低时，pH值对Cr(Ⅵ) 的还原率影响更加明显。例如，当土壤pH值为3.35或7.89时，Cr(Ⅵ) 的还原率分别为54%和20%。

当含铬废水进入农田时，其中的 Cr(Ⅲ) 被土壤胶体吸附固定；Cr(Ⅵ) 迅速被有机质还原成 Cr(Ⅲ)，再被土壤胶体吸附；导致铬的迁移能力及生物有效性降低，同时使铬在土壤中积累起来。然而，在一定条件下，Cr(Ⅲ) 可转化为 Cr(Ⅵ)；如pH=6.5～8.5时，土壤中的 Cr(Ⅲ) 能被氧化为 Cr(Ⅵ)，其反应为：

$$4Cr(OH)^{2+} + 3O_2 + 2H_2O \longrightarrow 4CrO_4^{2+} + 12H^+$$

此外，土壤中的氧化锰也能使 Cr(Ⅲ) 转化为 Cr(Ⅵ)。因此，Cr(Ⅲ) 存在着潜在危害。

植物在生长发育过程中，可从外界环境中吸收铬，铬可以通过根和叶进入植物体内。植物体内含铬量随植物种类及土壤类型的不同有很大差别，植物中铬的残留量与土壤含铬量呈正相关。植物从土壤中吸收的铬绝大部分积累在根中，其次是茎叶，籽粒里积累的铬量最少。

微量元素铬是植物所必需的。植物缺少铬就会影响其正常发育，低浓度的铬对植物生长有刺激作用，但植物体内累积过量铬又会引起毒害作用，直接或间接地给人类健康带来危害。例如，土壤中 Cr(Ⅲ) 为20～40μg/g时，对玉米苗生长有明显的刺激作用；当 Cr(Ⅲ) 为320μg/g时，则有抑制作用；又如，土壤中 Cr(Ⅵ) 为20μg/g时，对玉米苗生长有刺激作用；Cr(Ⅵ) 为80μg/g时，则有显著的抑制作用。

高浓度铬不仅对植物产生危害，而且会影响植物对其他营养元素的吸收。例如，当土壤含铬量大于5μg/g时会干扰植株上部对钙、钾、磷、硼、铜的吸收，受害的大豆最终表现为植株顶部严重枯萎。

土壤中的铬对植物的毒性与下列因素有关。

① 铬的化学形态。如 Cr(Ⅵ) 的毒性比 Cr(Ⅲ) 强。

② 土壤性质。土壤胶体对 Cr(Ⅲ) 有强烈的吸附固定作用，在酸性或中性条件下对 Cr(Ⅵ) 也有很强的吸附作用；土壤有机质具有吸附或螯合作用，还能使可溶性 Cr(Ⅵ) 还原成难溶的 Cr(Ⅲ)；因此，土壤黏粒和有机质的含量会影响铬对植物的毒性。

③ 土壤氧化还原电位。如在同一 Cr(Ⅲ) 浓度下，旱地土壤中有效态铬比在水田中高得多。

④ 土壤 pH 值。Cr(Ⅵ) 在中性和碱性土壤中的毒性要比在酸性土壤中大；而 Cr(Ⅲ) 对植物的毒性在酸性土壤中较大。

总的说来，铬对植物生长的抑制作用较弱，其原因是铬在植物体内迁移性很低。水稻栽培试验结果表明，重金属在植物体内的迁移顺序为 Cd>Zn>Ni>Cu>Cr。可见，铬是金属元素中较难被吸收的元素之一，其可能的原因是：①三价铬还原成二价铬再被植物吸收的过程在土壤-植物体系中难以发生；②六价铬是有效性铬，但植物对六价铬的吸收受到硫酸根等阴离子的强烈抑制作用。

(5) 铜

地壳中铜的平均值为 70μg/g。土壤中铜的含量为 2～200μg/g。我国土壤含铜量为 3～300μg/g，大部分土壤含铜量在 15～60μg/g，平均为 20μg/g。

土壤铜污染的主要来源是铜矿山和冶炼厂排出的废水。此外，工业粉尘、城市污水以及含铜农药，都能造成土壤的铜污染。如我国华南某铜矿附近受污染土壤的铜含量为 1730～2630μg/g，为对照土壤的 91～138 倍。日本被铜污染的土地面积约为 30430hm^2，占重金属污染总面积的 80%左右，其中渡良濑川流域土壤平均含铜达 1000μg/g，最高达 2020 μg/g，可溶性铜 250μg/g。

土壤中的铜的存在形态可分为：①可溶性铜，约占土壤总铜量的 1%；主要是可溶性铜盐，如 $Cu(NO_3)_2 \cdot 3H_2O$、$CuCl_2 \cdot 2H_2O$、$CuSO_4 \cdot 5H_2O$ 等；②代换性铜，被土壤有机、无机胶体所吸附，可被其他阳离子代换出来；③非代换性铜，指被有机质紧密吸附的铜和原生矿物、次生矿物中的铜，不能被中性盐所代换；④难溶性铜，大多是不溶于水而溶于酸的盐类，如 CuO、Cu_2O、$Cu(OH)_2$、$Cu(OH)^+$、$CuCO_3$、Cu_2S、$Cu_3(PO_4)_2 \cdot 3H_2O$等。

土壤中的腐殖质能与铜形成螯合物。土壤有机质及黏土矿物对铜离子有很强的吸附作用，吸附强弱与其含量及组成有关。黏土矿物及腐殖质吸附铜离子的强度为：腐殖质>蒙脱石>伊利石>高岭石。我国几种主要土壤对铜的吸附强度为：黑土>褐土>红壤。

土壤 pH 值对铜的迁移及生物效应有较大的影响。游离铜与土壤 pH 值呈负相关；在酸性土壤中，铜易发生迁移，其生物效应较强。

铜是生物必需元素，广泛地分布在一切植物中。在缺铜的土壤中施用铜肥，能显著提高作物产量。例如，硫酸铜是常用的铜肥，可以用作基肥、种肥、追肥，还可用来处理种子。但过量铜会对植物生长发育产生危害。如当土壤含铜量达 200μg/g 时，小麦枯死；当含铜量达 250μg/g 时，水稻也将枯死。又如，用含铜 0.06μg/mL 的溶液灌溉农田，水稻减产 15.7%；浓度增至 0.6μg/mL，减产 45.1%；若铜浓度增至 3.2μg/mL，水稻无收获。研究表明，铜对植物的毒性还受其他元素的影响。在水培液中只要有 1μg/mL 的硫酸铜，即可使大麦停止生长；然而加入其他营养盐类，即使铜浓度达 4μg/mL，也不至于使大麦停止生长。

生长在铜污染的土壤中的植物，其体内会发生铜的累积。植物中铜的累积与土壤中的总铜量无明显的相关性，而与有效态铜的含量密切相关。有效态铜包括可溶性铜和土壤胶体吸附的代换性铜，土壤中有效态铜量受土壤 pH 值、有机质含量等的直接影响。不同植物对铜的吸收累积是有差异的，铜在同种植物不同部位的分布也是不一样的。

(6) 锌

土壤中锌的总含量为 10～300μg/g，平均值为 50μg/g，我国土壤含锌量为 3～70μg/g，平均值为 100μg/g。

用含锌废水污灌时，锌以 Zn^{2+}，也可以络离子 $Zn(OH)^+$、$ZnCl^+$、$Zn(NO_3)^+$ 等形态进入土壤，并被土壤胶体吸附累积；有时则形成氢氧化物、碳酸盐、磷酸盐和硫化物沉淀，或与土壤中的有机质结合。锌主要被富集在土壤表层。

根据 L. M. Shuman 的研究，土壤中各部分的含锌量为：黏土＞氧化铁＞有机质＞粉砂＞砂＞交换态。土壤中大部分锌是以结合状态存在的，或为有机复合物及各种矿物，一般不易被植物吸收。植物只能吸收可溶性或代换态锌。锌的迁移能力及有效性主要取决于土壤的酸碱性，其次是土壤吸附和固定锌的能力。总体而言，土壤中有效态锌的浓度比其他重金属的有效浓度高，有效态锌平均占总锌量的 5%～20%。

土壤中锌的迁移主要取决于 pH 值。当土壤为酸性时，被黏土矿物吸附的锌易解吸，不溶性氢氧化锌可和酸作用，转化为 Zn^{2+}。因此，酸性土壤中锌容易发生迁移。当土壤中以 Zn^{2+} 为主存在时，容易流失迁移或被植物吸收。故缺锌现象常常发生在酸性土壤中。

由于稻田淹水，处于还原状态，硫酸盐还原菌将 SO_4^{2-} 转化为 H_2S，土壤中 Zn^{2+} 与 S^{2-} 形成溶度积小的 ZnS，土壤中的锌发生累积。锌与有机质相互作用，可以形成可溶性的或不溶性的络合物。可见，土壤中的有机质对锌的迁移会产生较大的影响。

锌是植物生长发育不可缺少的元素。常把硫酸锌用作微量元素肥料，但过量的锌会伤害植物的根系，从而影响作物的产量和质量。土壤酸度的增加会加重锌对植物的危害。例如，在中性土壤里加入 100μg/mL 的锌溶液，洋葱生长正常；当加入 500μg/mL 锌时，洋葱茎叶变黄；但在酸性土壤中，加入 100μg/mL 的锌溶液，洋葱生长发育受阻，加入 500μg/mL 锌时，洋葱几乎不生长。

植物对锌的忍耐浓度大于其他元素。各种植物对高浓度锌毒害的敏感性也不同。

一般来说，锌在土壤中富集，必然导致在植物体内累积，植物体内累积的锌与土壤含锌量密切相关。如水稻糙米中锌的含量与土壤的含锌量呈线性相关。土壤中其他元素可影响植物对锌的吸收。如施用过多的磷肥，可使锌形成不溶性磷酸锌而被固定，植物吸收的锌就减少，甚至引起锌缺之症。温度和阳光对植物吸收锌也有影响。不同植物对锌的吸收累积差异很大，一般植物体内自然含锌量为 10～160μg/g，但有些植物对锌的吸收能力很强，植物体内累积的锌可达 0.2～10mg/g。锌在植物体各部位的分布也是不均匀的。如在水稻、小麦中含锌量分布为：根＞茎＞果实。

(7) 砷

地壳中砷的平均含量为 2μg/g，一般土壤含砷量约为 6μg/g，我国部分土壤平均含砷量为 10μg/g 左右。

砷是变价元素。土壤中砷以三价或五价状态存在，其存在形态分为可溶性砷，吸附、代换态砷及难溶态砷。可溶性砷主要为 AsO_4^{3-}、AsO_3^{3-} 等阴离子，一般只占总砷量的 5%～10%。我国土壤中可溶性砷低于 1%，其总量低于 1μg/g。因此，即使以可溶性砷进入土壤，也容易转化为难溶性砷累积于土壤表层里。

土壤中砷的迁移转化与其中铁、铝、钙、镁及磷的含量有关，还和土壤 pH 值、氧化还原电位、微生物的作用有关。

土壤胶体对 AsO_4^{3-} 和 AsO_3^{3-} 有吸附作用。如带正电荷的氢氧化铁、氢氧化铝和铝硅酸盐黏土矿物表面的铝离子都可吸附含砷的阴离子，但有机胶体对砷无明显的吸附作用。不同黏土矿物或不同的阴离子组成对砷的吸附作用有差异。研究表明，用 Fe^{3+} 饱和的黏土矿物对砷的吸附量为 620～1172μg/g；吸附强度为：蒙脱石＞高岭石＞白云石；用 Ca^{2+} 饱和的黏土矿物的吸附量为：75～415μg/g；吸附强度依次为：高岭石＞蒙脱石＞白云石。

砷可以和铁、铝、钙、镁等离子形成难溶的砷化合物，还可以和无定形的铁、铝等氢氧化物产生共沉淀，故砷可被土壤中的铁、铝、钙及镁等固定，使之难以迁移。含砷（Ⅴ）化合物的溶解度为：$Ca_3(AsO_4)_2$＞$Mg_3(AsO_4)_2$＞$AlAsO_4$＞$FeAsO_4$，故 Fe^{3+}固定 AsO_4^{3-} 的能力最强。几种土壤对砷的吸附能力顺序如下：红壤＞砖红壤＞黄棕壤＞黑土＞碱土＞黄土。

土壤中吸附态砷可转化为溶解态的砷化物，这个过程与土壤 pH 值和氧化还原条件有关。如土壤 E 降低，pH 值升高，砷溶解度显著增加。在碱性条件下，土壤胶体的正电荷减少，对砷的吸附能力也就降低，可溶性砷含量增加。由于 AsO_4^{3-} 比 AsO_3^{3-} 容易被土壤吸附固定，如果土壤中砷以 AsO_3^{3-} 状态存在，砷的溶解度相对增加。土壤中 AsO_4^{3-} 与 AsO_3^{3-} 之间的转化取决于氧化还原条件。旱地土壤处于氧化状态，AsO_3^{3-} 可氧化成 AsO_4^{3-}；而水田土壤处于还原状态，大部分砷以 AsO_3^{3-} 形态存在，砷的溶解度及有效性相对增加，砷害也就增加。此外，AsO_3^{3-} 对作物的危害比 AsO_4^{3-} 更大。

土壤微生物也能促进砷的形态变化。有人分离出 15 个系的异养细菌，它们可把 AsO_3^{3-} 氧化为 AsO_4^{3-}。土壤微生物还可起气化逸脱砷的作用。盆栽实验发现，施砷量和水稻吸收砷及土壤残留量之和有一个很大差值，认为由于砷霉菌对砷化合物有气化作用，使这部分砷还原为 AsH_3 等形式，从土壤中气化逸脱。此外，土壤微生物还可使无机砷转化为有机砷化物。

磷化合物和砷化合物的特性相似，因此土壤中磷化合物的存在将影响砷的迁移能力和生物效应。一般土壤吸附磷的能力比砷强，致使磷能夺取土壤中固定砷的位置，砷的可溶性及生物有效性相对增加。Gile 就砷的土壤吸附问题指出，磷可被土壤胶体中的铁、铝所吸附，而砷的吸附主要是铁起作用；另外，铝对磷的亲和力远远超过对砷的亲和力，被铝吸附的砷很容易被磷交换取代。

由此可见，砷与镉、铬等的性质相反；当土壤处于氧化状态时，它的危害比较小；当土壤处于淹水还原状态时，AsO_4^{3-} 还原为 AsO_3^{3-}，加重了砷对植物的危害。因此，在实践中，对砷污染的水稻土，常采取措施提高土壤氧化还原电位或加入某些物质，以减轻砷对作物生长的危害。

一般认为砷不是植物必需的元素。低浓度砷对许多植物生长有刺激作用，高浓度砷则有危害作用。砷中毒可阻碍作物的生长发育。研究表明：土壤含砷量为 25μg/g 或 50μg/g 时，可使小麦分别增产 8.7%和 20%；含砷达 100μg/g 时，则严重影响小麦生长；含砷200～1000μg/g 时，小麦全部死亡。不同砷化物对作物生长发育的影响是有差别的。如有机砷化物易被水稻吸收，其毒性比无机砷大得多，即使是无机砷，AsO_3^{3-} 对作物的危害比 AsO_4^{3-} 大。

作物对砷的吸收累积与土壤含砷量有关，不同植物吸收累积砷的能力有很大的差别，植物的不同部位吸收累积的砷量也是不同的。砷进入植物的途径主要是根、叶吸收。植物的根系可从土壤中吸收砷，然后在植株内迁移运转到各个部分；有机态砷被植物吸收后，可在体内逐渐降解为无机态砷。同重金属一样，砷可以通过土壤-植物系统，经由食物链最终进入人体。

综上所述，土壤重金属污染主要来自废水污灌、污泥的施用及大气降尘；废渣及城市垃圾的任意堆放也可造成土壤重金属污染。土壤中高浓度的重金属会危害植物的生长发育，影响农产品的产量和质量。重金属对植物生长发育的危害程度取决于土壤中重金属的含量，特别是有效态的含量。影响土壤中重金属迁移转化及生物效应的主要因素有：胶体对重金属的吸附，各种无机及有机配体的配合或螯合作用，土壤的氧化还原状态，土壤的酸碱性及共存离子的作用，还有土壤-微生物的作用等。由此可见，影响土壤中重金属迁移转化及生物效应的因素是多方面的。

重金属可通过土壤植物系统及食物链最终进入人体，影响人类健康。重金属不能被微生物所降解，同时由于胶体对重金属离子有强烈的吸附作用等，使其不易迁移。因此，土壤一旦遭受重金属污染，就很难予以彻底消除。可以认为，土壤是重金属污染的“汇”，故应积极防治土壤的重金属污染。

2. 有机污染物

包括化学农药、除草剂、石油类有机物、洗涤剂及酚类等。其中农药是土壤的主要有机物，常用的农药约有50种。

3. 放射性物质

我国28个省市土壤中放射性核素^{238}U、^{226}Ra、^{232}Th和^{40}K的含量按面积加权平均值分别为39.5Bq/kg，36.5Bq/kg，49.1Bq/kg和580.0Bq/kg；土壤中^{238}U、^{226}Ra、^{232}Th的含量随地域的变化规律基本一致，^{40}K含量随地域的分布规律则不显著。土壤中的天然放射性核素含量与成土母岩和铀资源分布有着明显的相关性。这些放射性污染物除可直接危害人体外，还可以通过生物链和食物链进入人体，在人体内产生内照射，损伤人体组织细胞，引发肿瘤、白血病和遗传障碍等疾病。

4. 病原微生物

土壤中的病原微生物，如肠道细菌、炭疽杆菌、肠寄生虫、结核杆菌等，主要来自含病原体的人畜粪便、生活污水、垃圾、工业废水、医院污水和废渣的污染。病原微生物一旦从外界进入土壤后就会大量繁殖而引起土壤质量下降，使生态平衡遭到破坏，植物直接病变或死亡。人类若直接接触含有病原微生物的土壤，可能会给健康带来影响。若食用被土壤污染的蔬菜、水果等则会受到间接污染而染病。

第五节 土壤污染防治

污染物可以通过多种途径进入土壤，引起土壤正常功能的变化，从而影响植物的正常生长和发育。然而，土壤对污染物也能起净化作用，特别是进入土壤的有机污染物可经过扩散、稀释、挥发及光化学降解、生物化学降解、化学降解等作用而得到净化。如果进入土壤中的污染物在数量和速度上超过土壤的净化能力，即超过土壤的环境容量，最终将导致土壤正常功能的失调，阻碍作物正常生长。

土壤环境容量又称土壤负载容量，是一定土壤环境单元在一定时限内遵循环境质量标准，既维持土壤生态系统的正常结构与功能，保证农产品的生物学产量与质量，又不使环境系统污染超过土壤环境所能容纳污染物的最大负荷量。

从某种程度上来说，土壤是环境中污染物的“汇”。土壤与植物的生命活动紧密相连；污染物可通过土壤-植物系统及食物链，最终影响人体健康。因此，土壤污染的防治十分重要。首先要控制和消除污染源，对已经污染的土壤，要采取一切有效措施，消除土壤中的污染物，或控制土壤污染物的迁移转化，使其不能进入食物链。

一、控制和消除土壤污染源

控制和消除土壤污染源是防止污染的根本措施。控制土壤污染源，即控制进入土壤中污染物的数量和速度，使其在土体中缓慢地自然降解，以免产生土壤污染。

1. 控制和消除工业“三废”的排放

应大力推广清洁工艺，以减少或消除污染源，对工业“三废”及城市废弃物必须处理与回收，即进行废弃物资源化。对排放的“三废”要净化处理，控制污染物的排放数量和浓度。

我国水资源短缺，分布又不均匀，近几年来水体污染日益严重，故我国的水资源日益匮缺，农业用水甚为紧张。因此，我国许多地方已发展了污水灌溉。这一方面解决了部分农田的用水问题；另一方面，污水中含有相当多的肥料成分，但也可以导致土壤污染。因此利用污水灌溉和施用污泥时，首先要根据土壤的环境容量，制定区域性农田灌溉水质标准和农用污泥施用标准，要经常了解污水中污染物质的成分、含量及动态。必须控制污灌水量及污泥施用量，避免盲目滥用污水灌溉引起土壤污染。此外，工业废渣不能任意堆放。

2. 控制化学农药的使用

禁止或限制使用剧毒、高残留农药，如有机氯农药；发展高效、低毒、低残留农药，如除虫菊酯、烟碱等植物体天然成分的农药；大力开展微生物与激素农药的研究。微生物可使昆虫引起感染而死亡，如核角体病毒防治桑毛虫，效果较好。激素农药有昆虫内激素（昆虫体内腺体分泌物）、蜕皮激素（蜕皮激素固酮防治蛾类幼虫）、保幼激素（天蚕油保幼激素使昆虫无法成活）和外激素，如果蝇、午毒蛾及棉红铃虫等性引诱剂。另外，可采用含有自然界中构成生物体的氨基酸、脂肪酸、核酸等成分的农药，它们易被降解。探索和推广生物防治病虫害的途径，开展生物上的天敌防治法，如应用昆虫、细菌、霉、病毒等微生物作为病虫害的天敌。还应开展害虫不孕化防治法。

随着农药的长期大量使用，各种问题逐渐暴露出来，主要体现在以下几个方面。

（1）难降解，污染范围广

主要是有机氯农药，尤其是 DDT，它已经渗透进了地球上包括人类在内的几乎所有的生物体内，对全球生态和人类健康都构成了严重的危害。

(2）利用率低

据估算，农药试用后大约 10％～20％的部分附着在作物的茎、叶和果实上，80％以上的部分在喷洒过程中散落在土壤和空气中，经风吹雨淋广泛传播，严重污染了周围的水体，甚至造成鱼类死亡。

（3）选择性差

化学农药在杀灭害虫的同时，也杀伤了天敌及其他有益生物，破坏了生态平衡，引起害虫再增猖獗。

(4) 危害鸟类

为了保护农作物的种子和幼芽不受害虫侵袭，农民常用杀虫剂拌种，结果使不少鸟类因啄食种子而丧命。同时，鸟类啄食了喷洒过农药的粮食后也会导致死亡和病变。

在自然界中鸟类与昆虫是一对矛盾关系，鸟类是昆虫的天敌，可以起到调节昆虫数量和改变其分布区域的作用。昆虫的生产和繁殖是相当迅速的，如果没有鸟类对它们的控制作用，地球恐怕早就被昆虫淹没了，因此，鸟类其实是人类防治害虫的得力助手，大部分鸟类对林业、牧业都有益处。人类在用化学农药防治害虫的同时，使得一些鸟类也跟着受害，鸟类数目减少，害虫数目反而多起来，于是加大了农药使用量，形成了恶性循环，使自然界的生态平衡遭到了严重破坏。

(5) 害虫产生抗药性

害虫的抗药性不断加深和扩大，全世界至少已有504种害虫对农药产生了抗性，有的害虫抗性高达几万倍，可能导致目标害虫再猖獗以及次要害虫爆发。这样，不得不加大农药的施用量和频度，从而造成化学农药应用的恶性循环，进而导致由于使用化学农药造成的环境污染更加严重。

(6) 通过食物链危害人类健康

生物体对农药的富集作用相当显著。农药可以积蓄在农作物、粮食和禽蓄体内，经食品危害人体健康。

另外，农药引起的投毒和自杀事件也屡屡发生，造成了一定的社会影响。据不完全统计，全世界每年约有100万人因农药中毒而发病，至少有2万人死于农药污染。据卫生部食品卫生监督检验所资料显示，2000年全国因农药引起的中毒41起，中毒1300人，死亡35人，超过全年总死亡人数的1/5，触目惊心的数字，足以引起我们的警惕！从20世纪60年代开始，政府和公众开始关注化学农药及其所带来的上述危害，并开始寻找控制害虫的更为安全有效的方法。进入21世纪以后，化学农药残留的危害已经引起了人们的高度重视，并开始寻求解决方案。

3. 大力发展生物农药

我国虽然加入了WTO，但发达国家针对中国入世纷纷调整和提高了产品进口的技术门槛，形成了“绿色壁垒”。美国食品与药物管理局常常以农药残留不符合FDA标准而扣留从中国进口的食品，欧盟、日本、韩国也紧随其后，给我国农产品出口设置了重重障碍，而生物农药则是实现农产品安全生产与顺利出口的重要保障，这一现实给生物农药的发展带来了机遇和挑战。

苏云金杆芽孢菌（Bacillus thuringiensis，简称Bt）作为一种生物农药，以其绿色和环保而备受青睐，主要用于粮食、棉花、蔬菜、瓜果、绿地及林木上200多种害虫的防治，它具有选择性强、不杀死害虫天敌、对环境友好、对人畜无害的特点，能满足人们对绿色食品的需求，符合农业可持续发展的方向，因此，得到了迅猛的发展。

20世纪70年代，Bt开始大规模生产。由于当时的发酵水平落后，导致成本偏高，再加上人们的环保意识不强，结果Bt生物农药在化学农药的强大竞争压力下萎靡低落，许多Bt生物农药生产厂家纷纷倒闭。然而，科学的脚步没有停止，国家从“七五”开始投入资金进行Bt生物农药攻关，经过30多年的发展，Bt生物农药已经在生产成本、科技含量、生产技术、理论研究等方面取得了长足的进步。今天，Bt生物农药的面目已经焕然一新了。随着人们环保意识的增强和对绿色食品的需求，Bt生物农药迎来了快速发展时期。

Bt 生物农药具有以下优点。

① 防治虫害效果好，能保持农产品的优良品质。

② 对病虫害特异性强，不杀伤害虫天敌和有益动物，对人畜安全无毒。

③ 多种因素和成分发挥作用，害虫和病原难以产生抗药性。

④ 能保持生态平衡。

⑤ 无残留，有利于可持续发展。

目前，Bt 生物农药仍然存在一些问题，主要表现在以下几个方面。

① 价格与化学农药持平，还需要进一步降低成本，提高市场竞争力。

② 光稳定性差，在太阳光照射下容易失去活性，田间有效期只有 5～7 天，新型抗晒 Bt 生物农药已经问世，亟待推广。

③ 一般只能防治食叶性害虫。为了克服 Bt 的缺点，科学工作者在不断地进行创新，预计在未来的 5 年里，Bt 的价格将低于化学农药，表现出明显的市场竞争力。

二、土壤改良与修复

增加土壤有机质含量，砂掺黏和改良砂性土壤，可以增加或改善土壤胶体的性质，增加土壤对有毒物质的吸附能力和吸附量，从而增加土壤环境容量，提高土壤的净化能力。

分析、分离或培养新的微生物品种以增加微生物对有机污染的降解作用，也是提高土壤净化能力极为重要的一环，这方面目前已取得了一些进展。

1. 施用化学改良剂

化学改良剂包括抑制剂和强吸附剂。一般施用的抑制剂有石灰、磷酸盐和碳酸盐等，它们能与重金属发生化学反应而生成难溶化合物以阻碍重金属向作物体内转移。在酸性污染土壤中施用石灰，可提高土壤 pH 值，使镉、铜、锌、汞等重金属形成氢氧化物沉淀。据试验表明，施用石灰后，稻米的含镉量可降低 30%。施用钙铁磷肥也能有效地抑制 Cd、Hg、Pb、Cu、Zn 等重金属的活性，如 Cd^{2+} 和 Hg^{2+} 与磷酸盐分别形成难溶 $Cd_3(PO_4)_2$ 沉淀和 $Hg_3(PO_4)_2$ 沉淀，这对消除土壤中的 Cd、Hg 污染具有重要意义。

施用强吸附剂可使农药分子失去活性，也可减轻农药对作物的危害。如加入 0.4% 的活性炭，豌豆从土壤中吸收的艾氏剂量可降低 96%。有机质、绿肥、蒙脱土等都具有类似的缓解效果。

2. 生物改良措施

通过植物的富集而排除部分污染物，包括种植对重金属吸收能力极强的作物，如黄颔蛇草对重金属的吸收量比水稻高 10 倍；种植这些非食用性作物，在一定程度上可排除土壤中的重金属。另外，种植抗性植物，即种植对重金属吸收能力极差的植物，也是生物改良的方法之一。

为了减少污染物对作物生长等的危害，也可采用耕翻土层，即采用深耕，将上下土层翻动混合，使表层土壤污染物含量降低。这种方法动土量较少，但在污染严重的地区不宜采用。

近年来，土壤及地下水污染的化学与生物修复已成为研究热点，并取得了一定进展。土壤有机污染的化学修复是用表面活性剂或有机溶剂清洗土壤中的有机物；生物修复是利用微生物将土壤中有毒有害的有机污染物降解为无害的有机物质（CO_2 和 H_2O）的过程。降解过程可以由改变土壤理化条件（包括 pH 值、湿度、温度、通气条件及添加营养物）来完成，也可接种特殊驯化与构建的工程微生物提高降解速率。

污染土壤生物修复的特点如下：成本低于热处理及物理化学方法；不破坏植物生长所需要的土壤环境；污染物氧化比较完全，不会产生二次污染；处理效果好，对低分子量的污染物去除率可达99%以上；可原地处理，操作简单。

三、发展绿色食品，促进人类健康

发展绿色食品，促进人类健康是人们追求的目标。为了实现这一目标，常用的措施如下。

① 使用高效低毒化学农药替代高毒高残留的化学农药。

② 使用绿色农药控制田间害虫。

③ 利用害虫天敌捕杀田间害虫。

④ 采用诱捕技术捕捉田间害虫。

⑤ 使用水覆盖灭虫卵技术清除田间害虫。

⑥ 使用蒸汽杀虫卵技术清除田间害虫。

⑦ 利用杂草控制田间害虫数量。

⑧ 采用氮气窒息法杀死害虫技术。

⑨ 紫外灯除虫技术。

思考题

1. 试述土壤的形成和组成。
2. 土壤具有哪些基本特性？
3. 什么是土壤环境容量？
4. 什么是土壤污染，如何判别？
5. 土壤污染物主要有哪些？它们是通过何种途径进入土壤的？
6. 土壤中化学农药的降解方式有哪些？
7. 常见的土壤生物修复技术有哪些？

第五章 灾害与环境保护

本章要点：地震成因分析及预警；火山的形成机制、喷发条件及环境影响；泥石流成因分析及灾害防治；厄尔尼诺现象发生的原因、成因及危害；飓风的形成过程及分级；生物多样性的主要组成、生物多样性的保护措施及重要意义；树皮蝎子、红火蚁、蜱虫、黑寡妇蜘蛛、非洲杀人蜂五种致命害虫的特征、生活习性、毒性及预防措施；食品安全、食品质量、食品安全事件、原因分析及解决方案；城市化相伴带来诸多环境问题、环境污染背后的深层原因、积极探索环保新道路及环境保护措施。

第一节 地质灾害

一、地震

1. 地震的定义与危害

地震是地球内部介质局部发生急剧破裂产生的震波，是地球表层的快速震动，进而在一定范围内引起地面震动的现象。地震是地壳构造运动的一种表现，也是地球地壳岩体能量集中释放的一种表现形式。全球每年发生地震约 500 万次，其中大约 95%的地震由于发生在地下深处，或者其能量很小，因而人们无从感觉，只有用专用的仪器才能记录下来。震级达到三级以上时人类开始有感觉，但是小的震级对人类的生命安全与健康并无危害。能造成严重灾害的地震，全世界每年会发生十几次。

地震是威胁人类安全及生态环境最大的自然灾害。强烈地震破坏性很强，瞬时可使很大范围的城市和乡村沦为废墟，产生严重灾害。我国台湾地区和日本列岛位于环太平洋地震带，我国的西部和西南地区位于地中海-喜马拉雅地震带。由于我国很大一部分领土都位于比较活跃的地震带，所以造成我国地震比较频发。例如 1966 年 3 月的河北邢台地震、1975 年 2 月的辽宁海城地震、1976 年 7 月的河北唐山地震、1999 年 9 月的台湾南投地震、2008 年 5 月的汶川地震、2010 年 4 月的玉树地震、2013 年 4 月的雅安地震，这些强烈的地震都给震区人民带来了严重的伤害。

2. 地震成因

地震的成因目前有三种学说：第一种学说是 1911 年理德提出的“弹性回跳说”，即地球内部不断积累的应变能超过岩石强度时产生断层，断层形成后，岩石弹性回跳，趋向于恢复原来状态，于是把积累的能量突然释放出来而引起地震；第二种学说是 1955 年日本的松泽武雄提出的“岩浆冲击说”，即地下岩石导热不均，部分熔融体积膨胀，挤压围岩，导致围岩破裂产生地震；第三种学说是美国学者布里奇曼提出的“相变说”，即地下物质在一定临界温度和压力下，从一种结晶状态转化为另一种结晶状态，体积突然变化从而发生地震。三种学说的想象力都很丰富，遗憾的是不能合理地解释全球每年约 500 万次的地震。其他各种学说层出不穷，但得到学术界公认的学说尚未出现。

3. 地震预警

按照传播方式的不同，地震波可分为三种类型：纵波、横波和面波。其中纵波在地壳中的传播速度为5.5～7.0km/s，它最先从震中向外扩散，使地面上下震动，破坏性较弱；横波在地壳中的传播速度为3.2～4.0km/s，它在纵波扩散之后开始向外扩散，使地面发生前后、左右抖动，破坏性较强。纵波和横波叠加形成面波，面波最后到达，它是由纵波与横波在地表相遇后激发产生的混合波，其波长长、振幅大，只能沿地表传播，是造成建筑物被强烈破坏的主要因素。

地震预警是在地震已经发生的前提下，利用电波比地震波快的原理，在地震波未到达前发出的警告。地震预警可以使人们提前逃离成灾现场和处理有可能形成次生灾的隐患，从而达到减少人员伤亡及次生灾害的目的。地震预警关键是利用地震波前几秒的数据准确估计震级、震中位置以及快速估计地震对预警目标的影响等。

美国、日本、墨西哥、罗马尼亚等国家都在积极开展地震预警系统的研发，并建成了一定规模和数量的地震预警系统或地震紧急自动处置系统。中国是个多地震国家，由地震造成的人员伤亡与经济损失巨大，但除大亚湾核电站建立了一个由地震监测网络和人工决策相结合的地震预警系统外，尚未建设过其他重大工程地震预警系统，有关研究工作也才刚刚开始。

二、火山

1. 火山的形成机制

火山的形成涉及一系列物理化学过程。地壳上地幔岩石在一定温度压力条件下产生部分熔融体。熔融体与母岩分离，并通过孔隙或裂隙向上运移，在一定部位逐渐富集形成岩浆囊。随着岩浆的不断补给，岩浆囊的岩浆过剩压力逐渐增大。当地壳覆盖层的强度不足以阻止岩浆继续向上运动时，岩浆通过薄弱带向地表攀升。在攀升过程中由于压力的降低，溶解在岩浆中的挥发分逐渐溶出，形成气泡。当气泡占有的体积分数超过75%时，禁锢在液体中的气泡会迅速释放出来，导致爆炸性喷发。气体释放后岩浆黏度急剧下降，形成湍流。若岩浆黏滞系数较高或挥发分较少，则形成宁静式溢流。从地幔岩石熔融到爆炸性喷发的物理化学过程多种多样，因而形成了形形色色的火山活动。

2. 火山喷发条件

一个地方能否形成火山主要在于是否具备以下5个条件：①必须有部分熔融体的形成；②必须有岩浆在地壳中富集；③必须有岩浆囊中的岩石熔融、气体析出等物理化学过程；④有构造活动产生的引张应力场；⑤有压力梯度与浮力的双重驱动力。

3. 火山喷发对环境的影响

火山爆发时喷出的大量火山灰和火山气体，对气候造成极大的影响。火山喷发的气体中含有硫黄和有毒气体二氧化硫。火山灰和火山气体被喷到高空中去，它们就会随风散布到很远的地方。火山灰会遮住阳光，导致气温下降，农作物减产，严重的将导致物种灭绝。二氧化硫在大气中被氧化，形成酸雨，对植被构成伤害。火山爆发喷出的大量火山灰和暴雨结合形成泥石流能冲毁道路、桥梁，淹没附近的乡村和城市，使得无数人无家可归。泥土、岩石碎屑形成的泥浆可像洪水一般淹没整座城市。火山喷发的火山灰不同于烟灰，它是坚硬的小颗粒，不溶于水。吸入火山灰会导致人和动物的呼吸道和肺部受损，导致呼吸疾病。火山灰对机械设备会产生影响，使汽车、飞机等交通工具受损。

4. 火山喷发造成的灾害

火山造成的灾害主要是类似土石流一样的泥流堆积。热的火山碎屑流、熔岩流和火山涌浪流在流动的过程中，碰到大量的地表水，形成了热的火山泥流堆积。这些堆积物受到大雨冲刷或地震引起的崩塌，流入河流或湖泊，形成冷的火山泥流堆积。火山泥流堆积不仅能淹没广大区域，也能冲垮建筑物。如1985年哥伦比亚的内瓦多·德·鲁伊斯火山和皮纳吐坡火山的喷发，前者因喷发所引起的火山泥流，把整个河谷和城镇淹没，造成近2500人死亡。后者喷发后所堆积的未固结火山灰受到暴雨的冲刷侵蚀，引发泥流堆积，淹没广大区域，造成重大财产损失。

5. 火山喷发的益处

火山喷发的益处主要包括火山产生的地热、温泉和肥沃的土壤。如果岩浆能留在地表下，就会产生很好的地热资源。火山附近常有温泉或热泉，这是因为岩浆散发出的热量使地下水变热的缘故。这种热源称为地热，规模大的可形成“地热田”。火山作用的另一个好处是制造陆地。地球表面大约有71%被海水所覆盖，海底火山经年累月不断地冒出岩浆，冷凝成岩石。如此长期堆积，直到有一天岩石高出水面形成岛屿。夏威夷群岛与冰岛就是这么形成的。至今，岛上还有活动火山不时喷出岩浆。此外，火山爆发所形成的火山灰云层会在爆发后一段时间内阻挡太阳光，导致该区的平均温度下降。因此，火山爆发可以自发地调节地球的气候平衡。

三、泥石流

1. 泥石流的定义和特点

泥石流是山区或沟谷深壑、地形险峻的地区，由暴雨、冰雪融水或冰湖溃决等水源激发，使山谷中积存的松散岩土体向下游开阔地倾泻，携带大量泥沙、石块的特殊洪流。泥石流常常发生在某些区域，预知性差，有很强的隐蔽性、突发性和破坏性。泥石流发生的全过程一般只有几个小时，个别情况下只有几分钟。由于突然暴发、历时短暂、冲击力大、来势凶猛，泥石流对土地资源、生态环境及人类社会等具有极大的破坏力。泥石流是山地环境退化、地表侵蚀作用加剧、水土泥沙流失的产物，其形成和发展主要受到地质、地貌、水文、气候等大尺度环境因素的综合影响。人类不合理、不明智的破坏活动，更是加剧了泥石流发生的频率和强度。

2. 泥石流的成因

(1) 自然因素

地形地貌是形成泥石流的基础条件。在地形上具备山高沟深、地形陡峻、沟床落差大、流域形状便于水流汇集、“两山夹谷”的地区容易发生泥石流现象。泥石流的地貌一般可分为形成区、流通区和堆积区三部分。上游形成区的地形多为“三面环山，一面出口”的瓢状或漏斗状，这种结构能够积蓄能量。中游流通区的地形多为狭窄陡深的峡谷，谷床纵坡降大，这种结构能够提高流速，使泥石流能迅猛直泻。下游堆积区的地形为开阔平坦的山前平原或河谷阶地，使堆积物有堆积场所。这些开阔地域往往是农田或农舍，一旦发生泥石流则损失惨重。

松散物质来源是形成泥石流的增害条件。泥石流常发生于地质构造复杂、断裂褶皱发育，新构造活动强烈，地震烈度较高的地区。地表岩石破碎、崩塌、错落、滑坡等地质活动为泥石流的形成提供了丰富的固体物质来源。另外，岩层结构松散、软弱、易于风化、节理发育或软硬相间成层的地区也能为泥石流提供碎屑物来源。

水源是形成泥石流的诱发条件。水既是泥石流的重要组成部分，又是泥石流的激发条件、搬运介质和动力来源。泥石流的水源有暴雨、强降雨、长时间的连续降雨、雪融水和水库（池）溃崩水等类型。

（2）人为因素

人类不合理的活动是诱发泥石流的内在因素，它在一定程度上加剧了泥石流发生的频率和危害程度。受到利益的驱动，某些地区的人们会做出一些不明智的活动，如滥伐林木、过度放牧、破坏植被、陡坡开荒、随意兴修交通和水利设施、随意丢弃废渣废料等，这些活动破坏了地表径流的下垫面条件，为泥石流提供了碎屑物，加剧了水土流失，增加了发生泥石流的风险。

3. 泥石流的危害

（1）泥石流灾害猛于洪水

包含着大量泥、沙、石块的泥石流，从高而下，在运行过程中又不断增加物质，能量远远大于洪水。由于其流程较短，人们几乎没有时间撤离，往往造成重大损失。

（2）泥石流冲击力大，破坏性强

泥石流来势凶猛，所到之处房屋倒塌、道路中断，公路、铁路、车站被埋，桥涵、路基被摧毁，耕地毁损，大树被拦腰砸断。甘肃舟曲县特大泥石流，半个县城瞬间被吞没，250万平方米被夷为平地。

4. 泥石流灾害防治

不同地区泥石流有不同的特点，应采取相应的治理措施。常用的治理措施分为工程治理措施和非工程治理措施。工程治理措施主要包括蓄水、引水工程，拦挡、排导工程和清淤工程等。非工程治理措施主要包括退耕还林，种植植被，涵养水源，减小地表径流，防止水土流失，恢复生态系统功能等。在崩塌、滑坡强烈活动的泥石流发生区，应以工程措施为主，兼用非工程措施。在以坡面侵蚀及沟谷侵蚀为主的泥石流地区，应以非工程措施为主，辅以工程措施。而在坡面侵蚀和重力侵蚀兼有的泥石流地区，则以综合治理效果最好。

第二节　天气灾害

一、厄尔尼诺

1. 什么是厄尔尼诺现象

厄尔尼诺是指在南美洲的秘鲁和厄瓜多尔沿海地带，某些年份圣诞节前后海水突然出现增暖的现象。具体来说，流经南美沿岸的秘鲁海流，在几乎与秘鲁海岸平行的东南信风吹送下，表层海水离岸外流，深层海水上涌补充，同时将营养盐类挟至上层，因而浮游生物繁盛，吸引大量秘鲁沙丁鱼等冷水性鱼类在此繁衍、栖息，使该地区成为著名的东南太平洋渔场。可是在某些年份，东南信风暂时减弱，太平洋赤道逆流的南支流越过赤道沿厄瓜多尔沿岸南下，使厄瓜多尔和秘鲁沿岸水温迅速升高，冷水性浮游生物和鱼类因不适应新的环境而大量死亡。由于沿海水温上升在圣诞节即圣子耶稣诞辰前后最为激烈，秘鲁居民将这种海水温度季节性上升的现象称为厄尔尼诺（厄尔尼诺为西班牙文音译，意为圣婴或圣子）。

2. 厄尔尼诺现象发生的原因

地球是一个由地核、地幔、地壳组成的椭球体，地核在不断地进行热核反应积累热量，同时地壳、地幔又在不停地释放热量，有两种主要释放。一种是火山喷发、地震、造山运动；另一种是地幔热对流，即在海洋中，热岩浆从中央海岭流出并带出热量，表层热量传给海水后，冷凝成洋壳向两边推移，移至大陆架边沿海沟处完全变冷，俯冲下沉到大陆壳下面熔于地幔中，然后再开始新一轮的循环。和沸腾的水锅中的液体流动很相似。此理论即为洋底扩张说。目前古地磁学已证实了该学说的正确性，并用其成功地解释了沿大洋边沿的火山、地震现象。太平洋中央海岭南北向分布，自然把太平洋分成东西两部分。

太平洋的中央海岭并不正好在太平洋的中央，而是偏在太平洋的东部。从中央海岭到美洲大陆的大洋东部只是狭窄的一片，其洋底构造简单，洋壳薄，最后俯冲到美洲大陆以下，其速度较快，约为10cm/年。而赤道西太平洋地域宽广，构造复杂，洋壳最后俯冲到地幔中，其速度约4～6cm/年，较东部慢。洋壳向地幔俯冲的构造不同，速度不同，就造成了其上洋面的水温差异。

由于地球自转速度加速而地壳鼓起时，地球内部的软流圈的物质就要向赤道方向流动。此时物质是向距地球旋转轴运动的方向上流动，所以按照角动量守恒，地球自转就要转慢。这种事件是在海水增温后一段时间出现，由于海水增温要传到海面有一个滞后过程，所以厄尔尼诺现象多出现在地球转速变慢之时或与其同时。西传型易于在秘鲁和厄瓜多尔西边海域出现，还与日月引潮力在这一地区较大有关。因此，当地球内部不稳定时，这个不大的固体潮就可以触发深部热物质涌至海底，并加温海水形成厄尔尼诺现象。

3. 厄尔尼诺的危害

厄尔尼诺所带来的影响是显著而广泛的。首先，厄尔尼诺现象就是海洋异常事件，它的出现直接影响到了当地的渔业。其次，厄尔尼诺会通过海-气间的相互作用影响到当地的气候，并通过流动的大气将这种异常扩展到更大范围。同时，气候异常又会引发其他一系列与之有关的问题，例如：暴雨、洪涝灾害、干旱、森林大火、沙尘暴、暖冬和多种疾病。所以，厄尔尼诺影响之广泛充分地表现在两方面，即空间范围的广泛性（不仅对低纬度地区有影响，而且对中高纬度地区也有影响）和影响领域的广泛性（大气、海洋、生态、农业、人体健康等）。

厄尔尼诺是气候出现异常的强烈信号。但厄尔尼诺对全球不同地区的影响是不同的。根据其影响程度将全球分为三类地区：第一类地区为基本不受或很少受厄尔尼诺影响的地区，如欧洲特别是欧洲北部地区；第二类地区为有时受厄尔尼诺影响且影响较严重，但厄尔尼诺影响尚不确定的地区，如中国、南非等；第三类地区为经常受厄尔尼诺影响的地区，如南美的一些国家、西太平洋和印度洋地区、澳大利亚等。

二、飓风

1. 飓风定义和分级

西洋、墨西哥湾、加勒比海和北太平洋东部的热带气旋称为飓风。飓风通常发生在夏季和早秋，它来临时常常电闪雷鸣。在仅仅一天内，飓风就能够释放出大量的能量，而这些能量足以满足整个美国约六个月电的需要量。世界气象组织对热带气旋的定义和分类标准是，按热带气旋中心附近最大平均风力将热带气旋划分为四级：风力＜8 级为热带低压；风力 8～9 级为热带风暴；风力 10～11 级为强热带风暴；风力 12 级为台风或飓风。

飓风是指风速达到 33m/s 以上的热带气旋，只是因发生的地域不同，才有了不同的名称。出现在北太平洋西部和我国南海的强烈热带气旋称为“台风”；发生在大西洋、墨西哥湾、加勒比海和北太平洋东部的则称为“飓风”。飓风与龙卷风也不能混淆。后者的时间很短暂，属于瞬间爆发，最长也不超过数小时。此外，龙卷风一般是伴随着飓风而产生的。龙卷风最大的特征在于它出现时，往往有一个或数个如同“大象鼻子”一样的漏斗状云柱，同时伴随狂风暴雨、雷电或冰雹。龙卷风经过水面时，能吸水上升形成水柱，然后同云相接，俗称“龙取水”。经过陆地时，常会卷倒房屋，甚至把人吸卷到空中。

飓风分为 5 个级别：一级飓风中心持续风速 64～82 海里/h，或 119～153km/h；二级飓风中心持续风速 83～95 海里/h，或 154～177km/h；三级飓风中心持续风速 96～113 海里/h，或 178～209km/h；四级飓风中心持续风速 114～135 海里/h，或 210～249km/h；五级飓风中心持续风速超过 135 海里/h，或 249km/h。根据联合国国际气象组织的定义，持续风速指 10min 平均风速。

2. 飓风的形成过程

飓风形成需要的三个条件是温暖的水域、潮湿的大气和海洋洋面上的风。开始时形成一连串的雷雨风暴，继而形成一股暖而潮湿的空气在热带洋面上徘徊。当热空气上升时，水蒸气凝结变成水滴，释放出潜热加热周围的空气，并将热空气送往大气高层（潜热是热带飓风的形成因素之一）。如果近旁有足量的风暴云存在，暖湿的空气就会形成“空气柱”，从洋面垂直通往高空同温层的边界。与此同时，空气柱在上升过程中其底部的低压吸收了周围许多空气，并以水汽凝结释放潜热的形式使空气柱获得更多的能量在洋面上游走。随着空气柱的一再上升，其底部则进一步降压。如果这种现象发生在赤道南北纬 4°以上，一种复合的加速度则作用于流入柱内的空气，且越转越快最终形成旋风。当条件有利时，热带旋风在形成中迅速获得能量，从低气压转为风暴，最后变为飓风，其能量之大是惊人的。美国国家海洋和大气管理局计算过，典型的飓风一天持续能释放 2×10^{14} J 的能量，相当于 100 万颗投在日本广岛的原子弹。幸运的是，飓风内大部分能量是向其上空中散发的，而不是向侧面散发。面对飓风所带来的破坏，气候转暖只能使情况更糟。初始形成的湿空气柱一般只在洋面温度超过 26℃时形成。当海洋温暖时，大部分洋面超过这个温阈。每超过温阈 1℃，都能提升飓风的强度。“卡特琳娜”飓风发生时，此时墨西哥湾洋面温度为 30℃左右，这也为伴随着温度的上升将会发生更多、更强飓风的观点提供了佐证。

第三节 生物多样性及致命害虫

一、生物多样性

1. 定义

生物多样性是指生物及其环境形成的生态复合体以及与此相关的各种生态过程的综合，包括动物、植物、微生物和它们所拥有的基因以及它们与其生存环境形成的复杂的生态系统，通常包括遗传多样性、物种多样性和生态系统多样性三个组成部分。

随着时间的推移，人们在开展自然保护的实践中逐渐认识到，自然界中各个物种之间、生物与周围环境之间都存在着十分密切的联系，因此自然保护仅仅着眼于对物种本身进行保护是远远不够的，往往也难以取得理想的效果。要拯救珍稀濒危物种，不仅要对所涉及的物种的野生种群进行重点保护，而且还要保护好它们的栖息地。或者说，需要对物种所在的整个生态系统进行有效的保护。在这样的背景下，生物多样性的概念便应运而生了。

2. 保护生物多样性势在必行

20 世纪后期生命科学各领域取得了巨大的进展，特别是分子生物学的突破性成就，使生命科学在自然科学中的地位起了革命性的变化，很多科学家认为在未来的自然科学中生物科学将成为带头学科，甚至预言21世纪是生物学的世纪。从事生命科学研究的专业人员也越来越多。在生物科学诸多的分支中，保护生物多样性是生物科学较紧迫的任务之一，也是全球生物学界共同关心的焦点问题之一。据可靠数据说明，每天约有100多种生物在地球上灭绝，很多生物在没有被人类认识前就消亡了，这对人类来说无疑是一种悲哀和灾难。保护生物多样性的行动势在必行、迫在眉睫。

3. 生物多样性主要组成

(1) 遗传多样性

遗传多样性是生物多样性的重要组成部分。广义的遗传多样性是指地球上生物所携带的各种遗传信息的总和，这些遗传信息储存在生物个体的基因之中。因此，遗传多样性也就是生物的遗传基因的多样性。任何一个物种或一个生物个体都保存着大量的遗传基因，因此，可被看做是一个基因库。一个物种所包含的基因越丰富，它对环境的适应能力越强，对生命进化和物种分化的贡献越大。

狭义的遗传多样性主要是指生物种内基因的变化，包括种内显著不同的种群之间以及同一种群内的遗传变异。此外，遗传多样性可以表现在多个层次上，如分子、细胞、个体等。在自然界中，对于绝大多数有性生殖的物种而言，种群内的个体之间往往没有完全一致的基因型，而种群就是由这些具有不同遗传结构的多个个体组成的。

在生物的长期演化过程中，遗传物质的改变（或突变）是产生遗传多样性的根本原因。遗传物质的突变主要有两种类型，即染色体数目和结构的变化以及基因位点内部核苷酸的变化。前者称为染色体的畸变，后者称为基因突变（或点突变）。此外，基因重组也可以导致生物产生遗传变异。

(2) 物种多样性

物种多样性是指地球上动物、植物、微生物等生物种类的丰富程度。物种多样性包括两个方面：其一是指一定区域内的物种丰富程度，可称为区域物种多样性；其二是指生态学方面的物种分布的均匀程度，可称为生态多样性或群落物种多样性。物种多样性是衡量一定地区生物资源丰富程度的一个客观指标。区域物种多样性是衡量一个国家或地区生物多样性丰富程度的指标。区域物种多样性包括：①物种总数，即特定区域内所拥有的特定类群的物种数目；②物种密度，指单位面积内的特定类群的物种数目；③特有物种比例，指在一定区域内某个特定类群特有种占该地区物种总数的比例。

(3) 生态系统多样性

生态系统是指各种生物与其周围环境所构成的自然综合体。所有的物种都是生态系统的组成部分。生态系统的多样性主要是指地球上生态系统组成、功能的多样性以及各种生态过程的多样性，包括生态环境的多样性、生物群落和生态过程的多样化等多个方面。其中，生态环境的多样性是生态系统多样性形成的基础，生物群落的多样化可以反映生态系统类型的多样性。

4. 重要意义

生物多样性是人类社会赖以生存和发展的物质基础。生物多样性为我们提供了食物、纤维、木材、药材和多种工业原料。生物多样性还在保持土壤肥力、保证水质以及调节气候等方面发挥了重要作用。生物多样性在大气层成分、地球表面温度、地表沉积层氧化还原电位以及pH值等方面的调控方面发挥着重要作用。生物多样性的维持，将有益于一些珍稀濒危物种的保存。

任何一个物种一旦灭绝，便永远不可能再生，人类将永远丧失这些宝贵的生物资源。保护生物多样性对科学研究具有重大的战略意义，迫在眉睫。

5. 保护措施

第一，就地保护。为了保护生物多样性，把包含保护对象在内的一定面积的陆地或水体划分出来，进行保护和管理。比如，建立自然保护区实行就地保护。自然保护区是有代表性的自然系统、珍稀濒危野生动植物种的天然分布区，包括自然遗迹、陆地、陆地水体、海域等不同类型的生态系统。自然保护区还具备科学研究、科普宣传、生态旅游的重要功能。

第二，迁地保护。迁地保护是在生物多样性分布的异地，通过建立动物园、植物园、树木园、野生动物园、种子库、基因库、水族馆等不同形式的保护设施，对那些比较珍贵的物种、具有观赏价值的物种或其基因实施由人工辅助的保护。迁地保护目的只是使即将灭绝的物种找到一个暂时生存的空间，待其元气得到恢复、具备自然生存能力的时候，还是要让被保护者重新回到生态系统中。

第三，建立基因库。人们已经开始建立基因库，来实现保存物种的愿望。比如，为了保护作物的栽培种及其会灭绝的野生亲缘种，建立全球性的基因库网。大多数基因库储藏着谷类、薯类和豆类等主要农作物的种子。

第四，构建法律体系。人们还必须运用法律手段，完善相关法律制度，来保护生物多样性。比如，加强对外来物种引入的评估和审批，实现统一监督管理。建立基金制度，保证国家专门拨款，争取个人、社会和国际组织的捐款和援助，为实践工作的开展提供强有力的经济支持等。

二、致命害虫

昆虫，是地球上历史较悠久的生物之一，它们已演化了数百万年，数量胜过世上其他各种生物，昆虫会飞、会爬、会蜿蜒滑行、会成群聚集、会蜇人、会叮咬，它们会在毫无预警的情况下突袭，对人类造成极大威胁，至今已有数百万人死于昆虫所散播的潜伏疾病，它们是致命的害虫。

1. 树皮蝎子

（1）形态特征

亚利桑那树皮蝎有 7～8cm 长，外形好似琵琶，全身表面都是高度角质的硬皮，身体分节明显。头胸部由六节组成，是梯形，背面覆有半透明硬甲，背部中央有一对黑色中眼，前端两侧各有 3 个侧眼，有附肢 6 只，第一对为有助食作用的螯肢，第二对为长而粗的形似蟹螯的角须，有捕食、触觉及防御功能，其余四对为步足。口位于腹面前腔的底部。雌雄异体，外形略有差异。前腹部较宽，由 7 节组成。后腹部为易弯曲的狭长部分，由 5 个体节及一个尾刺组成，第一节有一生殖器，生殖器覆盖着生殖孔。

（2）生活习性

这些喜欢潮湿空气和偏暖温度的蝎子，经常光顾人们居住的地方，室内的水槽和浴缸、壁橱、储藏室暗区都可以发现被困的亚利桑那树皮蝎。也可以看到它们在攀爬墙壁，或附着在天花板上。在室外，亚利桑那树皮蝎经常生活在木材或砖柱缝里。它们冬眠时甚至集大群生活在树皮下。由于亚利桑那树皮蝎喜欢生活在杨木河岸地区，梧桐园，这些地方都具有足够的水分和湿度，以支持昆虫和其他猎物物种生存。灌溉草坪的广布增加了住宅区环境湿度，这些已经导致了亚利桑那树皮蝎在一些地区的爆炸式增长。树皮蝎夜间活动，喜欢伏击猎物，喜吃小蟋蟀、蟑螂、蜘蛛等中小型昆虫。分布于亚利桑那州西南部、新墨西哥州和科罗拉多河的加利福尼亚沿岸地区、索诺拉州北部、犹他州东南部。

(3) 毒性

亚利桑那树皮蝎的螯刺通常仅引起局部疼痛和轻微肿胀，淋巴管炎和区域性淋巴结肿，伤口周围皮肤温度上升和触痛。墨西哥雕像木蝎毒液可立即引起疼痛和受累部位的麻木和刺麻感，但通常无肿胀。儿童可出现紧张不安和头、颈、眼异常而胡乱地晃动，成人主要表现为心动过速、高血压、呼吸加快、体弱无力和运动障碍。成人和儿童均可发生呼吸困难并常伴有流涎过多。褐尾蝎螯刺 6 岁以下儿童和高血压患者可致死。

2. 蜱虫

(1) 形态特征

蜱 (pí) 也叫壁虱，鳖吃，俗称草扒子、狗鳖、草别子、牛虱、隐翅虫、草蜱虫、狗豆子、牛鳖子、草爬子（东北某些地区把蚰蜒也称为“草爬子”）。蜱蛰伏在浅山丘陵的草丛、植物上，或寄宿于牲畜等动物皮毛间。不吸血时，小的才干瘪绿豆般大小，也有极细如米粒的；吸饱血液后，有饱满的黄豆大小，大的可达指甲盖大。蜱叮咬的无形体病属于传染病，人对此病普遍易感，与危重患者有密切接触、直接接触病人血液等体液的医务人员或其陪护者，如不注意防护，也可能感染。

该寄生虫极其喜欢皮毛丛密的动物，尤其喜欢黄牛，经常可以在黄牛的脖子下方、四腿内侧发现其身影，多时会聚集成群，并且非常不容易剔除。在四川、云南、贵州等地农村极为常见。

成虫在躯体背面有壳质化较强的盾板，通称为硬蜱 (pí)，属硬蜱科；无盾板者，通称为软蜱，属软蜱科。全世界已发现的约 800 余种，硬蜱科约 700 多种，软蜱科约 150 种，纳蜱科 1 种（仅存于欧洲）。中国已记录的硬蜱科约 100 种，软蜱科 10 种。蜱是许多种脊椎动物体表的暂时性寄生虫，是一些人兽共患病的传播媒介和储存宿主。

(2) 吸血习性

蜱的幼虫、雌雄成虫都吸血。宿主包括陆生哺乳类、鸟类、爬行类和两栖类，有些种类侵袭人体。多数蜱种的宿主很广泛，例如全沟硬蜱的宿主包括哺乳类 200 种、鸟类 120 种和少数爬行类，并可侵袭人体。这在流行病学上有重要意义。硬蜱多在白天侵袭宿主，吸血时间较长，一般需要数天。软蜱多在夜间侵袭宿主，吸血时间较短，一般数分钟到 1h。蜱的吸血量很大，各发育期饱血后可胀大几倍至几十倍，雌硬蜱甚至可达 100 多倍。

(3) 宿主关系

蜱在宿主的寄生部位常有一定的选择性，一般在皮肤较薄，不易被搔动的部位。例如全沟硬蜱寄生在动物或人的颈部、耳后、腋窝、大腿内侧、阴部和腹股沟等处。微小牛蜱多寄生于牛的颈部肉垂和乳房，次为肩胛部。波斯锐缘蜱多寄生在家禽翅下和腿腋部。

蜱的嗅觉敏锐，对动物的汗臭和二氧化碳很敏感，当与宿主相距 15m 时，即可感知，由被动等待到活动等待，一旦接触宿主即攀登而上。如栖息在森林地带的全沟硬蜱，成虫寻觅宿主时，多聚集在小路两旁的草尖及灌木枝叶的顶端等候，当宿主经过并与之接触时即爬附宿主；栖息在荒漠地带的亚东璃眼蜱，多在地面活动，主动寻觅宿主；栖息在牲畜圈舍的蜱种，多在地面或爬上墙壁、木柱寻觅宿主。

(4) 致死蜱虫

蜱虫一般都是生长在树林或者长久腐烂的枯叶堆中，所以一般情况下大城市森林覆盖率不是很高的情况下、一般不会见到蜱虫的出现。软蜱虫和硬蜱虫有一点区别很明显：就是硬壳蜱虫用来吸血的嘴比较大，而在吸满血后背壳是硬的，而不像另外的软体蜱虫它在吸血之前体积是很小的，即使给自己吸得圆滚滚的它的背后依然是软的。硬壳的蜱虫它的体积比软壳的小，但是它给人们带来的恐惧却超过了软壳蜱虫，硬壳蜱虫它在吸血的过程中是将整个头部都埋藏在肉里面，即使吸满了血它也不会自动脱落，人工去取也不是一件容易的事情，更有甚者将它（蜱虫）的身体一分为二，却依然取不掉钻进肉内的那一部分，所以硬壳蜱虫是最可怕的。

（5）疾病种类

① 森林脑炎：是一种由森林脑炎病毒引起的神经系统急性传染病，为森林区的自然疫源性疾病。中国主要的病媒蜱种为全沟硬蜱，病毒在蜱体内可长期保存，并可传代。本病多发生在5～8月，在中国主要分布于黑龙江和吉林两省林区，患者主要是伐木工人。此外，四川、河北、新疆、云南等省和自治区也有病例发生。

② 新疆出血热：是一种蜱媒急性传染病，是荒漠牧场的自然疫源性疾病。病原为一种蜱媒RNA病毒。传播媒介主要为亚东璃眼蜱，病原体可在蜱体内保存数月，并经卵传递。本病除经蜱传播外，羊血经皮肤伤口，以及医务人员接触急性期病人新鲜血液后，也可感染发病。

③ 蜱媒回归热：又称地方性回归热，是由钝缘蜱传播的自然疫源性螺旋体病，不规则间歇发热为其主要临床特征。中国新疆有该病流行，其病原体，在南疆村镇型的为伊朗包柔氏螺旋体，乳突钝缘蜱为传播媒介；北疆荒野型的为拉氏包柔氏螺旋体，特突钝缘蜱为传播媒介。动物传染源主要是鼠类，病人也可作为本病的传染源。

④ 莱姆病：中国于1985年夏天在黑龙江海林县林区首次发现。病原体是伯氏包柔螺旋体。它是一种由硬蜱传播的自然疫源性疾病，好发于春、夏季。中国主要媒介是全沟硬蜱，某些野生小型啮齿动物为储存宿主。本病分布广泛，在五大洲20多个国家都有病例报告。中国已证实有20个省、市、自治区有本病流行。

⑤ Q热：病原体为贝氏立克次体。本病临床特点为起病急骤。常在野生动物（啮齿类）与家畜之间传播流行，牛、羊为人体Q热的主要传染源。感染方式主要由呼吸道吸入传播，也可通过消化道及蜱的叮咬、粪便污染伤口而感染。病原体能在蜱体内长期存在，并经卵传递，如乳突钝缘蜱可储存病原体2～10年。本病分布遍及世界各地，在中国已有十几个省、市、自治区证实有Q热存在。

（6）防治方法

① 环境防治：草原地带采用牧场轮换和牧场隔离办法灭蜱。结合垦荒，清除灌木杂草，清理禽畜圈舍，堵洞嵌缝以防蜱类孳生；捕杀啮齿动物。

② 化学防治：蜱类栖息及越冬场所可喷洒敌敌畏、马拉硫磷、杀螟硫磷等。牲畜可定期药浴杀蜱。

③ 药物预防：皮肤涂抹或居室喷洒罗浮山百草油，能有效预防蜱虫叮咬。

④ 生物防治：由于蜱虫主要栖息在草地、树林中，用生物农药狂扫喷洒主要发生蜱虫的地面，持效期特长，对人畜无害；也可用藻盖杀（0.12%藻酸丙二醇酯）物理防治，无色无味，对人体无毒、无害、无污染。

⑤ 个人防护：进入有蜱地区要穿五紧服，长袜长靴，戴防护帽。外露部位要涂布驱避剂，离开时应相互检查，勿将蜱带出疫区。

3. 红火蚁

(1) 红火蚁的形态特征

红火蚁的体型长度为3～6mm，与其他的蚂蚁差不多。它们体内没有骨骼，但却拥有强硬的外骨架，以坚固的外壳来保护自己。它们以气管来呼吸。它们的躯体分为头、胸、腹三部分，拥有三对足及一对触角。一如其他蚁种，它们的工蚁和兵蚁全为没有生育能力的雌蚁，蚁后负责生产和孵化蚁卵。在众多火蚁中，由于在每次生育过程中只需要一只雄蚁，因此雄蚁的数量远比雌蚁少。红火蚁不能在极为寒冷的环境下生存，在17℃的环境下便会死亡。

(2) 红火蚁的生活习性

成虫食性广泛，其捕杀昆虫、蚯蚓、青蛙、蜥蜴、鸟类和小哺乳动物，也采集植物种子。其进攻体型相对大的鸟类等的眼等要害器官。入侵红火蚁给被入侵地往往带来严重的生态灾难，是生物多样性保护和农业生产的大敌。红火蚁为单或多后制群体，后蚁每天可最高产卵800枚，一个几只蚁后的巢穴每天共可以产生2000～3000枚卵。当食物充足时产卵量即可达到最大，一个成熟的蚁巢可以达到24万头工蚁。

红火蚁有“五凶”：喜攻击、爱咬人、吃种子、蜇动物、钻电箱。红火蚁入侵住房、学校、草坪等地，与人接触的机会较大，叮咬现象时有发生。红火蚁扩散途径有主动扩散和被动扩散。主动扩散（自然扩散）：红火蚁会自然迁飞3～5km，也会随雨水、臭水沟扩散。被动扩散（人为扩散）：园艺植栽污染、草皮污染、土壤废土移动、堆肥、园艺农耕机具设备污染、空货柜污染、车辆污染等。

(3) 红火蚁的危害

对动植物的危害：红火蚁取食作物种子、幼芽、根系、果实，造成作物减产；红火蚁破坏生态环境，大量捕食其他小动物，如刚刚孵化的小鸟、节肢动物和蚯蚓等。

对基础设施的危害：红火蚁损坏基础设施，可危害供电设备、电信设备和堤坝等。啃咬电线会经常造成电线短路甚至引发小型火灾。

对人体的危害：其尾刺排放的毒液可引起过敏反应，人被红火蚁叮咬后，被叮咬部位会有持续灼热样疼痛，局部皮肤形成红斑、水泡、硬肿，有痒感，水泡破裂后又常会引起细菌性二次感染。少数体质敏感的人可能发生严重的过敏性休克，主要表现为全身过敏反应，还有喉头或支气管水肿与痉挛、呼吸困难，面色苍白，四肢厥冷，血压下降等循环衰竭症状。此外被叮咬的人还会因脑缺氧和脑水肿，出现头晕、乏力、烦躁不安、昏迷、抽搐、大小便失禁等现象。

(4) 红火蚁的防治

红火蚁防治可分为季节性防治和技术性防治两大方面。主要有物理防治、化学防治和生物防治三大类。

① 季节性防治：在春、夏、秋湿度比较高的时候，红火蚁活动能力强，诱饵诱杀的办法效果较好；在冬季温度较低、红火蚁活动较弱或基本不活动的时期，药剂灌巢效果较好。

② 技术性防治：物理防治主要为沸水处理蚁巢和水淹法对红火蚁进行控制。沸水处理一般采取向可见蚁巢内直接灌入沸水的做法，防除效果接近60%。沸水必须灌注达到蚁巢所有区域。单次的处理成功率较低，必须每隔5～10天处理1次，连续处理3～4次，但处理区域很容易再发生。处理过程中应注意安全防护，避免被沸水烫伤，或伤害周围的植物。在热水中加入一点肥皂或洗衣液，可增加防效。水淹法则是用水淹没蚁巢将蚂蚁淹死的方法，需挖掘整个蚁丘并放入水中浸泡24h以上。

红火蚁化学防治技术是目前研究得最多、最深入，应用最为广泛的一类防治方法。目前在国内外红火蚁防治中主要使用的农药剂型有：乳油、悬浮剂、微乳剂、可湿性粉剂、粉剂和饵粒。防治红火蚁的农药可用于喷洒、灌穴、撒布等。触杀性的农药有：联苯菊酯、氯菊酯、氰戊菊酯、氯氰菊酯、溴氰菊酯、氯氟氢菊酯、七氟菊酯、除虫菊素等。用于投放诱杀性饵剂的农药有：氟虫胺、氟虫腈、氟蚁腙、多杀霉素、乙酰甲胺磷、烯虫酯、除虫菊素等。这类方法一般分为两大类：一类是对蚁丘进行喷洒、灌穴和撒布等方法，这些方法多采用触杀性农药，见效快，但由于红火蚁是社会性昆虫，它可传递信息，产生移窝现象；另一类重要方法是采用饵剂诱杀的防治方法，虽然见效相对较慢，但却是各国青睐的产品。

抑制红火蚁发生的生物因子种类很多。在其原产地南美洲，红火蚁的天敌包括至少 18 种寄生性蚤蝇，多于 10 种病原微生物，至少 3 种寄生性线虫，1 种寄生蜂，1 种寄生蚁以及数十种共生生物。在该地区，红火蚁之所以无法爆发成灾，主要的原因是存在强有力的竞争物种和各种捕食性、寄生性天敌。

4. 非洲杀人蜂

(1) 生活习性

多年以前，26 只坦桑尼亚女王蜂被意外地带入巴西，并与当地蜜蜂杂交，形成了今日凶残的非洲杀人蜂。这是一种因为人为事故而繁育出的杂交蜂种，该种昆虫毒性剧烈，可以在短时间内致人死命，每年都有成千上万人因此受害。这些蜜蜂在野外大量繁殖起来，攻击人畜，科学家认为，杀人蜂生活在非洲，那里的天敌很多，如果不主动发起进攻，就会被其他动物消灭。在艰难的生涯中，经过自然选择，那些富有进攻性的群蜂得以保存下来，繁殖后代。它们成群结队，来势凶猛，许多动物见了，闻风而逃，就连狮子也无法对付它们。非洲杀人蜂的致命天性在于其对外界极为敏感的群集防御体系。更可怕的是，即使受害者早已被逼退，远离蜂巢，杀人蜂也照样穷追不舍，赶尽杀绝。

非洲杀人蜂种群很大，变异能力也是很强的，假如长期使用一些化学农药，它本身的抗性增强，可能会影响它的毒性。这些杀人蜂本身，也是对自然环境和生态建设有益的昆虫，由于它们往往以森林中的各种害虫为捕食对象，因此对防治森林病虫害有很大作用。伤害人们的凶手，同时也是森林农田的卫士，人们应当辩证地认识杀人蜂。

(2) 致毒机理

“杀人蜂”的致命物质是心脏毒素，毒蜂身体中究竟是什么物质让人丧命？蜂毒是由工蜂的毒腺分泌出的一种淡黄色透明液体，其化学成分极其复杂，除了含有大量水分外，还含有多种多肽、酶、生物胺、胆碱、甘油等物质和 19 种游离氨基酸等。在组成蜂毒的多肽类物质中，蜂毒肽的含量最高，约占干蜂毒的 50%。正是这一成分成为了患者致病致死的罪魁祸首。因为蜂毒肽是一种强烈的心脏毒素，具有收缩血管的作用，同时蜂毒的血溶性又极强，因此对心脏的损害也就极大。

(3) 预防对策

蜂王会分泌出一种叫弗罗蒙的物质，群蜂一闻到这种气味，会顿时变得温顺起来，就会停止战斗。人工合成这种物质，可用于人类的自我保护。

5. 黑寡妇蜘蛛

(1) 特征

黑寡妇蜘蛛又名红斑寇蛛，常称做黑寡妇，是一种具强烈神经毒素的蜘蛛，人若被咬，可能致命。成年雌性黑寡妇蜘蛛腹部呈亮黑色，并有一个红色的沙漏状斑记。这个斑记通常是红色的，有些可能是介于白色和黄色之间或是某种红色到橘黄色之间的颜色。对某些物种，斑记可能是分开的两个点。雌性黑寡妇蜘蛛腿展大约 38mm 长，躯体大约 13mm 长。雄性黑寡妇蜘蛛的大小约只有雌性蜘蛛的一半，甚至更小。它们相对于躯体大小具有更长的腿和较小的腹部。它们通常呈黑褐色，并具有黄色条纹以及一个黄色的沙漏斑记。成年雄性蜘蛛可以通过更纤细的躯体和更长的腿以及更大的须肢，与未成年雌性蜘蛛区别开来。成体雄性没有毒腺。

黑寡妇的毒液比草原响尾蛇还厉害 15 倍，在美洲市场，它的价值比黄金贵千百倍，一盎司最少 230 万美元。此外，它的蛛丝强度超过所有已知的天然纤维，甚至连以坚韧著称的钢丝和凯夫拉纤维都望尘莫及，它的韧性，在拉断前足可以延伸 27%，强度则超过普通蛛丝的 2 倍。

(2) 生活习性

黑寡妇蜘蛛通常生活在温带或热带地区，分布在前苏联南方、欧洲东南部与中亚细亚、美洲，以及我国的海南岛、新疆等地。它们一般以各种昆虫为食，不过偶尔它们也捕食虱子、马陆、小型蜈蚣、蝎子和其他蜘蛛。当猎物缠在网上后，黑寡妇蜘蛛就会迅速从栖所出击，用坚韧的网将猎物稳妥地包裹住，然后刺穿猎物并将毒素注入。毒素 10min 左右起效，此间猎物始终由蜘蛛紧紧把持着。当猎物的活动停止后，蜘蛛将消化酶注入伤口。随后黑寡妇蜘蛛将猎物带回栖所待用。

(3) 毒理学特征

尽管其毒素的毒性很强，但注射剂量很小，且寡妇蜘蛛咬人导致死亡的案例非常少。黑寡妇蜘蛛的毒液会促进神经递质乙酰胆碱（acetylcholine）的释放。乙酰胆碱作用于肌肉，能引起肌肉收缩。一般而言，黑寡妇蜘蛛的毒素对儿童和体弱者威胁较大。

(4) 预防

黑寡妇蜘蛛在没有受到惊吓的情况下，一般不会主动攻击人类。黑寡妇蜘蛛常躲在茂密的草丛中，因此人们在草丛中劳作、休息时，应该看看周围有没有黑寡妇蜘蛛的踪迹，而且最好不要赤脚。俄罗斯有 6 个种类的黄蜂是黑寡妇蜘蛛的天敌，可以引进这种黄蜂控制黑寡妇蜘蛛的数量。

第四节　食品安全

随着经济和社会的不断发展，食品安全问题已受到全球关注。保障食品安全已成为保护人类生命健康，提高人类生活质量，促进食品贸易和维护世界和平与发展的基础。各国政府及学术界都在不断探寻能有效解决日益严峻的食品安全问题的途径和方法。

一、食品安全与食品质量

目前，较为公认的是 FAO/WHO 于 2003 年在《保障食品质量与安全——强化国家食品控制体系指南》中提出的概念，即食品安全指的是有损于消费者健康的急性或慢性危害，这些危害因素必须消除，食品安全具有不可协商性。

在现实生活中，食品质量和食品安全经常被混淆。食品质量是指影响食品价值的所有特性的总和，包括：①外观、质地、风味等感官品质；②蛋白质、脂肪、维生素等营养品质；③化学残留物、重金属含量、微生物毒素等安全品质；④动物福利、环境友好等伦理品质。

可以看出，食品安全仅指食品中可能对人体健康造成损害的属性，仅仅是食品质量的一个组成部分。另外，食品安全（food safety）也不同于“粮食安全”或“食物数量安全”(food security)，后者是指一个国家或社会的食物数量保障。

动物福利一般指动物（尤其是受人类控制的）不应受到不必要的痛苦，即使是供人用作食物、工作工具、友伴或研究需要。这个立场是建基于人类所做的行为需要有相当的道德情操，而并非像一些动物权益者将动物的地位提升至与人类相当，并在政治及哲学方面追寻更大的权益。动物福利概念由五个基本要素组成：生理福利，即无饥渴之忧虑；环境福利，也就是要让动物有适当的居所；卫生福利，主要是减少动物的伤病；行为福利，应保证动物表达天性的自由；心理福利，即减少动物恐惧和焦虑的情绪。

二、食品安全事件

1. 三聚氰胺污染奶粉事件

2008年9月11日，有媒体报道，甘肃14名婴儿患肾结石，省卫生厅介入调查。当晚，三鹿集团承认婴幼儿奶粉受到三聚氰胺污染。截至2008年9月21日上午8时，全国因食用含三聚氰胺的奶粉导致住院的婴幼儿1万余人，官方确认4例患儿死亡。2008年9月16日，三鹿集团党委书记被免职。同时，石家庄市分管农业的副市长、市畜牧水产局局长、市食品药品监督管理局局长和市质量技术监督局局长也被免职。2008年9月22日，国家质量监督检验检疫总局局长因“毒奶粉”事件引咎辞职。2008年10月8日，卫生部等五部门公布了乳及乳制品中三聚氰胺临时限量标准。其中1000g婴幼儿配方乳粉中允许存在1mg三聚氰胺。

2. 苏丹红事件

某食品公司2002年从印度进口5t红辣椒粉，将其用于生产各类品牌的伍斯特调料，为众多下游食品商或者饭店所使用。后被英国食品标准管理局查出此种辣椒含致癌物质，因此受到通缉。2005年2月，英国食品标准局在官方网站上公布了一份通告：亨氏、联合利华等30家企业的产品中可能含有致癌性的工业染色剂苏丹红一号。随后，一场声势浩大的查禁苏丹红一号的行动席卷全球。

就在英国食品标准局把这份通告发出的十多天之后，北京市食品安全办向社会通报北京市情况：红心鸭蛋等。苏丹红（一号）是一种红色染料，用于为溶剂、油、蜡、汽油增色以及鞋、地板等的增光。有关研究表明，苏丹红（一号）具有致癌性，我国和欧盟都禁止其用于食品生产。

3. 其他食品安全事件

2011年3·15特别行动中，央视曝光了双汇“瘦肉精”养猪一事。瘦肉精可以增加动物的瘦肉量使肉品提早上市、降低成本。但瘦肉精有着较强的毒性，长期使用有可能导致染色体畸变，诱发恶性肿瘤。2011年11月9日，国家质检总局抽查结果显示，联合利华有限公司“立顿”铁观音稀土超标3倍多。稀土对人体健康的作用好坏取决于其浓度的高低，过量摄入将对人体造成危害。2011年10月19日“思念”三鲜水饺检出金黄色葡萄球菌；11月7日，三全白菜猪肉水饺也检出金黄色葡萄球菌；11月17日，湾仔码头上汤小云吞也检出金黄色葡萄球菌。三大冷冻食品品牌陷“细菌门”。2011年4月，合肥、南京等多地的一些熟食店、面馆为牟利而用牛肉膏将猪肉变成牛肉。专家指出，食品添加剂在一定安全剂量内食用，并无危害，但若违规超量和长期食用，则对人体有危害，甚至可能致癌。2011年6

月底，“新华视点”初步揭开了京津冀“地沟油”产业链的黑幕。调查发现天津、河北甚至北京都存在“地沟油”加工窝点，其加工工艺科技含量高，产业链庞大并以小包装的形式进入超市。2011年8月，浙工商在流通领域食品质量例行抽检中发现，血燕中亚硝酸盐的含量严重超标350倍之多，这些血燕产品多从广东、厦门等地进入，主要源自马来西亚等国家。2011年4月初，有媒体爆出在上海市浦东区的一些华联超市和联华超市的主食专柜都在销售同一个公司生产的三种馒头：高粱馒头、玉米馒头和黑米馒头。这些馒头都是回收馒头中加香精和色素加工而成的。2011年4月，沈阳市公安局皇姑分局端掉6个黄豆芽黑加工点，查获掺入非食品添加剂豆芽25余吨，据了解，这些豆芽中被检测出亚硝酸钠、尿素、恩诺沙星、6-苄基腺嘌呤激素等有害物质。2011年12月，有关媒体报道，在南京六合大厂丁家山路上，藏匿着一处黑作坊非法生产鸭血。仅三四斤（1斤＝0.5kg）原料，加入添加剂后即可生产多达20斤的鸭血，“膨大”近5倍。

随着中国食品工业的发展，食品安全问题也日益凸显，其主要出现在贯穿食品源头、加工、生产、流通销售及消费即从农田到餐桌的各个环节。我国的食品安全问题频频发生，食品安全事件的原因在哪里？

三、食品安全事件原因分析

1. 表征因素

从表观上看，食品安全事件的原因包括：①水、土壤和空气等农业环境资源的污染；②种植业和养殖业生产过程中使用化肥、农药、生长激素致使有害化学物质在农产品中残留；③农产品加工和储藏过程中违规或超量使用食品添加剂（防腐剂）；④微生物污染引起的食源性疾病；⑤新原料、新工艺带来的食品安全风险，如转基因食品的安全性；科技进步对食品安全控制带来新的挑战；⑥生产经营者在食品中掺杂使假。

2. 法规与管理方面的原因

中国食品安全管理存在的制度缺陷主要包括：法律制度不健全。虽然涉及食品安全的相关法律法规已有几十部之多，但是法律条款相对分散，调整范围较窄，没有充分适应新形势下消费者对食品安全的需求，其对食品安全问题的惩罚力度也较低；“分段监管”导致监管制度碎片化（八个部门管不了一头猪）；地方政府职责定位偏差；缺乏专业的管理部门；生产者和消费者之间存在严重的信息不对称；食品安全制度缺乏创新激励；食品消费者只具有有限理性，食品生产者机会主义特征明显；制度执行不力等。

3. 深层次分析

化学污染物在食品生产、加工、流通、销售、消费等环节中起着重要的作用。化学品的物理学性质、化学性质、生物学性质、毒理学性质、毒性、临床表现、潜在风险、迁移转化规律、分析检测技术、损伤修复技术、消除技术等方面的学术研究亟待加强。

四、解决方案

1. 构建食品安全体系

关于食品安全体系研究的理论与实践表明：构建一个系统、强大的国家食品安全控制体系具有非常重要的意义，国家食品安全体系应包括法规体系、管理体系、监控体系、科学支撑体系等单元；食品安全体系的建立应涉及所有利益相关者，为了共同的目标一起工作，共同建立一套确保食品安全的方法，允许不同风险的比较和平衡，在现有资源上取得最大效果。

2. 完善政府监管与规制

关于食品安全政府监管与规制的研究表明，由于外部性、人的有限理性及生产者与消费者之间关于食品安全认识、成因及对策问题的研究综述存在严重的信息不对称等原因，市场本身无法解决当前存在的各种食品安全问题，必须通过政府管制来解决。政府规制从制度供给方面为解决食品安全管理方面的问题提供了有益思路，我国已经意识到政府规制对于解决食品安全管理问题的重要性，在消除信息不对称、强化外部规制资源的作用、实现政府规制成本最小化等方面做了大量的工作，收到了一定的成效。因此，此项工作需要高度重视。

3. 食品安全控制管理模式

关于食品安全控制管理模式研究方面，国际上目前的研究主要集中在针对危害因素进行控制的 HACCP 模式方面；国内则在对 HACCP 模式进行系统研究和应用的同时，也开始从供应链角度研究有关的食品安全控制模式，另外对量化分级、信息追踪等具体模式也做了探讨。但作者认为，中国现有关于食品安全控制模式研究的文献非常零散，主题不明确，对于被广大理论和实践工作者看好的 HACCP 模式过于偏重企业生产加工环节的食品安全控制，实际上并不能解决“从农田到餐桌”整个过程的食品安全问题。因此，对于食品安全控制管理模式的研究还很不系统，针对食品供应链整体进行控制的模式或方法尚需进一步深入研究。

第五节　环境保护

一、城市化相伴带来诸多环境问题，当前我国环境形势依然严峻

发达国家一两百年出现的环境问题，在我国近 30 多年来的快速发展中集中显现出来。形成了结构型、压缩型、复合型的环境问题。老的环境问题尚未解决，新的环境问题接踵而至，环境质量改善与人民群众的期待还有不小差距。

1. 大气污染和大气环境劣化

自从 2013 年 1 月 1 日起，我国 74 个城市按空气质量新标准开展监测并实时发布 $PM_{2.5}$ 等数据以来，我国第一批 74 个率先实现空气质量新标准监测的城市中，京津冀区域城市的 80 个国家网监测点位中半数以上出现空气质量连续超标现象。长三角区域城市的 129 个国家网监测点位中约有 1/3 出现空气质量连续超标现象其他直辖市及省会城市的监测点位中也有不同程度的空气质量超标。2014 年年初，受极端不利气象条件影响，我国中东部地区出现长时间、大范围、重污染雾霾天气。按照新的环境空气质量标准，全国 70%左右的城市不达标。$PM_{2.5}$ 问题引起公众普遍关注。事实上，雾霾天气持续、空气质量下降并不是今年的新现象。这几年，每到秋冬特别是入冬以后，我国中东部地区不时会遇到这样的情况。其中既有气象原因，也有污染排放原因。雾是自然，霾是人为。持续的雾霾其实是工业文明对自然损害程度的直观展示，是人类必须承担污染后果的现身说法。燃煤、机动车、工业、扬尘，这些污染源排放量大是造成严重污染的根本原因。

2. 水质污染和水环境劣化

中国水资源总量虽然居世界第六位，但人均占有量却位居世界第 109 位，被列为世界 13

个贫水国之一。尤其西北、华北等地区缺水严重，水资源问题已严重影响到地区经济的可持续发展。水污染问题依然突出。目前工业废水经处理后仍有一半达不到排放标准。在全国受到监测的1200条河流中，850条受到了不同程度的污染，流经城市的河段污染更加严重，有近86%的城市河段水质污染超标，近50%的重点城镇水源不符合饮用水标准。2012年，10大流域中劣Ⅴ类水质比例占10.2%，61个重点湖（库）中24个劣于Ⅲ类标准。地下饮用水水源地水质10%不达标，农村仍有2.98亿人饮水不安全。生活污水的处理能力则更低。

3. 地面塌陷

频繁出现的地面塌陷事件成了城市里的潜在陷阱，这也与城市化进程的推进不无关系。每个城市的地下都有水电、煤燃气、石油以及通信等不同的管网走廊，它们分别由自来水公司、电力部门、煤气公司、天然气公司、石油以及通信公司进行管道以及电缆的铺设，形成错综复杂的地下网路。然而，随着科技的进步，这些管网铺设有着时间上的错位，以致同条道路要多次经历开挖、填埋再开挖的循环过程，一旦遭遇大雨浸泡，路面便会出现空洞，以致容易发生塌陷，对城市安全构成隐性威胁。

4. 生态保护和农村环保亟待加强

生态系统退化，水土流失严重。全国近60万个行政村大部分没有环保基础设施，每年产生生活污水90多亿吨，生活垃圾2.8亿吨。污染源普查结果显示，农业源排放的化学需氧量、总氮、总磷分别占全国排放总量的44%、57%和67%。环境风险不断凸显。

二、环境污染背后的深层原因

1. 对可持续发展战略认识不足

当前在GDP考核的指挥棒下，一些地方政府热衷于吸引大项目、快项目，以期对拉动经济、增加就业等产生立竿见影的效果。然而，大干快上的发展模式，使得不少地方在项目的选择上，重发展、轻环境成为一个突出问题。环境问题的背后，还是发展方向、手段、思维之积弊的累积。对于环境保护，特别考验政府及政府官员的“战略眼光”，牺牲环境或许能够收获短暂的经济增长，但种下的苦果及所造成的长期危害可能需要用几代人的健康乃至生命来偿还。

2. 环保法执法不力、投入过低

政出多门，有令不行，有令难行。法律通常被认为是政府能够动用的最有效的规制手段。早在1979年，全国人大常委会就通过了环境保护法（试行）。这是改革开放后首批颁布的法律之一。治国需重典，处罚要到人。我国环境保护十多年来的艰难历程表明，如果环境执法只停留在罚款上、只局限在“公对公”上，就只能是隔靴搔痒、难儆效尤。如果不把那些保护污染的“幕后黑手”绳之以法、晓以利害，就免不了故伎重演、污染反弹。超前立法未能落实的原因还在于我国的环保投入占经济总量的比例太低。如果要维持现状，环保的投入应该占GDP的2%。如果要改善污染现状，投入要达到3%。

3. 官员重经济增长，轻环境保护

对于国家专门的环保法规，地方政府和官员也每每“高度重视”。而另一方面当环境保护与地方经济发展相矛盾时，环境保护显然又处于弱势。当有上级部门、媒体记者、当地群众对地方企业排污超标进行查处或提出质疑时，地方政府及官员那种对经济增长的“重视”

和对环境保护的“轻视”之意却又暴露无遗。其实，地方官员追求GDP和上级对基层GDP政绩考核本身并没有错。因为社会进步与民生改善都离不开经济发展所提供的财力支撑。错的是什么呢？错的是GDP至上的评价机制，错的是以牺牲土地、环境、百姓利益为代价的政绩冲动。在近乎丧失理智地追求GDP指标与相应评价制度下，社会很难实现统筹协调、科学发展。城市政府作为城市的管理者没有经营城市的先进理念，仅片面追求经济利益，忽视或根本不重视城市生态环境保护和污染的防治，其结果导致城市环境问题日益严重，城市形象大受影响，反而阻碍了城市经济的发展。

4. 缺乏有效的环保问责监督机制

环保问责监督机制是保证环保问责能够顺利进行的保障。建立长效的环保问责机制，并且对其进行有效的监督是环保问责可持续发展的基础。目前，我国长效的环保问责机制尚不健全，环保有责不问、大责小问现象十分普遍，让人感觉环保问责流于形式，使环保问责失去权威性和震慑性，从另一个角度助长了环保违法的泛滥。

三、积极探索环保新道路

我国环境政策的演变历程，从环境政策理念和政策目标上看，经历了局部环境治理—先污染后治理—边污染边治理—环境优先、保护优先的可持续发展的演变过程。从环境政策工具上看，经历了从单一使用行政管制工具到行政管制工具、市场化工具并用阶段，再到行政管制工具、市场化工具和公众参与工具综合运用阶段的演变过程。

我国已经到了以环境保护优化经济发展的新阶段。西方发达国家曾走过的“先污染后治理、牺牲环境换取经济增长”的老路，在我国走不通，也走不起。我国一些地方把经济发展与环境保护割裂开来，甚至重蹈“先污染后治理”的覆辙，付出了过大的生态环境代价。教训告诫我们，千万不能走老路，必须积极探索环保新道路。探索环保新道路要遵循代价小、效益好、排放低、可持续的基本要求。以最小的资源环境代价支撑更大规模的经济社会发展，使经济社会活动对生态环境的损害降低到最小限度，实现经济效益、社会效益和生态环境效益多赢。探索环保新道路的着力点是加快推进环境管理战略转型，以改善环境质量为目标导向，加快构建与我国国情相适应的生态环境保护宏观战略体系、全面高效的污染防治体系、健全的环境质量评价体系、完善的环境保护法规政策和科技标准体系、完备的环境管理和执法监督体系、全民参与的社会行动体系。

四、严格治理$PM_{2.5}$污染

$PM_{2.5}$污染是当前环保工作的热点之一。国务院常务会议已审议通过的《大气污染防治行动计划》提出了十条35项具体措施，主要包括以下六个方面。

1. 推进产业结构优化升级，压缩过剩产能

严格控制高耗能、高污染行业新增产能，钢铁、电解铝、焦炭等行业新改扩建项目实行产能等量或减量置换。加快淘汰落后产能，提前一年完成21个重点行业“十二五”淘汰任务。压缩过剩产能，认真清理产能严重过剩行业违规在建项目。

2. 加快调整能源结构，控制煤炭消费总量

增加天然气、煤制甲烷、煤层气供应，发展核电、水电、风能和生物质能，提高清洁能源比例。强化能源消费总量控制，京津冀、长三角、珠三角等区域及山东省实现煤炭消费总量负增长，耗煤建设项目实行煤炭减量替代。

3. 严格治理机动车污染，提升燃油品质

实施机动车国Ⅴ标准，加快淘汰黄标车。加快车用燃油低硫化步伐，实行优质优价。实施优先发展公共交通战略，加强步行、自行交通系统建设。

4. 强化综合治理，实施多污染物协同控制

采取燃煤电厂脱硫脱硝除尘、工业锅炉窑炉污染治理、挥发性有机物综合整治、扬尘环境管理、餐饮油烟污染治理、北方采暖季节污染控制等综合措施。

5. 建立监测预警应急体系，妥善应对重污染天气

尽快建立重污染天气监测预警体系，及时发布监测预警信息。抓紧制订和完善应急预案。根据预警等级，采取重污染企业限产限排停产、停止建筑施工土方作业、机动车限行等应对措施。

6. 加强组织领导，切实落实保障措施

建立区域协调机制。充分运用价格税收杠杆，加大财政资金投入，引导社会资金治理大气污染。加快修改环境保护法和大气污染防治法，严厉打击环境违法行为。明确政府社会企业职责，构建全民行动的大格局。

五、环境保护“功在当代”、“利在千秋”

1. 产业结构调整、能源结构调整是必由之路

调整工业布局和产业结构。依靠科技进步，推行清洁生产，压缩、淘汰技术含量低、能耗高、污染重的产业和产品，提升和优化产业产品层次。结合城区改造，逐步调整工业布局。将污染严重的企业搬出城区，坚决实行关、停、并、转，大力提倡清洁生产，提高生态农业、绿色工厂在农业和工业产业中的比例。强化和推动城市经营的先进理念，科学规划，严格管理，将整个城市化过程纳入到国家和地区的可持续发展战略之中，实施可持续发展城市化战略。

2. 调整中国社会的各项重要功能在不同地区的配置

要改善城市环境，首先就要从调整中国社会的各项重要功能在不同地区的配置着手，从功能过于集中的部分地区，移出部分功能。这样才能釜底抽薪，避免人文效应与地理气候条件的叠加，缓解污染及其对人类生活的负面影响。在这一点上，新一轮的城镇化担负着特殊的重任。如果能够实现人口和社会活动的优化分布，可以为治理环境污染奠定良好的基础。

3. 财政政策助力环境保护

首先，通过信贷杠杆来调整企业，防治环境污染。企业防治环境污染的资金或综合利用废弃物的技改项目，可向银行申请低息、贴息贷款。而对污染严重的企业进行信贷控制，不接受贷款申请或提高贷款利率。其次，由国家和地方财政提供专项资金对污染治理做得好的企业给予财政补贴。再者，由排污企业向保险公司交纳一定量的保证金。出现污染环境事故后，由保险公司向排污企业提供污染赔偿和污染治理资金。

4. 进一步完善环保立法

要取得理想的环保执法效果，没有完善的环保立法不行。当前，在加大环保执法力度的同时，应加快进一步完善我国环保立法的步伐。通过立、改、废等方式，把那些不适应当前环保形势发展需要的规定及时进行调整，使我国的环保立法与时俱进。消除民法通则与环境保护法中不协调的相关内容，是当前需及时解决的立法问题之一。

5. 问责必须体系化

依法从上到下进行问责。问责不仅是为了平息事端，更不能掩盖问题，而是要彻底地解决问题。所以，问责机制应该是系统的。合理的问责制度应该让所有的关系人对自己的行为后果有预期。知道哪些事情、哪些行为是不能做的，做了一定会被追究责任的。

六、未来十年我国环境管理政策发展趋势

到 2025 年，我国经济发展水平将迈入中上等收入国家行列，并逐步迈进工业化后期，城市化率将达到 60%左右，资源、能源等消费高峰有望陆续到来。这些因素在为我国环境保护工作提供强大驱动力的同时，也成为我国环境保护工作压力的基本动因。随着生活水平的不断提高，老百姓对环境质量的要求也相应提高。实现与 2020 年中国小康社会相适应的环境质量水平的难度将加大，重视并加快环境政策创新与实践将成为新形势下我国环保工作的迫切要求。用全局视野和战略思维去制定环境政策，以环境承载力优化区域布局，可以判断未来 10 年我国将在以下 10 个方面加大推进力度：①强化污染物总量减排政策，将碳减排逐步纳入日程；②环境政策将由污染控制型政策逐步转向以环境质量改善和环境风险防控为目标的环境政策；③加大环境保护投入，提升环境经济政策的执行力度；④实施区域差异化的环境政策；⑤推进环境污染协同控制政策；⑥认识环境保护的国家责任，提高环境政策的执行力；⑦支持公众参与，架构“政府-企业-公众”的共同治理模式；⑧改进绩效考核指标体系，提高地方政府的监管能力；⑨全面深化经济与环境源头控制制度，提高我国经济发展的质量；⑩全面推进生态文明建设。

思考题

1. 试述地震、火山、泥石流的形成原因。
2. 厄尔尼诺和飓风是如何形成的？
3. 常见的致命生物有哪些？
4. 如何避免食品安全事件发生？
5. 最新的环境保护措施有哪些？

参考文献

[1] Ronald A. Bailey，Herbert M. Clark，et. al. Chemistry of the Environment，Academic Press，2005.
[2] 刘兆荣，陈忠明，赵广荣等．环境化学教程．北京：化学工业出版社，2004.
[3] Thomas G Spiro，William M Stigliani. Chemistry of the Environment，Second Edition. 北京：清华大学出版社，2003.
[4] Eldon D Enger，Bradley F Smith. Environmental Science—A Study of Interrelationships (Seventh Edition)，Mc Graw-Hill. 北京：清华大学出版社，2000.
[5] 王晓蓉．环境化学．南京：南京大学出版社，1993.
[6] 戴树桂．环境化学（面向21世纪课程教材）．北京：高等教育出版社，2002.
[7] 俞誉福，叶明吕，郑志坚．环境化学导论．上海：复旦大学出版社，1997.
[8] 邓南圣，吴峰．环境化学教程．武汉：武汉大学出版社，2000.
[9] 叶常明，王春霞，金龙珠．21世纪的环境化学．北京：科学出版社，2004.